Bayesian Cognitive Modeling

A Practical Course

Bayesian inference has become a standard method of analysis in many fields of science. Students and researchers in experimental psychology and cognitive science, however, have failed to take full advantage of the new and exciting possibilities that the Bayesian approach affords. Ideal for teaching and self study, this book demonstrates how to do Bayesian modeling. Short, to-the-point chapters offer examples, exercises, and computer code (using WinBUGS or JAGS, and supported by Matlab and R), with additional support available online. No advance knowledge of statistics is required and, from the very start, readers are encouraged to apply and adjust Bayesian analyses by themselves. The book contains a series of chapters on parameter estimation and model selection, followed by detailed case studies from cognitive science. After working through this book, readers should be able to build their own Bayesian models, apply the models to their own data, and draw their own conclusions.

Michael D. Lee is a professor in the Department of Cognitive Sciences at the University of California, Irvine.

Eric-Jan Wagenmakers is a professor in the Department of Psychological Methods at the University of Amsterdam.

Bayesian Cognitive Modeling

A Practical Course

MICHAEL D. LEE

ERIC-JAN WAGENMAKERS

CAMBRIDGE
UNIVERSITY PRESS

University Printing House, Cambridge CB2 8BS, United Kingdom

Cambridge University Press is part of the University of Cambridge.

It furthers the University's mission by disseminating knowledge in the pursuit of education, learning and research at the highest international levels of excellence.

www.cambridge.org
Information on this title: www.cambridge.org/9781107603578

First published 2013

A catalogue record for this publication is available from the British Library

ISBN 978-1-107-60357-8 Paperback

Contents

Part IV Case studies 143

For Colleen and David, and Helen and Mitchell — Michael

Preface

This book, together with the code, answers to questions, and other material at `www.bayesmodels.com`, teaches you how to do Bayesian modeling. Using modern computer software—and, in particular, the WinBUGS program—this turns out to be surprisingly straightforward. After working through the examples provided in this book, you should be able to build your own models, apply them to your own data, and draw your own conclusions.

This book is based on three principles. The first is that of *accessibility*: the book's only prerequisite is that you know how to operate a computer; you do not need any advanced knowledge of statistics or mathematics. The second principle is that of *applicability*: the examples in this book are meant to illustrate how Bayesian modeling can be useful for problems that people in cognitive science care about. The third principle is that of *practicality*: this book offers a hands-on, "just do it" approach that we feel keeps students interested and motivated.

In line with these three principles, this book has little content that is purely theoretical. Hence, you will not learn from this book why the Bayesian philosophy to inference is as compelling as it is; neither will you learn much about the intricate details of modern sampling algorithms such as Markov chain Monte Carlo, even though this book could not exist without them.

The goal of this book is to facilitate and promote the use of Bayesian modeling in cognitive science. As shown by means of examples throughout this book, Bayesian modeling is ideally suited for applications in cognitive science. It is easy to construct a basic model, and then add individual differences, add substantive prior information, add covariates, add a contaminant process, and so on. Bayesian modeling is flexible and respects the complexities that are inherent in the modeling of cognitive phenomena.

We hope that after completing this book, you will have gained not only a new understanding of statistics (yes, it can make sense), but also the technical skills to implement statistical models that professional but non-Bayesian cognitive scientists dare only dream about.

MICHAEL D. LEE
Irvine, USA
ERIC-JAN WAGENMAKERS
Amsterdam, The Netherlands

Acknowledgements

The plan to produce this book was hatched in 2006. Since then, the core material has undergone a steady stream of additions and revisions. The revisions were inspired in part by students and colleagues who relentlessly suggested improvements, pointed out mistakes, and attended us to inconsistencies and inefficiencies. We would especially like to thank Ryan Bennett, Adrian Brasoveanu, Eddy Davelaar, Joram van Driel, Wouter Kruijne, Alexander Ly, John Miyamoto, James Negen, Thomas Palmeri, James Pooley, Don van Ravenzwaaij, Hedderik van Rijn, J. P. de Ruiter, Anja Sommavilla, Helen Steingroever, Wolf Vanpaemel, and Ruud Wetzels for their constructive comments and contributions. We are particularly grateful to Dora Matzke for her help in programming and plotting. Any remaining mistakes are the sole responsibility of the authors. A list of corrections and typographical errors will be available on `www.bayesmodels.com`. When you spot a mistake or omission that is not on the list please do not hesitate to email us at `BayesModels@gmail.com`.

The material in this book is not independent of our publications in the cognitive science literature. Sometimes, an article was turned into a book chapter; at other times, a book chapter spawned an article. Here we would like to acknowledge our published articles that contain text and figures resembling, to varying degrees, those used in this book. These articles often may be consulted for a more extensive and formal exposition of the material at hand.

Chapter 1: The basics of Bayesian analysis

- Wagenmakers, E.-J., Lodewyckx, T., Kuriyal, H., & Grasman, R. (2010). Bayesian hypothesis testing for psychologists: A tutorial on the Savage–Dickey method. *Cognitive Psychology, 60*, 158–189.

Chapter 6: Latent-mixture models

- Ortega, A., Wagenmakers, E.-J., Lee, M. D., Markowitsch, H. J., & Piefke, M. (2012). A Bayesian latent group analysis for detecting poor effort in the assessment of malingering. *Archives of Clinical Neuropsychology, 27*, 453–465.

Chapter 7: Bayesian model comparison

- Scheibehenne, B., Rieskamp, J., & Wagenmakers, E.-J. (2013). Testing adaptive toolbox models: A Bayesian hierarchical approach. *Psychological Review, 120*, 39–64.

- Wagenmakers, E.-J., Lodewyckx, T., Kuriyal, H., & Grasman, R. (2010). Bayesian hypothesis testing for psychologists: A tutorial on the Savage–Dickey method. *Cognitive Psychology, 60*, 158–189.

Chapter 8: Comparing Gaussian means

- Wetzels, R., Raaijmakers, J. G. W., Jakab, E., & Wagenmakers, E.-J. (2009). How to quantify support for and against the null hypothesis: A flexible WinBUGS implementation of a default Bayesian t test. *Psychonomic Bulletin & Review, 16*, 752–760.

Chapter 9: Comparing binomial rates

- Wagenmakers, E.-J., Lodewyckx, T., Kuriyal, H., & Grasman, R. (2010). Bayesian hypothesis testing for psychologists: A tutorial on the Savage-Dickey method. *Cognitive Psychology, 60*, 158–189.

Chapter 10: Memory retention

- Shiffrin, R. M., Lee, M. D., Kim, W., & Wagenmakers, E.-J. (2008). A survey of model evaluation approaches with a tutorial on hierarchical Bayesian methods. *Cognitive Science, 32*, 1248–1284.

Chapter 11: Signal detection theory

- Lee, M. D. (2008). BayesSDT: Software for Bayesian inference with signal detection theory. *Behavior Research Methods, 40*, 450–456.
- Lee, M. D. (2008). Three case studies in the Bayesian analysis of cognitive models. *Psychonomic Bulletin & Review, 15*, 1–15.

Chapter 13: Extrasensory perception

- Wagenmakers, E.-J. (2012). Can people look into the future? Contribution in honor of the University of Amsterdam's 76th lustrum.
- Wagenmakers, E.-J., Wetzels, R., Borsboom, D., van der Maas, H. L. J., & Kievit, R. A. (2012). An agenda for purely confirmatory research. *Perspectives on Psychological Science, 7*, 627–633.

Chapter 14: Multinomial processing trees

- Matzke, D., Dolan, C. V., Batchelder, W. H., & Wagenmakers, E.-J. (in press). Bayesian estimation of multinomial processing tree models with heterogeneity in participants and items. *Psychometrika*.

Chapter 15: The SIMPLE model of memory

- Shiffrin, R. M., Lee, M. D., Kim, W., & Wagenmakers, E.-J. (2008). A survey of model evaluation approaches with a tutorial on hierarchical Bayesian methods. *Cognitive Science, 32*, 1248–1284.

Chapter 16: The BART model of risk taking

- van Ravenzwaaij, D., Dutilh, G., & Wagenmakers, E.-J. (2011). Cognitive model decomposition of the BART: Assessment and application. *Journal of Mathematical Psychology, 55*, 94–105.

Chapter 17: Generalized context model

- Lee, M. D. & Wetzels, R. (2010). Individual differences in attention during category learning. In R. Catrambone & S. Ohlsson (Eds.), *Proceedings of the 32nd Annual Conference of the Cognitive Science Society*, pp. 387–392. Austin, TX: Cognitive Science Society.
- Bartlema, A., Lee, M. D., Wetzels, R., & Vanpaemel, W. (2012). Bayesian hierarchical mixture models of individual differences in selective attention and representation in category learning. Manuscript submitted for publication.

Chapter 18: Heuristic decision-making

- Lee, M. D. & Newell, B. R. (2011). Using hierarchical Bayesian methods to examine the tools of decision-making. *Judgment and Decision Making, 6*, 832–842.

Chapter 19: Number concept development

- Lee, M. D. & Sarnecka, B. W. (2010). A model of knower-level behavior in number-concept development. *Cognitive Science, 34*, 51–67.
- Lee, M. D. & Sarnecka, B. W. (2011). Number knower-levels in young children: Insights from a Bayesian model. *Cognition, 120*, 391–402.

PART I

GETTING STARTED

[T]he theory of probabilities is basically just common sense reduced to calculus; it makes one appreciate with exactness that which accurate minds feel with a sort of instinct, often without being able to account for it.

Laplace, 1829

1 The basics of Bayesian analysis

1.1 General principles

The general principles of Bayesian analysis are easy to understand. First, uncertainty or "degree of belief" is quantified by probability. Second, the observed data are used to update the *prior* information or beliefs to become *posterior* information or beliefs. That's it!

To see how this works in practice, consider the following example. Assume you are given a test that consists of 10 factual questions of equal difficulty. What we want to estimate is your ability, which we define as the rate θ with which you answer questions correctly. We cannot directly observe your ability θ. All that we can observe is your score on the test.

Before we do anything else (for example, before we start to look at your data) we need to specify our prior uncertainty with respect to your ability θ. This uncertainty needs to be expressed as a probability distribution, called the *prior distribution*. In this case, keep in mind that θ can range from 0 to 1, and that we do not know anything about your familiarity with the topic or about the difficulty level of the questions. Then, a reasonable "prior distribution," denoted by $p(\theta)$, is one that assigns equal probability to every value of θ. This uniform distribution is shown by the dotted horizontal line in Figure 1.1.

Now we consider your performance, and find that you answered 9 out of 10 questions correctly. After having seen these data, the updated knowledge about θ is described by the *posterior distribution*, denoted $p(\theta \mid D)$, where D indicates the observed data. This distribution expresses the uncertainty about the value of θ, quantifying the relative probability that each possible value is the true value. Bayes' rule specifies how we can combine the information from the data—that is, the likelihood $p(D \mid \theta)$—with the information from the prior distribution $p(\theta)$, to arrive at the posterior distribution $p(\theta \mid D)$:

$$p(\theta \mid D) = \frac{p(D \mid \theta)\, p(\theta)}{p(D)}. \tag{1.1}$$

This equation is often verbalized as

$$\text{posterior} = \frac{\text{likelihood} \times \text{prior}}{\text{marginal likelihood}}. \tag{1.2}$$

Note that the marginal likelihood (i.e., the probability of the observed data) does not involve the parameter θ, and is given by a single number that ensures that

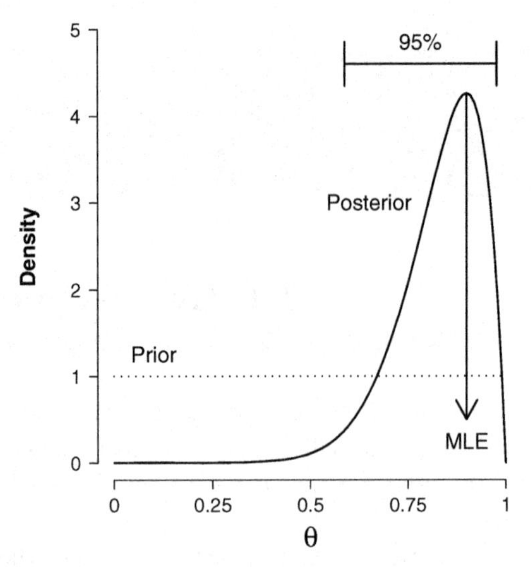

Fig. 1.1 Bayesian parameter estimation for rate parameter θ, after observing 9 correct responses and 1 incorrect response. The mode of the posterior distribution for θ is 0.9, equal to the maximum likelihood estimate (MLE), and the 95% credible interval extends from 0.59 to 0.98.

the area under the posterior distribution equals 1. Therefore, Equation 1.1 is often written as

$$p(\theta \mid D) \propto p(D \mid \theta)\, p(\theta)\,, \tag{1.3}$$

which says that the posterior is proportional to the likelihood times the prior. Note that the posterior distribution is a combination of what we knew before we saw the data (i.e., the information in the prior distribution), and what we have learned from the data. In particular, note that the new information provided by the data has reduced our uncertainty about the value of θ, as shown by the posterior distribution being narrower than the prior distribution.

The solid line in Figure 1.1 shows the posterior distribution for θ, obtained when the uniform prior is updated with the data. The central tendency of a posterior distribution is often summarized by its mean, median, or mode. Note that with a uniform prior, the mode of a posterior distribution coincides with the classical maximum likelihood estimate or MLE , $\hat{\theta} = k/n = 0.9$ (Myung, 2003). The spread of a posterior distribution is most easily captured by a Bayesian $x\%$ credible interval that extends from the $(100 - x)/2^{\text{th}}$ to the $(100 + x)/2^{\text{th}}$ percentile of the posterior distribution. For the posterior distribution in Figure 1.1, a 95% Bayesian credible interval for θ extends from 0.59 to 0.98. In contrast to the orthodox confidence interval, this means that one can be 95% confident that the true value of θ lies in between 0.59 and 0.98.

Exercises

Exercise 1.1.1 The famous Bayesian statistician Bruno de Finetti published two big volumes entitled *Theory of Probability* (de Finetti, 1974). Perhaps surprisingly, the first volume starts with the words "probability does not exist." To understand why de Finetti wrote this, consider the following situation: someone tosses a fair coin, and the outcome will be either heads or tails. What do you think the probability is that the coin lands heads up? Now suppose you are a physicist with advanced measurement tools, and you can establish relatively precisely both the position of the coin and the tension in the muscles immediately before the coin is tossed in the air—does this change your probability? Now suppose you can briefly look into the future (Bem, 2011), albeit hazily. Is your probability still the same?

Exercise 1.1.2 On his blog, prominent Bayesian Andrew Gelman wrote (March 18, 2010): "Some probabilities are more objective than others. The probability that the die sitting in front of me now will come up '6' if I roll it ... that's about 1/6. But not exactly, because it's not a perfectly symmetric die. The probability that I'll be stopped by exactly three traffic lights on the way to school tomorrow morning: that's well, I don't know exactly, but it is what it is." Was de Finetti wrong, and is there only one clearly defined probability of Andrew Gelman encountering three traffic lights on the way to school tomorrow morning?

Exercise 1.1.3 Figure 1.1 shows that the 95% Bayesian credible interval for θ extends from 0.59 to 0.98. This means that one can be 95% confident that the true value of θ lies between 0.59 and 0.98. Suppose you did an orthodox analysis and found the same confidence interval. What is the orthodox interpretation of this interval?

Exercise 1.1.4 Suppose you learn that the questions are all true or false questions. Does this knowledge affect your prior distribution? And, if so, how would this prior in turn affect your posterior distribution?

1.2 Prediction

The posterior distribution θ contains all that we know about the rate with which you answer questions correctly. One way to use the knowledge is *prediction*.

For example, suppose you are confronted with a new set of 5 questions, all of the same difficulty as before. How can we formalize our expectations about your performance on this new set? In other words, how can we use the posterior distribution $p(\theta \mid n = 10, k = 9)$—which, after all, represents everything that we know about θ from the old set—to *predict* the number of correct responses out of the new set of $n^{\text{rep}} = 5$ questions? The mathematical solution is to integrate over the posterior,

$\int p(k^{\text{rep}} \mid \theta, n^{\text{rep}} = 5)\, p(\theta \mid n = 10, k = 9)\ d\theta$, where k^{rep} is the predicted number of correct responses out of the additional set of 5 questions.

Computationally, you can think of this procedure as repeatedly drawing a random value θ_i from the posterior, and using that value to every time determine a single k^{rep}. The end result is $p(k^{\text{rep}})$, the posterior predictive distribution of the possible number of correct responses in the additional set of 5 questions. The important point is that by integrating over the posterior, all predictive uncertainty is taken into account.

Exercise

Exercise 1.2.1 Instead of "integrating over the posterior," orthodox methods often use the "plug-in principle." In this case, the plug-in principle suggests that we predict $p(k^{\text{rep}})$ solely based on $\hat{\theta}$, the maximum likelihood estimate. Why is this generally a bad idea? Can you think of a specific situation in which this may not be so much of a problem?

1.3 Sequential updating

Bayesian analysis is particularly appropriate when you want to combine different sources of information. For example, assume that you are presented with a new set of 5 questions of equal difficulty. You answer 3 out of 5 correctly. How can we combine this new information with the old? Or, in other words, how do we update our knowledge of θ? Consistent with intuition, Bayes' rule entails that the prior that should be updated based on your performance for the new set is the posterior that was obtained based on your performance for the old set. Or, as Lindley put it, "today's posterior is tomorrow's prior" (Lindley, 1972, p. 2).

When all the data have been collected, however, the order in which this was done is irrelevant. The results from the 15 questions could have been analyzed as a single batch; they could have been analyzed sequentially, one-by-one; they could have been analyzed by first considering the set of 10 questions and next the set of 5, or vice versa. For all these cases, the end result, the final posterior distribution for θ, is identical. Given the same available information, Bayesian inference reaches the same conclusion, independent of the order in which the information was obtained. This again contrasts with orthodox inference, in which inference for sequential designs is radically different from that for non-sequential designs (for a discussion, see, for example, Anscombe, 1963).

Thus, a posterior distribution describes our uncertainty with respect to a parameter of interest, and the posterior is useful—or, as a Bayesian would have it, necessary—for probabilistic prediction and for sequential updating. To illustrate, in the case of our binomial example the uniform prior is a beta distribution with parameters $\alpha = 1$ and $\beta = 1$, and when combined with the binomial likelihood this yields a posterior that is also a beta distribution, with parameters $\alpha + k$ and $\beta + n - k$. In simple *conjugate* cases such as these, where the prior and the posterior belong to the same distributional family, it is possible to obtain analytical solutions for the posterior distribution, but in many interesting cases it is not.

1.4 Markov chain Monte Carlo

In general, the posterior distribution, or any of its summary measures, can only be obtained analytically for a restricted set of relatively simple models. Thus, for a long time, researchers could only proceed easily with Bayesian inference when the posterior was available in closed-form or as a (possibly approximate) analytic expression. As a result, practitioners interested in models of realistic complexity did not much use Bayesian inference. This situation changed dramatically with the advent of computer-driven sampling methodology, generally known as Markov chain Monte Carlo (MCMC: e.g., Gamerman & Lopes, 2006; Gilks, Richardson, & Spiegelhalter, 1996). Using MCMC techniques such as Gibbs sampling or the Metropolis–Hastings algorithm, researchers can directly sample sequences of values from the posterior distribution of interest, forgoing the need for closed-form analytic solutions. The current adage is that *Bayesian models are limited only by the user's imagination.*

In order to visualize the increased popularity of Bayesian inference, Figure 1.2 plots the proportion of articles that feature the words "Bayes" or "Bayesian," according to Google Scholar (for a similar analysis for specific journals in statistics and economics see Poirier, 2006). The time line in Figure 1.2 also indicates the introduction of WinBUGS, a general-purpose program that greatly facilitates Bayesian analysis for a wide range of statistical models (Lunn, Thomas, Best, & Spiegelhalter, 2000; Lunn, Spiegelhalter, Thomas, & Best, 2009; Sheu & O'Curry, 1998). MCMC methods have transformed Bayesian inference to a vibrant and practical area of modern statistics.

For a concrete and simple illustration of Bayesian inference using MCMC, consider again the binomial example of 9 correct responses out of 10 questions, and the associated inference problem for θ, the rate of answering questions correctly. Throughout this book, we use WinBUGS to do Bayesian inference, saving us the effort of coding the MCMC algorithms ourselves.[1] Although WinBUGS does not work for every research problem application, it will work for many in cognitive sci-

[1] At this point, some readers want to know how exactly MCMC algorithms work. Other readers feel the urge to implement MCMC algorithms themselves. The details of MCMC sampling are covered in many other sources and we do not repeat that material here. We recommend the relevant chapters from the following books, listed in order of increasing complexity: Kruschke (2010a), MacKay (2003), Gilks et al. (1996), Ntzoufras (2009), and Gamerman and Lopes (2006). An introductory overview is given in Andrieu, De Freitas, Doucet, and Jordan (2003). You can also browse the internet, and find resources such as `http://www.youtube.com/watch?v=4gNpgSPal_8` and `http://www.learnbayes.org/`.

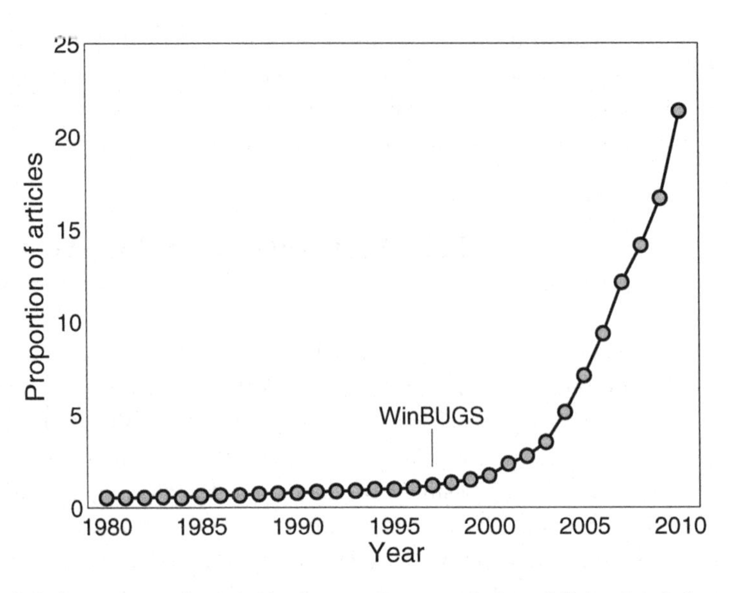

Fig. 1.2 A Google Scholar perspective on the increasing popularity of Bayesian inference, showing the proportion of articles matching the search "bayes OR bayesian -author: bayes" for the years 1980 to 2010.

ence. WinBUGS is easy to learn and is supported by a large community of active researchers.

The WinBUGS program requires you to construct a file that contains the model specification, a file that contains initial values for the model parameters, and a file that contains the data. The model specification file is most important. For our binomial example, we set out to obtain samples from the posterior of θ. The associated WinBUGS model specification code is two lines long:

```
model{
  theta ~ dunif(0,1) # the uniform prior for updating by the data
  k ~ dbin(theta,n)  # the data; in our example, k = 9 and n = 10
}
```

In this code, the "$\sim$" or twiddle symbol denotes "is distributed as", `dunif(a,b)` indicates the uniform distribution with parameters a and b, and `dbin(theta,n)` indicates the binomial distribution with rate θ and n observations. These and many other distributions are built in to the WinBUGS program. The "#" or hash sign is used for comments. As WinBUGS is a declarative language, the order of the two lines is inconsequential. Finally, note that the values for k and n are not provided in the model specification file. These values constitute the data and they are stored in a separate file.

When this code is executed, you obtain a sequence of MCMC samples from the posterior $p(\theta \mid D)$. Each individual sample depends only on the one that immediately preceded it, and this is why the entire sequence of samples is called a *chain*.

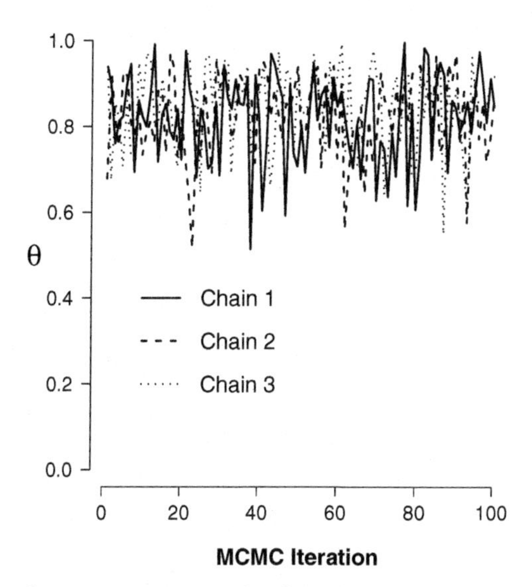

Fig. 1.3 Three MCMC chains for rate parameter θ, after observing 9 correct responses and 1 incorrect response.

In more complex models, it may take some time before a chain converges from its starting value to what is called its stationary distribution. To make sure that we only use those samples that come from the stationary distribution, and hence are unaffected by the starting values, it is good practice to diagnose convergence. This is an active area of research, and there is an extensive set of practical recommendations regarding achieving and measuring convergence (e.g., Gelman, 1996; Gelman & Hill, 2007).

A number of worked examples in this book deal with convergence issues in detail, but we mention three important concepts now. One approach is to run multiple chains, checking that their different initial starting values do not affect the distributions they sample from. Another is to discard the first samples from each chain, when those early samples are sensitive to the initial values. These discarded samples are called *burn-in* samples. Finally, it can also be helpful not to record every sample taken in a chain, but every second, or third, or tenth, or some other subset of samples. This is known as *thinning*, a procedure that is helpful when the chain moves slowly through the parameter space and, consequently, the current sample in the MCMC chain depends highly on the previous one. In such cases, the sampling process is said to be autocorrelated.

For example, Figure 1.3 shows the first 100 iterations for three chains that were set up to draw values from the posterior for θ. It is evident that the three chains are "mixing" well, suggesting early convergence. After assuring ourselves that the chains have converged, we can use the sampled values to plot a histogram, construct a density estimate, and compute values of interest. To illustrate, the three chains from Figure 1.3 were run for 3000 iterations each, for a total of 9000 samples from the posterior of θ. Figure 1.4 plots a histogram for the posterior. To visualize how the

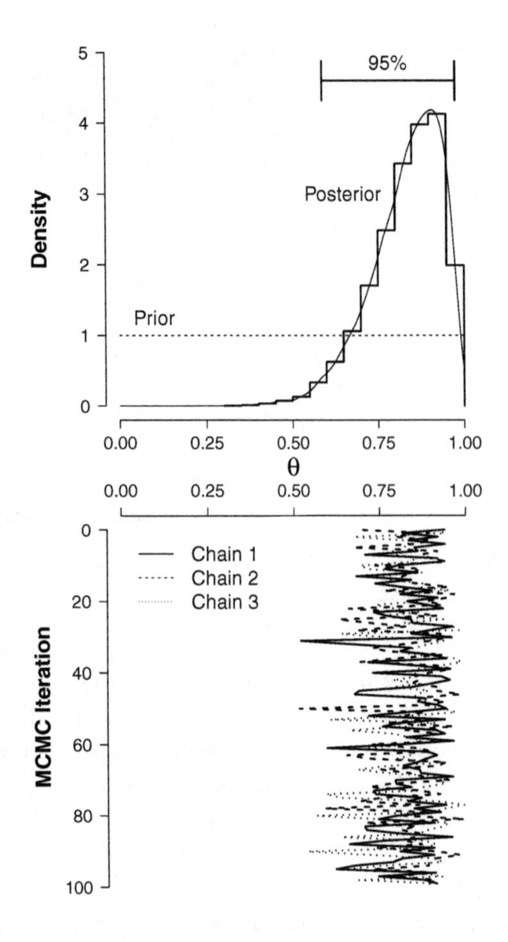

Fig. 1.4 MCMC-based Bayesian parameter estimation for rate parameter θ, after observing 9 correct responses and 1 incorrect response. The thin solid line indicates the fit of a density estimator. Based on this density estimator, the mode of the posterior distribution for θ is approximately 0.89, and the 95% credible interval extends from 0.59 to 0.98, closely matching the analytical results from Figure 1.1.

histogram is constructed from the MCMC chains, the bottom panel of Figure 1.4 plots the MCMC chains sideways; the histograms are created by collapsing the values along the "MCMC iteration" axis and onto the "θ" axis.

In the top panel of Figure 1.4, the thin solid line represent a density estimate. The mode of the density estimate for the posterior of θ is 0.89, whereas the 95% credible interval is $(0.59, 0.98)$, matching the analytical result shown in Figure 1.1.

The key point is that the analytical intractabilities that limited the scope of Bayesian parameter estimation have now been overcome. Using MCMC sampling, posterior distributions can be approximated to any desired degree of accuracy. This

Box 1.1 Why isn't every statistician a Bayesian?

"The answer is simply that statisticians do not know what the Bayesian paradigm says. Why should they? There are very few universities in the world with statistics departments that provide a good course in the subject. Only exceptional graduate students leave the field of their advisor and read for themselves. A secondary reason is that the subject is quite hard for someone who has been trained in the sampling-theory approach to understand. ... The subject is difficult. Some argue that this is a reason for not using it. But it is always harder to adhere to a strict moral code than to indulge in loose living. ... Every statistician would be a Bayesian if he took the trouble to read the literature thoroughly and was honest enough to admit that he might have been wrong." (Lindley, 1986, pp. 6–7).

book teaches you to use MCMC sampling and Bayesian inference to do research with cognitive science models and data.

Exercises

Exercise 1.4.1 Use Google and list some other scientific disciplines that use Bayesian inference and MCMC sampling.

Exercise 1.4.2 The text reads: "Using MCMC sampling, posterior distributions can be approximated to any desired degree of accuracy." How is this possible?

1.5 Goal of this book

The goal of this book is to show, by working through concrete examples, how Bayesian inference can be applied to modeling problems in cognitive science. Bayesian data analysis has received increasing attention from cognitive scientists, and for good reason.

1. Bayesian inference is *flexible*. This means that Bayesian models can respect the complexity of the data, and of the processes being modeled. For example, data analysis may require the inclusion of a contaminant process, a multi-level structure, or an account of missing data. Using the Bayesian approach, these sorts of additions are relatively straightforward.
2. Bayesian inference is *principled*. This means that all uncertainty is accounted for appropriately, and no useful information is discarded.

Box 1.2 Common sense expressed in numbers

"The Bayesian approach is a common sense approach. It is simply a set of techniques for orderly expression and revision of your opinions with due regard for internal consistency among their various aspects and for the data. Naturally, then, much that Bayesians say about inference from data has been said before by experienced, intuitive, sophisticated empirical scientists and statisticians. In fact, when a Bayesian procedure violates your intuition, reflection is likely to show the procedure to have been incorrectly applied." (Edwards et al., 1963, p. 195).

3. Bayesian inference yields *intuitive conclusions*. This reflects the fact that Bayesian inference is normative, stipulating how rational agents should change their opinion in the light of incoming data. Of course, it can nevertheless happen that you occasionally find a Bayesian conclusion to be surprising or counter-intuitive. You are then left with one of two options—either the analysis was not carried out properly (e.g., errors in coding, errors in model specification) or your intuition is in need of schooling.
4. Bayesian inference is *easy to undertake*. This means that with the software packages used in this book, Bayesian inference is often (but not always!) a trivial exercise. This frees up resources so more time can be spent on the substantive issues of developing theories and models, and interpreting results when they are applied to data.

At this point you may be champing at the bit, eager to apply the tools of Bayesian analysis to the kinds of cognitive models that interest you. But first we need to cover the basics and this is why Parts I, II, and III prepare you for the more complicated case studies presented in Part IV. This is not to say that the "elementary" material in Parts I, II, and III are devoid of cognitive context. On the contrary, we have tried to highlight how even the binomial model finds meaningful application in cognitive science.

Perhaps the material covered in this first chapter is still relatively abstract for you. Perhaps you are currently in a state of confusion. Perhaps you think that this book is too difficult, or perhaps you do not yet see clearly how Bayesian inference can help you in your own work. These feelings are entirely understandable, and this is why this book contains more than just this one chapter. Our teaching philosophy is that you learn the most by doing, not by reading. So if you still do not know exactly what a posterior distribution is, do not despair. The chapters in this book make you practice core Bayesian inference tasks so often that at the end you will know exactly what a posterior distribution is, whether you like it or not. Of course, we rather hope you like it, and we also hope that you will discover that Bayesian statistics can be exciting, rewarding, and, indeed, fun.

1.6 Further reading

This section provides some references for further reading. We first list Bayesian textbooks and seminal papers, then some texts that specifically deal with WinBUGS. We also note that Smithson (2010) presents a useful comparative review of six introductory textbooks on Bayesian methods.

1.6.1 Bayesian statistics

This section contains an annotated bibliography of Bayesian articles and books that we believe are particularly useful or inspiring.

- Berger, J. O. & Wolpert, R. L. (1988). *The Likelihood Principle* (2nd edn.). Hayward, CA: Institute of Mathematical Statistics. This is a great book if you want to understand the limitations of orthodox statistics. Insightful and fun.
- Bolstad, W. M. (2007). *Introduction to Bayesian Statistics* (2nd edn.). Hoboken, NJ: Wiley. Many books claim to introduce Bayesian statistics, but forget to state on the cover that the introduction is "for statisticians" or "for those comfortable with mathematical statistics." The Bolstad book is an exception, as it does not assume much background knowledge.
- Dienes, Z. (2008). *Understanding Psychology as a Science: An Introduction to Scientific and Statistical Inference*. New York: Palgrave Macmillan. An easy-to-understand introduction to inference that summarizes the differences between the various schools of statistics. No knowledge of mathematical statistics is required.
- Gamerman, D. & Lopes, H. F. (2006). *Markov Chain Monte Carlo: Stochastic Simulation for Bayesian Inference*. Boca Raton, FL: Chapman & Hall/CRC. This book discusses the details of MCMC sampling; a good book, but too advanced for beginners.
- Gelman, A. & Hill, J. (2007). *Data Analysis Using Regression and Multilevel/Hierarchical Models*. New York: Cambridge University Press. This book is an extensive practical guide on how to apply Bayesian regression models to data. WinBUGS code is provided throughout the book. Andrew Gelman also has an active blog that you might find interesting: `http://andrewgelman.com/`
- Gilks, W. R., Richardson, S., & Spiegelhalter, D. J. (1996). *Markov Chain Monte Carlo in Practice*. Boca Raton, FL: Chapman & Hall/CRC. A citation classic in the MCMC literature, this book features many short chapters on all kinds of sampling-related topics: theory, convergence, model selection, mixture models, and so on.
- Gill, J. (2002). *Bayesian Methods: A Social and Behavioral Sciences Approach*. Boca Raton, FL: CRC Press. A well-written book that covers a lot of ground. Readers need some background in mathematical statistics.

- Hoff, P. D. (2009). *A First Course in Bayesian Statistical Methods.* Dordrecht, The Netherlands: Springer. A clear and well-written introduction to Bayesian inference, with accompanying R code, requiring some familiarity with mathematical statistics.
- Jaynes, E. T. (2003). *Probability Theory: The Logic of Science.* Cambridge, UK: Cambridge University Press. Jaynes was one of the most ardent supporters of objective Bayesian statistics. The book is full of interesting ideas and compelling arguments, as well as being laced with Jaynes' acerbic wit, but it requires some mathematical background to appreciate all of the content.
- Jeffreys, H. (1939/1961). *Theory of Probability.* Oxford, UK: Oxford University Press. Sir Harold Jeffreys is the first statistician who exclusively used Bayesian methods for inference. Jeffreys also invented the Bayesian hypothesis test, and was generally far ahead of his time. The book is not always an easy read, in part because the notation is somewhat outdated. Strongly recommended, but only for those who already have a solid background in mathematical statistics and a firm grasp of Bayesian thinking. See `www.economics.soton.ac.uk/staff/aldrich/jeffreysweb.htm`
- Lee, P. M. (2012). *Bayesian Statistics: An introduction* (4th edn.). Chichester, UK: John Wiley. This well-written book illustrates the core tenets of Bayesian inference with simple examples, but requires a background in mathematical statistics.
- Lindley, D. V. (2000). The philosophy of statistics. *The Statistician, 49,* 293–337. One the godfathers of Bayesian statistics explains why Bayesian inference is right and everything else is wrong. Peter Armitage commented on the paper: "Lindley's concern is with the very nature of statistics, and his argument unfolds clearly, seamlessly and relentlessly. Those of us who cannot accompany him to the end of his journey must consider very carefully where we need to dismount; otherwise we shall find ourselves unwittingly at the bus terminus, without a return ticket."
- Marin, J.-M. & Robert, C. P. (2007). *Bayesian Core: A Practical Approach to Computational Bayesian Statistics.* New York: Springer. This is a good book by two reputable Bayesian statisticians. The book is beautifully typeset, includes an introduction to R, and covers a lot of ground. A firm knowledge of mathematical statistics is required. The exercises are challenging.
- McGrayne, S. B. (2011). *The Theory that Would not Die: How Bayes' Rule Cracked the Enigma Code, Hunted Down Russian Submarines, and Emerged Triumphant from Two Centuries of Controversy.* New Haven, CT: Yale University Press. A fascinating and accessible overview of the history of Bayesian inference.
- O'Hagan, A. & Forster, J. (2004). *Kendall's Advanced Theory of Statistics Vol. 2B: Bayesian Inference* (2nd edn.). London: Arnold. If you are willing to read only a single book on Bayesian statistics, this one is it. The book requires a background in mathematical statistics.
- Royall, R. M. (1997). *Statistical Evidence: A Likelihood Paradigm.* London: Chapman & Hall. This book describes the different statistical paradigms, and highlights

the deficiencies of the orthodox schools. The content can be appreciated without much background knowledge in statistics. The main disadvantage of this book is that the author is not a Bayesian. We still recommend the book, which is saying something.

1.6.2 WinBUGS texts

- Kruschke, J. K. (2010). *Doing Bayesian Data Analysis: A Tutorial Introduction with R and BUGS*. Burlington, MA: Academic Press. This is one of the first Bayesian books geared explicitly towards experimental psychologists and cognitive scientists. Kruschke explains core Bayesian concepts with concrete examples and OpenBUGS code. The book focuses on statistical models such as regression and ANOVA, and provides a Bayesian approach to data analysis in psychology, cognitive science, and empirical sciences more generally.
- Lee, S.-Y. (2007). *Structural Equation Modelling: A Bayesian Approach*. Chichester, UK: John Wiley. After reading the first few chapters from this book, you may wonder why not everybody uses WinBUGS for their structural equation modeling.
- Lunn, D., Jackson, C., Best, N., Thomas, A., & Spiegelhalter, D. (2012). *The BUGS Book: A Practical Introduction to Bayesian Analysis*. Boca Raton, FL: Chapman & Hall/CRC Press. Quoted from the publisher: "Bayesian statistical methods have become widely used for data analysis and modelling in recent years, and the BUGS software has become the most popular software for Bayesian analysis worldwide. Authored by the team that originally developed this software, The BUGS Book provides a practical introduction to this program and its use. The text presents complete coverage of all the functionalities of BUGS, including prediction, missing data, model criticism, and prior sensitivity. It also features a large number of worked examples and a wide range of applications from various disciplines."
- Ntzoufras, I. (2009). *Bayesian Modeling using WinBUGS*. Hoboken, NJ: John Wiley. Provides an accessible introduction to WinBUGS. The book also presents a variety of Bayesian modeling examples, with an emphasis on Generalized Linear Models. See `www.ruudwetzels.com` for a detailed review.
- Spiegelhalter, D., Best, N., & Lunn, D. (2003). *WinBUGS User Manual 1.4*. Cambridge, UK: MRC Biostatistic Unit. Provides an introduction to WinBUGS, including a useful tutorial and various tips and tricks for new users. The user manual has effectively been superseded by *The BUGS Book* mentioned above.

2 Getting started with WinBUGS

WITH DORA MATZKE

Throughout this book, you will use the WinBUGS (Lunn et al., 2000, 2009) software to work your way through the exercises. Although it is possible to do the exercises using the graphical user interface provided by the WinBUGS package, you can also use the Matlab or R programs to interact with WinBUGS.

In this chapter, we start by working through a concrete example using just WinBUGS. This provides an introduction to the WinBUGS interface, and the basic theoretical and practical components involved in Bayesian graphical model analysis. Completing the example will also quickly convince you that you do *not* want to rely on WinBUGS as your primary means for handling and analyzing data. It is not especially easy to use as a graphical user interface, and does not have all of the data management and visualization features needed for research.

Instead, we encourage you to choose either Matlab or R as your primary research computing environment, and use WinBUGS as an "add-on" that does the computational sampling part of analyses. Some WinBUGS interface capabilities will remain useful, especially in the exploratory stages of research. But either Matlab or R will be primary. Matlab and R code for every example in this book, as well as the scripts that implement the models in WinBUGS, are all available at `www.bayesmodels.com`.

This chapter first does a concrete example in WinBUGS, then re-works it in both Matlab and R. You should pay particular attention to the section that features your preferred research software. You will then be ready for the following chapters, which assume you are working in either Matlab or R, but understand the basics on the WinBUGS interface.

2.1 Installing WinBUGS, Matbugs, R, and R2WinBugs

2.1.1 Installing WinBUGS

WinBUGS is currently free software, and is available at `http://www.mrc-bsu.cam.ac.uk/bugs/`. Download the most recent version, including any patches, and make sure you download and apply the registration key. Some of the exercises in this book might work without the registration key, but some of them will not. You can download WinBUGS and the registration key directly from `http://www.mrc-bsu.cam.ac.uk/bugs/winbugs/contents.shtml`. A note to Windows 7 users: when you install the patch and the registration key, make sure that you have first opened WinBUGS using the "Run as administrator" option (right-click on the WinBUGS icon to make this option available); next, go to `File` → `New`, copy-paste the code (i.e., patches or key), and then select `Tools` → `Decode` → `Decode All`.

2.1.2 Installing Matlab and Matbugs

Matlab is a commercial software package, and is available at `http://www.mathworks.com/`. As far as we know, any reasonably recent version of Matlab should let you do the exercises in this book. Also, as faras we know, no toolboxes are required. To give Matlab the ability to interact with WinBUGS, download the freely available `matbugs.m` function and put it in your Matlab working directory. You can download `matbugs.m` directly from `https://code.google.com/p/matbugs`.

2.1.3 Installing R and R2WinBUGS

R is a free software package, and is available at `http://www.r-project.org/`: click "download R," choose your download location, and proceed from there. Alternatively, you can download the Windows version of R directly from `http://cran.xl-mirror.nl/`. To give R the ability to interact with WinBUGS, you have to install the `R2WinBUGS package`. To install the `R2WinBUGS` package, start R and select the `Install Package(s)` option in the `Packages` menu. Once you choose your preferred CRAN mirror, select `R2WinBUGS` in the `Packages` window and click on `OK`.

2.2 Using the applications

2.2.1 An example with the binomial distribution

We will illustrate the use of WinBUGS, Matbugs, and R2WinBUGS by means of the same simple example from Chapter 1, which involved inferring the rate of success for a binary process. A binary process is anything where there are only two possible outcomes. An inference that is often important for these sorts of processes is the underlying rate at which the process takes one value rather than the other. Inferences about the rate can be made by observing how many times the process takes each value over a number of trials.

Suppose that one of the outcomes (e.g., the number of successes) happens on k out of n trials. These are known, or observed, data. The unknown variable of interest is the rate θ at which the outcomes are produced. Assuming that what happened on one trial does not influence the other trials, the number of successes k follows a binomial distribution, $k \sim \text{Binomial}(\theta, n)$. This relationship means that by observing the k successes out of n trials, it is possible to update our knowledge

Box 2.1 **Our graphical model notation**

There is no completely agreed standard notation for representing graphical models visually. It is always the case that nodes represent variables, and the graph structure connecting them represents dependencies. And it is almost always the case that plates are used to indicate replication. Beyond that core, there are regularities and family resemblances in the approaches used by numbers of authors and fields, but not adherence to a single standard. In this book, we make distinctions between: *continuous* versus *discrete* valued variables, using circular and square nodes; *observed* and *unobserved* variables, using shaded and unshaded nodes; and *stochastic* versus *deterministic* variables, using single- and double-bordered nodes. Alternative or additional conventions are possible, and could be useful. For example, our notation does not distinguish between observed variables that are data (e.g., the decision a subject makes in an experiment) and observed variables that are known properties of an experimental design (e.g., the number of trials a subject completes). It is also possible to argue that, for deterministic variables, it is the functions that are deterministic, and so the arrows in the graph, rather than the nodes, should be double-bordered.

about the rate θ. The basic idea of Bayesian analysis is that what we know, and what we do not know, about the variables of interest is always represented by probability distributions. Data like k and n allow us to update prior distributions for the unknown variables into posterior distributions that incorporate the new information.

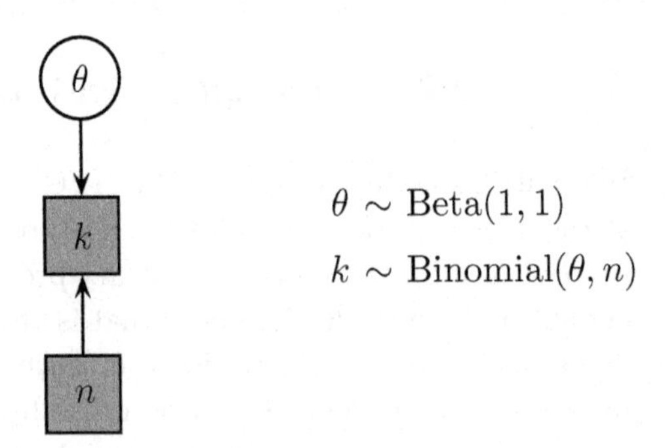

Fig. 2.1 Graphical model for inferring the rate of a binary process.

The graphical model representation of our binomial example is shown in Figure 2.1. The nodes represent all the variables that are relevant to the problem. The graph structure is used to indicate dependencies between the variables, with children depending on their parents. We use the conventions of representing unobserved variables without shading and observed variables with shading, and continuous variables with circular nodes and discrete variables with square nodes.

Thus, the observed discrete numbers of successes k and number of trials n are represented by shaded and square nodes, and the unknown continuous rate θ is represented by an unshaded and circular node. Because the number of successes k depends on the number of trials n and on the rate of success θ, the nodes representing n and θ are directed towards the node representing k. We will start with the prior assumption that all possible rates between 0 and 1 are equally likely. We will thus assume a uniform prior $\theta \sim \text{Uniform}(0,1)$, which can equivalently be written in terms of a beta distribution as $\theta \sim \text{Beta}(1,1)$.

One advantage of using the language of graphical models is that it gives a complete and interpretable representation of a Bayesian probabilistic model. Another advantage is that WinBUGS can easily implement graphical models, and its various built-in MCMC algorithms are then able to do all of the inferences automatically.

2.2.2 Using WinBUGS

WinBUGS requires the user to construct three text files: one that contains the data, one that contains the starting values for the model parameters, and one that contains the model specification. The WinBUGS model code associated with our binomial example is available at `www.bayesmodels.com`, and is shown below:

```
# Inferring a Rate
model{
   # Prior Distribution for Rate Theta
   theta ~ dbeta(1,1)
   # Observed Counts
   k ~ dbin(theta,n)
}
```

Note that the uniform prior on θ is implemented here as $\theta \sim \text{Beta}(1,1)$. An alternative specification may seem more direct, namely $\theta \sim \text{Uniform}(0,1)$, denoted `dunif(0,1)` in WinBUGS. These two distributions are mathematically equivalent, but in our experience WinBUGS has fewer computational problems with the beta distribution implementation.

Implementing the model shown in Figure 2.1, and obtaining samples from the posterior distribution of θ, can be done by following the sequence of steps outlined below. At the present stage, do not worry about some of the finer details, as these will be clarified in the remainder of this book. Right now, the best you can do is simply to follow the instructions below and start clicking away.

1. Copy the model specification text above and paste it in a text file. Save the file, for example as `Rate_1.txt`.
2. Start WinBUGS. Open your newly created model specification file by selecting the `Open` option in the `File` menu, choosing the appropriate directory, and double-clicking on the model specification file. Do not forget to select files of type "txt," or you might be searching for a long time. Now check the syntax

of the model specification code by selecting the **Specification** option in the **Model** menu. Once the **Specification Tool** window is opened, as shown in Figure 2.2, highlight the word "model" at the beginning of the code and click on **check model**. If the model is syntactically correct and all parameters are given priors, the message "model is syntactically correct" will appear in the status bar all the way in the bottom left corner of the WinBUGS window. (But beware: the letters are very small and difficult to see.)

3. Create a text file that contains the data. The content of the file should look like this:

```
list(
k = 5,
n = 10
)
```

Save the file, for example as `Data.Rate_1.txt`.

4. Open the data file and load the data. To open the data file, select the **Open** option in the **File** menu, select the appropriate directory, and double-click on the data file. To load the data, highlight the word "list" at the beginning of the data file and click on **load data** in the **Specification Tool** window, as shown in Figure 2.2. If the data are successfully loaded, the message "data loaded" will appear in the status bar.
5. Set the number of chains. Each chain is an independent run of the same model with the same data, although you can set different starting values for each chain.[1] Considering multiple chains provides a key test of convergence. In our binomial example, we will run two chains. To set the number of chains, type "2" in the field labelled **num of chains** in the **Specification Tool** window, shown in Figure 2.2.
6. Compile the model. To compile the model, click on **compile** in the **Specification Tool** window, shown in Figure 2.2. If the model is successfully compiled, the message "model compiled" will appear in the status bar.
7. Create a text file that contains the starting values of the unobserved variables (i.e., just the parameter θ for this model).[2] The content of the file should look like this:

```
list(
theta = 0.1
)
list(
theta = 0.9
)
```

[1] Running multiple chains is the best and easiest way to ensure WinBUGS uses different random number sequences in sampling. Doing a single-chain analysis multiple times can produce the same results because the random number sequence is identical.

[2] If you do not specify starting values yourself, WinBUGS will create them for you automatically. These automatic starting values are based on the prior and may occasionally result in numerical instability and program crashes. It is therefore safer to assign a starting value for all unobserved variables, and especially for variables at nodes "at the top" of the graphical model, which have no parents.

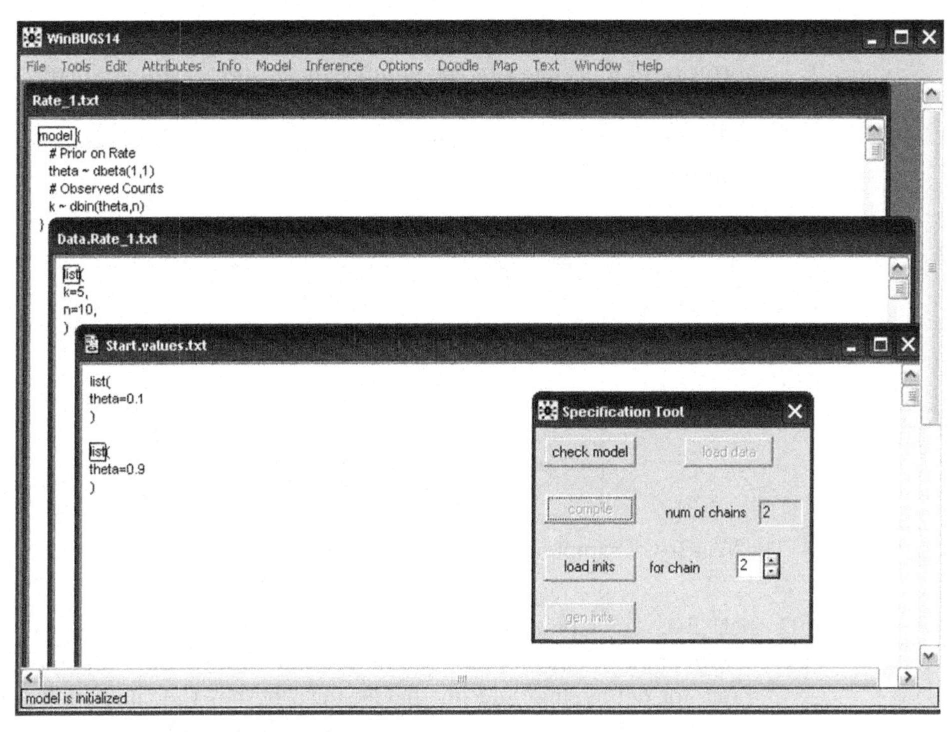

Fig. 2.2 The WinBUGS model specification tool.

Note that there are two initial values, one for each chain. Save the file, for example as `Start.values.txt`.

8. Open the file that contains the starting values by selecting the `Open` option in the `File` menu, selecting the appropriate directory, and double-clicking on the file. To load the starting value of θ for the first chain, highlight the word "list" at the beginning of the file and click on `load inits` in the `Specification Tool` window, shown in Figure 2.2. The status bar will now display the message "chain initialized but other chain(s) contain uninitialized variables." To load the starting value for the second chain, highlight the second "list" command and click on `load inits` once again. If all starting values are successfully loaded, the message "model is initialized" will appear in the status bar.
9. Set monitors to store the sampled values of the parameters of interest. To set a monitor for θ, select the `Samples` option from the `Inference` menu. Once the `Sample Monitor Tool` window, shown in Figure 2.3, is opened, type "theta" in the field labeled `node` and click on `set`.
10. Specify the number of samples you want to record. To do this, you first have to specify the total number of samples you want to draw from the posterior of θ, and the number of burn-in samples that you want to discard at the beginning of a sampling run. The number of recorded samples equals the total number of samples minus the number of burn-in samples. In our binomial example, we will

not discard any of the samples and will set out to obtain 20,000 samples from the posterior of θ. To specify the number of recorded samples, type "1" in the field labeled **beg** (i.e., WinBUGS will start recording from the first sample) and type "20000" in the field labeled **end** in the `Sample Monitor Tool` window, shown in Figure 2.3.

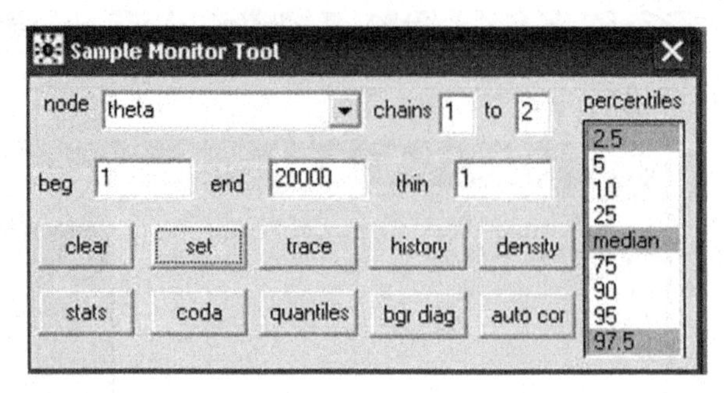

Fig. 2.3 The WinBUGS sample monitor tool.

11. Set "live" trace plots of the unobserved parameters of interest. WinBUGS allows you to monitor the sampling run in real-time. This can be useful on long sampling runs, for debugging, and for diagnosing whether the chains have converged. To set a "live" trace plot of θ, click on **trace** in the `Sample Monitor Tool` window, shown in Figure 2.3, and wait for an empty plot to appear on the screen. Once WinBUGS starts to sample from the posterior, the trace plot of θ will appear live on the screen.
12. Specify the total number of samples that you want to draw from the posterior. This is done by selecting the `Update` option from the `Model` menu. Once the `Update Tool` window, as in Figure 2.4, is opened, type "20000" in the field labeled **updates**. Typically, the number you enter in the `Update Tool` window will correspond to the number you entered in the **end** field of the `Sample Monitor Tool`.
13. Specify how many samples should be drawn between the recorded samples. You can, for example, specify that only every second drawn sample should be recorded. This ability to "thin" a chain is important when successive samples are not independent but autocorrelated. In our binomial example, we will record every sample that is drawn from the posterior of θ. To specify this, type "1" in the field labeled **thin** in the `Update Tool` window, shown in Figure 2.4, or in the `Sample Monitor Tool` window, shown in Figure 2.3. To record only every 10th sample, the **thin** field needs to be set to 10.
14. Specify the number of samples after which WinBUGS should refresh its display. To this end, type "100" in the field labeled **refresh** in the `Update Tool` window, shown in Figure 2.4.
15. Sample from the posterior. To sample from the posterior of θ, click on **update** in the `Update Tool` window, shown in Figure 2.4. During sampling, the message

"model is updating" will appear in the status bar. Once the sampling is finished, the message "updates took x s" will appear in the status bar.

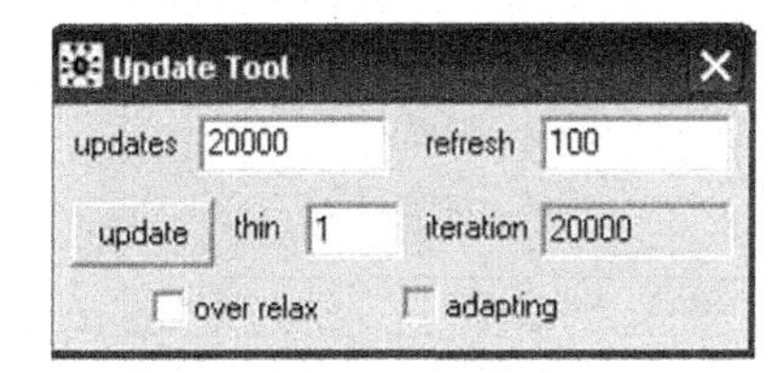

Fig. 2.4 Update Tool.

16. Specify the output format. WinBUGS can produce two types of output; it can open a new window for each new piece of output, or it can paste all output into a single log file. To specify the output format for our binomial example, select `Output options` from the `Options` menu, and click on `log` in the `Output options` window.
17. Obtain summary statistics of the posterior distribution. To request summary statistics based on the sampled values of θ, select the `Samples` option in the `Inference` menu, and click on `stats` in the `Sample Monitor Tool` window, shown in Figure 2.3. WinBUGS will paste a table reporting various summary statistics for θ in the log file.
18. Plot the posterior distribution. To plot the posterior distribution of θ, click on `density` in the `Sample Monitor Tool` window, shown in Figure 2.3. WinBUGS will paste the "kernel density" of the posterior distribution of θ in the log file.[3]

Figure 2.5 shows the log file that contains the results for our binomial example. The first five lines of the log file document the steps taken to specify and initialize the model. The first output item is the `Dynamic trace` plot that allows the θ variable to be monitored during sampling, and is useful for diagnosing whether the chains have reached convergence. In this case, we can be reasonably confident that convergence has been achieved because the two chains, shown in different colors, are overlapping one another.[4] The second output item is the `Node statistics` table that presents the summary statistics for θ. Among other things, the table shows the mean, the standard deviation, and the median of the sampled values of θ. The last output item is the `Kernel density` plot that shows the posterior distribution of θ.

How did WinBUGS produce the results in Figure 2.5? The model specification file implemented the graphical model from Figure 2.1, saying that there is a rate θ with a uniform prior, that generates k successes out of n observations. The data file supplied the observed data, setting $k = 5$ and $n = 10$. WinBUGS then sampled

[3] A kernel density is a fancy smoothed histogram. Here, it is a smoothed histogram for the sampled values of θ.

[4] Note that the `Dynamic trace` plot only shows 200 samples. To have the entire time series of sampled values plotted in the log file, click on `history` in the `Sample Monitor Tool` window.

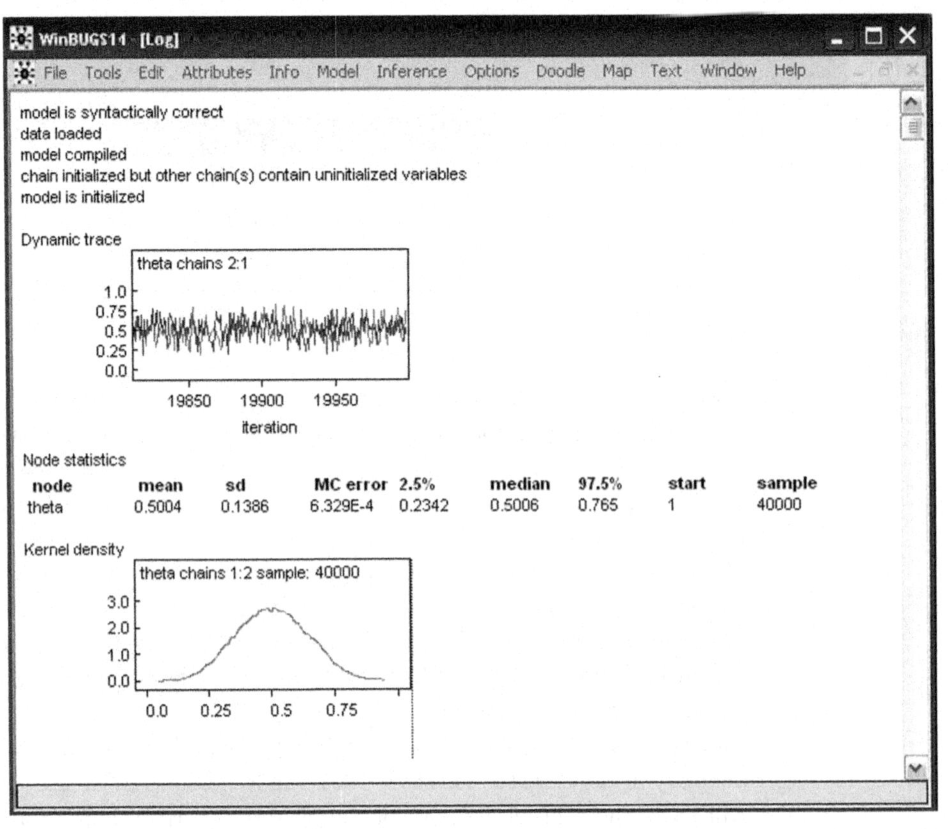

Fig. 2.5 Example of an output log file.

from the posterior of the unobserved variable θ. "Sampling" means drawing a set of values, so that the relative probability that any particular value will be sampled is proportional to the density of the posterior distribution at that value. For this example, the posterior samples for θ are a sequence of numbers like 0.5006, 0.7678, 0.3283, 0.3775, 0.4126, ... A histogram of these values is an approximation to the posterior distribution of θ.

Error messages

If the syntax of your model file is incorrect or the data and starting values are incompatible with your model specification, WinBUGS will balk and produce an error message. Error messages can provide useful information for debugging your WinBUGS code.[5] The error messages are displayed in the bottom left corner of the status bar, in very small letters.

Suppose, for example, that you mistakenly use the "assign" operator (`<-`) to specify the distribution of the prior on the rate parameter θ and the distribution of the observed data `k`:

[5] Although nobody ever accused WinBUGS of being user-friendly in this regard. Many error messages seem to have been written by the same people who did the Dead Sea Scrolls.

Box 2.2 Do I need or want to understand computational sampling?

Some people find the idea that WinBUGS looks after sampling, and that there is no need to understand the computational routines involved in detail, to be a relief. Others find it deeply disturbing. For the disturbed, there are many Bayesian texts that give detailed accounts of Bayesian inference using computational sampling. Start with the summary for cognitive scientists presented in Chapter 7 from Kruschke (2010a). Continue with the tutorial-style overview in Andrieu et al. (2003) or the relevant chapters in the excellent book by MacKay (2003), which is freely available on the Web, and move on to the more technical references such as Gilks et al. (1996), Ntzoufras (2009), and Gamerman and Lopes (2006). You can also browse the internet for more information; for example, there is an instructive applet at `http://www.lbreyer.com/classic.html`, and an excellent YouTube tutorial at `http://www.youtube.com/watch?v=4gNpgSPal_8`.

```
model{
  #Prior Distribution for Rate Theta
  theta <- dbeta(1,1)
  #Observed Counts
  k <- dbin(theta,n)
}
```

As WinBUGS requires you to use the tilde symbol "~" to denote the distributions of the prior and the data, it will produce the following error message: `unknown type of logical function`, as shown in Figure 2.6. As another example, suppose that you mistype the distribution of the observed counts `k`, and you mistakenly specify the distribution of `k` as follows:

```
k ~ dbon(theta,n)
```

WinBUGS will not recognize `dbon` as an existing probability distribution, and will produce the following error message: `unknown type of probability density`, as shown in Figure 2.7.[6]

With respect to errors in the data file, suppose that your data file contains the following data: `k` = -5 and `n` = 10. Note, however, that `k` is the number of successes in the 10 trials and it is specified to be binomially distributed. WinBUGS therefore expects the value of `k` to lie between 0 and `n` and it will produce the following error message: `value of binomial k must be between zero and order of k`.

[6] On Windows machines, the error message is accompanied by a penetrating "system beep." After experiencing a few such system beeps you will want to turn them off. Browse the web for information on how to do this, or go straight to `http://www.howtogeek.com/howto/windows/turn-off-the-annoying-windows-xp-system-beeps/`. The people sitting next to you will be grateful too.

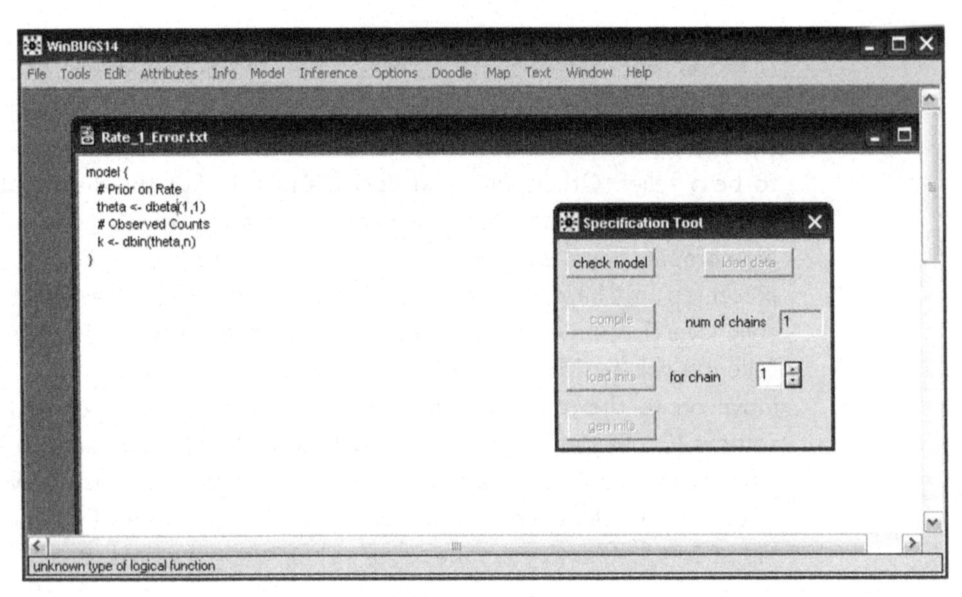

Fig. 2.6 WinBUGS error message as a result of incorrect logical operators. Note the small letters in the bottom left corner of the status bar.

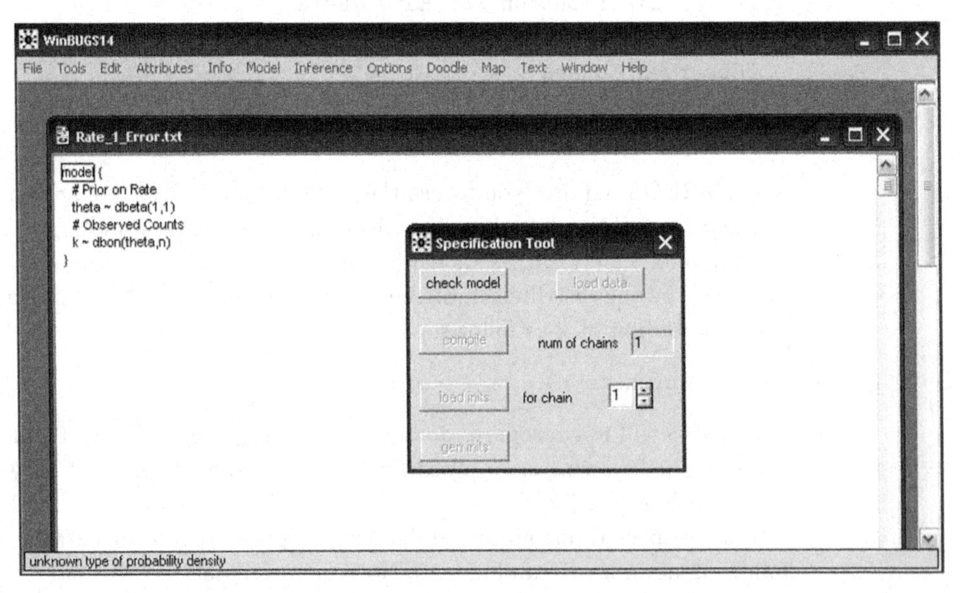

Fig. 2.7 WinBUGS error message as a result of a mis-specified probability density. Note the small letters in the bottom left corner of the status bar.

Finally, with respect to erroneous starting values, suppose that you chose 1.5 as the starting value of θ for the second chain. Because θ is the *probability* of getting 5 successes in 10 trials, WinBUGS expects the starting value for θ to lie between 0 and

Box 2.3 Changing the sampler

WinBUGS uses a suite of samplers, each fine-tuned to a particular class of statistical problems. Occasionally it may be worth the effort to change the default settings and edit the Updater/Rsrc/Methods.odc file. Any such editing should be done with care, and only after you have made a copy of the original Methods.odc file that contains the default settings. The advantage of changing the sampler is that it may circumvent traps or crashes. For example, the WinBUGS manual mentions that problems with the adaptive rejection sampler DFreeARS can sometimes be solved by replacing, for the log concave class, the method UpdaterDFreeARS by UpdaterSlice. Note for Windows 7 users: you may not be able to save any changes to files in the Updater/Rsrc directory. Work-around: copy the file to your desktop, edit it, save it, and copy it back to the Updater/Rsrc directory.

1. Therefore, specifying a value such as 1.5 produces the following error message: `value of proportion of binomial k must be between zero and one.`

2.2.3 Using Matbugs

We will use the `matbugs` function to call the WinBUGS software from within Matlab, and to return the results of the WinBUGS sampling to a Matlab variable for further analysis. The code we are using to do this is shown below:

```
% Data
k = 5;
n = 10;

% WinBUGS Parameters
nchains = 2; % How Many Chains?
nburnin = 0; % How Many Burn in Samples?
nsamples = 2e4;  % How Many Recorded Samples?
nthin = 1; % How Often is a Sample Recorded?

% Assign Matlab Variables to the Observed WinBUGS Nodes
datastruct = struct('k',k,'n',n);

% Initialize Unobserved Variables
start.theta= [0.1 0.9];

for i=1:nchains
    S.theta = start.theta(i); % An Intial Value for the Success Rate
    init0(i) = S;
end

% Use WinBUGS to Sample
[samples, stats] = matbugs(datastruct, ...
    fullfile(pwd, 'Rate_1.txt'),
```

```
'init', init0, 'view', 1, ...
'nChains', nchains, 'nburnin', nburnin, ...
'nsamples', nsamples, 'thin', nthin, ...
'DICstatus', 0, 'refreshrate',100, ...
'monitorParams', {'theta'}, ...
'Bugdir', 'C:/Program Files/WinBUGS14');
```

Some of the options in the Matbugs function control software input and output:

- `datastruct` contains the data that you want to pass from Matlab to WinBUGS.
- `fullfile` gives the name of the text file that contains the WinBUGS scripting of your graphical model (i.e., the model specification file).
- `view` controls the termination of WinBUGS. If `view` is set to 0, WinBUGS is closed automatically at the end of the sampling. If `view` is set to 1, WinBUGS remains open and it pastes the results of the sampling run in a log output file. To be able to inspect the results in WinBUGS, maximize the log output file and scroll up to the top of the page. Note that if you subsequently want WinBUGS to return the results to Matlab, you first have to close WinBUGS.
- `refreshrate` gives the number of samples after which WinBUGS should refresh its display.
- `monitorParams` gives the list of variables that will be monitored and returned to Matlab in the `samples` variable.
- `Bugdir` gives the location of the WinBUGS software.

Other options define the values for the computational sampling parameters:

- `init` gives the starting values for the unobserved variables.
- `nChains` gives the number of chains.
- `nburnin` gives the number of burn-in samples.
- `nsamples` gives the number of recorded samples that will be drawn from the posterior.
- `thin` gives the number of drawn samples between those that are recorded.
- `DICstatus` gives an option to calculate the Deviance Information Criterion (DIC) statistic (Spiegelhalter, Best, Carlin, & Van Der Linde, 2002). The DIC statistic is intended to be used for model selection, but is not universally accepted theoretically among Bayesian statisticians. If `DICstatus` is set to 0, the DIC statistic will not be calculated. If it is set to 1, WinBUGS will calculate the DIC statistic.

How did the WinBUGS script and Matlab work together to produce the posterior samples of θ? The WinBUGS model specification script defined the graphical model from Figure 2.1. The Matlab code supplied the observed data and the starting values for θ, and called WinBUGS. WinBUGS then sampled from the posterior of θ and returned the sampled values in the Matlab variable `samples.theta`. This flow of events is illustrated in Figure 2.9. You can plot the histogram of these sampled values using Matlab, in the way demonstrated in the script `Rate_1.m`. It should look something like the jagged line in Figure 2.8. Because the probability of any value

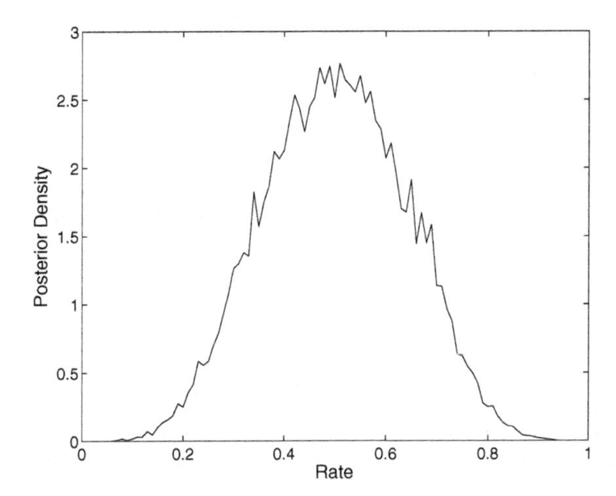

Fig. 2.8 Approximate posterior distribution of rate θ for $k = 5$ successes out of $n = 10$ trials, based on 20,000 posterior samples.

appearing in the sequence of posterior samples is decided by its relative posterior probability, the histogram is an approximation to the posterior distribution of θ.

Besides the sequence of posterior samples, WinBUGS also returns some useful summary statistics to Matlab. The variable `stats.mean` gives the mean of the posterior samples for each unobserved variable, which approximates its posterior expectation. This can often (but not always, as later exercises explore) be a useful point-estimate summary of all the information in the full posterior distribution. Similarly, `stats.std` gives the standard deviation of the posterior samples for each unobserved variable.

Finally, WinBUGS also returns the so-called $\hat{R}$ statistic in the `stats.Rhat` variable. This is a statistic about the sampling procedure itself, not about the posterior distribution. The $\hat{R}$ statistic is proposed by Brooks and Gelman (1998) and it gives information about convergence. The basic idea is to run two or more chains and measure the ratio of within-to-between-chain variance. If this ratio is close to 1, the independent sampling sequences are probably giving the same answer, and there is reason to trust the results.

Exercise

Exercise 2.2.1 Re-read the section on `view`. The Matlab code above specifies `view=1`. What does this do? Change the code to `view=0`. What has changed?

2.2.4 Using R2WinBUGS

We will use the `bugs()` function in the R2WinBUGS package to call the WinBUGS software from within R, and to return the results of the WinBUGS sampling to an

R variable for further analysis. Note for Windows 7 users: in order for the samples to be returned to R successfully, you may need to run R "as administrator" (right-click on the R icon to reveal this option). The R code we are using to obtain the WinBUGS samples is as follows:

```
setwd("D:/WinBUGS_Book/R_codes") #Set working directory, adjust as needed
library(R2WinBUGS) #Load the R2WinBUGS package
bugsdir <- "C:/Program Files/WinBUGS14" #Set WinBUGS directory, adjust as needed

k <- 5
n <- 10

data <- list("k", "n")
myinits <- list(
   list(theta = 0.1), #chain 1 starting value
   list(theta = 0.9)) #chain 2 starting value

parameters <- c("theta")

samples <- bugs(data, inits=myinits, parameters,
                model.file ="Rate_1.txt",
                n.chains=2, n.iter=20000, n.burnin=1, n.thin=1,
                DIC=T, bugs.directory=bugsdir,
                codaPkg=F, debug=F)
```

Note that lines 1 and 3 (i.e., the `setwd` line and the `bugsdir` line) specify the working directory and the WinBUGS directory, but only for the computer that runs the code. If you want to run the code on your own computer you need to modify these lines to match your setup.[7]

Some of the above options control software input and output:

- `data` contains the data that you want to pass from R to WinBUGS.
- `parameters` gives the list of variables that will be monitored and returned to R in the `samples` variable.
- `model.file` gives the name of the text file that contains the WinBUGS scripting of your graphical model (i.e., the model specification file). Avoid using non-alphanumeric characters (e.g., "&" and "*") in the directory and file names. Also, make sure that the name of the directory that contains the model file is not too long, otherwise WinBUGS will generate the following error message: `incompatible copy`. If WinBUGS fails to locate a correctly specified model file, try to include the entire path in the `model.file` argument.
- `bugs.directory` gives the location of the WinBUGS software.
- `codaPkg` controls the content of the variable that is returned from WinBUGS. If `codaPkg=F` (i.e., `codaPkg` is set to FALSE), WinBUGS returns a variable that contains the results of the sampling run. If `codaPkg=T` (i.e., `codaPkg` is set to TRUE), WinBUGS returns a variable that contains the file names of the

[7] When the code does not work immediately, check whether you have changed the directories correctly. The working directory should contain the model file, in this case `Rate_1.txt`, and the `bugsdir` variable should refer to the directory that contains the `WinBUGS14.exe` file.

WinBUGS outputs and the corresponding paths. You can access these output files by means of the R function `read.bugs()`.

- `debug` controls the termination of WinBUGS. If `debug` is set to FALSE, WinBUGS is closed automatically at the end of the sampling. If `debug` is set to TRUE, WinBUGS remains open and it pastes the results of the sampling run in a log output file. To be able to inspect the results in WinBUGS, maximize the log output file and scroll up to the top of the page. Note that if you subsequently want WinBUGS to return the results in the R `samples` variable, you first have to close WinBUGS. In general, you will not be able to use R again until after you terminate WinBUGS.

The other options define the values for the computational sampling parameters:

- `inits` assigns starting values to the unobserved variables. If you want WinBUGS to choose these starting values for you, replace `inits=myinits` in the call to bugs with `inits=NULL`.
- `n.chains` gives the number of chains.
- `n.iter` gives the number of samples that will be drawn from the posterior.
- `n.burnin` gives the number of burn-in samples.
- `n.thin` gives the number of drawn samples between those that are recorded.
- `DIC` gives an option to calculate the Deviance Information Criterion (DIC) statistic (Spiegelhalter et al., 2002). The DIC statistic is intended to be used for model selection, but is not universally accepted theoretically among Bayesian statisticians. If `DIC` is set to FALSE, the DIC statistic will not be calculated. If it is set to TRUE, WinBUGS will calculate the DIC statistic.[8]

WinBUGS returns the sampled values of θ in the R variable `samples`. You can access these values by typing `samples$sims.array` or `samples$sims.list`. The flow of events is illustrated in Figure 2.9.

You can use R to plot the histogram of sampled values of θ, as is demonstrated in the script `Rate_1.R`. In addition to the sequence of posterior samples, WinBUGS also returns to R some useful summary statistics. These summary statistics can be obtained by typing `samples` at the R prompt. When you run two or more chains, the `samples` command also provides the $\hat{R}$ statistic, introduced by Brooks and Gelman (1998). The $\hat{R}$ statistic provides information about the convergence of the sampling procedure, not about the posterior distribution. The basic idea is to run two or more chains and measure the ratio of within-to-between-chain variance. If this ratio is close to 1, the independent sampling sequences are probably giving the same answer, and there is reason to trust the results.

[8] For some reason, setting DIC equal to FALSE can lead to problems in the communication between R and WinBUGS. It is safest to set DIC equal to TRUE, even when you are not interested in the DIC.

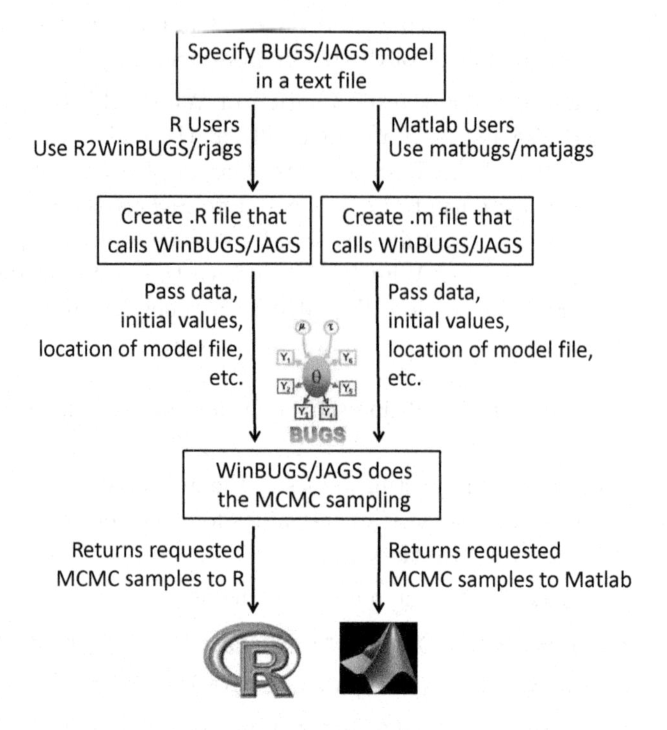

Fig. 2.9 Flowchart that illustrates the interaction between WinBUGS and R (left stream) or Matlab (right stream). JAGS is a program that is very similar to WinBUGS, described in the section on OpenBUGS and JAGS.

Exercise

Exercise 2.2.2 Re-read the section on `debug`. The R code above specifies `debug = F`; what does this do? Change the code to `debug = T`; what has changed?

2.3 Online help, other software, and useful URLs

2.3.1 Online help for WinBUGS

- The BUGS Project webpage `http://www.mrc-bsu.cam.ac.uk/bugs/weblinks/webresource.shtml` provides useful links to various articles, tutorial materials, and lecture notes about Bayesian modeling and the WinBUGS software.
- The BUGS discussion list `https://www.jiscmail.ac.uk/cgi-bin/webadmin?A0=bugs` is an online forum where WinBUGS users can exchange tips, ask questions, and share worked examples.

2.3.2 For Mac users

You can run WinBUGS on Macs using emulators, such as Darwine. As far as we know, you need a Dual Core Intel-based Mac and the latest stable version of Darwine to be able to use R2WinBUGS. Nevertheless, running WinBUGS on a Mac is not ideal. Mac users are encouraged to use JAGS instead. At the time of writing, the information below was useful for running WinBUGS on a Mac:

- The Darwine emulator is available at `www.kronenberg.org/darwine/`.
- The R2WinBUGS reference manual on the R-project webpage `cran.r-project.org/web/packages/R2WinBUGS/index.html` provides instructions on how to run R2winBUGS on Macs.
- Further information for running R2WinBUGS on Macs is available at `ggorjan.blogspot.com/2008/10/runnning-r2winbugs-on-mac.html` and `idiom.ucsd.edu/~rlevy/winbugsonmacosx.pdf`.
- Further information for running WinBUGS on Macs using a Matlab or R interface is available at `http://www.helensteingroever.com` and `www.ruudwetzels.com/macbugs`.

2.3.3 For Linux users

You can run WinBUGS under Linux using emulators, such as Wine and CrossOver.

- The BUGS Project webpage provides useful links to various examples on how to run WinBUGS under Linux `www.mrc-bsu.cam.ac.uk/bugs/faqs/contents.shtml` and how to run WinBUGS using a Matlab interface `www.mrc-bsu.cam.ac.uk/bugs/winbugs/remote14.shtml`.
- The R2WinBUGS reference manual on the R-project webpage `cran.r-project.org/web/packages/R2WinBUGS/index.html` provides instructions on how to run R2WinBUGS under Linux.

2.3.4 OpenBUGS, Stan, and JAGS

This book is designed primarily to work with WinBUGS. There are, however, alternative programs for generating MCMC samples from graphical models. Both OpenBUGS, Stan (Stan Development Team, 2013), and JAGS (Plummer, 2003) may be particularly attractive for Mac and Linux users, since they raise fewer issues than WinBUGS to install and run. The model code for OpenBUGS, Stan, and JAGS is very similar to WinBUGS, so that the transition from one program to the other is generally easy. An effort has been made to make most of the examples in this book compatible with JAGS. Often, in our experience, sampling is much faster in JAGS than it is in WinBUGS.

- OpenBUGS is available from `http://www.openbugs.info/w/`.
- Stan is available from `http://mc-stan.org/`.

- JAGS is available from `http://mcmc-jags.sourceforge.net/`.
- To give R the ability to interact with JAGS, you have to install the `rjags` package, and, optionally, the `R2jags` package. To ensure that you install the latest version of the `rjags` package, the safest procedure is to first Google the terms `rjags CRAN`, go to a website such as `http://cran.r-project.org/web/packages/rjags/index.html`, and—when using Windows—download the package zip file. Then start R, go to the `Packages` menu, choose `Install package(s) from local zip file...`, and select the package zip file you just downloaded. To check whether the installation was successful, type `library(rjags)` at the R prompt. To install the `R2jags` package, you can use the standard installation procedure: start R and select the `Install Package(s)` option in the `Packages` menu. After choosing your preferred CRAN mirror, select `R2jags` in the `Packages` window and click on `OK`.
- To give Matlab the ability to interact with JAGS, download the freely available `matjags.m` function and put it in your Matlab working directory. You can download `matjags.m` directly from `http://psiexp.ss.uci.edu/research/programs_data/jags/`.

PART II

PARAMETER ESTIMATION

Today's posterior is tomorrow's prior.

Lindley, 2000, p. 301

3 Inferences with binomials

3.1 Inferring a rate

Our first problem completes the introductory example in Chapter 2, and involves inferring the underlying success rate for a binary process. The graphical model is shown again in Figure 3.1. Recall that shaded nodes indicate known values, while unshaded nodes represent unknown values, and that circular nodes correspond to continuous values, while square nodes correspond to discrete values.

The goal of inference in the graphical model is to determine the posterior distribution of the rate θ, having observed k successes from n trials. The analysis starts with the prior assumption that all possible rates between 0 and 1 are equally likely. This corresponds to the uniform prior distribution $\theta \sim \text{Uniform}(0, 1)$, which can equivalently be written in terms of a beta distribution as $\theta \sim \text{Beta}(1, 1)$.

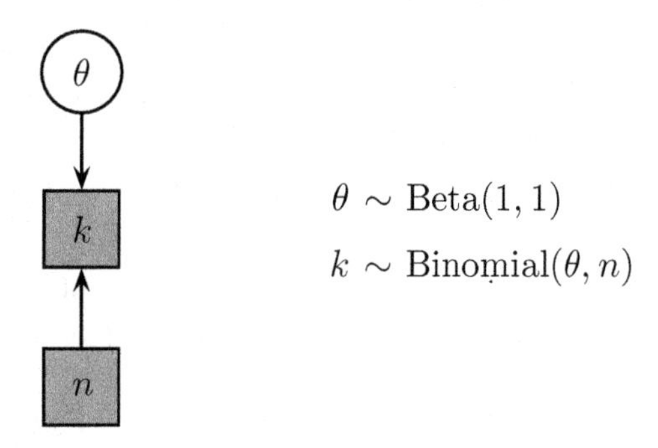

Fig. 3.1 Graphical model for inferring the rate θ of a binary process.

The script Rate_1.txt implements the graphical model in WinBUGS. The script is available at www.bayesmodels.com and is shown below:

```
# Inferring a Rate
model{
   # Prior Distribution for Rate Theta
   theta ~ dbeta(1,1)
   # Observed Counts
   k ~ dbin(theta,n)
}
```

The code Rate_1.m for Matlab or Rate_1.R for R, both available at www.bayesmodels.com, sets $k = 5$ and $n = 10$ and calls WinBUGS to sample from the graphical model. WinBUGS then returns to Matlab or R the posterior samples

Box 3.1 **Beta distributions as conjugate priors**

One of the nice properties of using the $\theta \sim \text{Beta}(\alpha, \beta)$ prior distribution for a rate θ is that it has a natural interpretation. The α and β values can be thought of as counts of, respectively, "prior successes" and "prior failures." This means that using a $\theta \sim \text{Beta}(3, 1)$ prior corresponds to having the prior information that 4 previous observations have been made, and 3 of them were successes. Or, more elaborately, starting with a $\theta \sim \text{Beta}(3, 1)$ is the same as starting with a $\theta \sim \text{Beta}(1, 1)$, and then seeing data giving two more successes (i.e., the posterior distribution in the second scenario will be the same as the prior distribution in the first). As always in Bayesian analysis, inference starts with prior information, and updates that information—by changing the probability distribution representing the uncertain information—as more information becomes available. When a type of likelihood function (in this case, the binomial) does not change the type of distribution (in this case, the beta) going from the prior to the posterior, they are said to have a "conjugate" relationship. This property is valued a lot in analytic approaches to Bayesian inference, because it makes for tractable calculations. It is not so important in the computational approaches emphasized in this book, because sampling methods can handle much more general relationships between parameter distributions and likelihood functions. But conjugacy is still useful in computational approaches because of the natural semantics it gives in setting prior distributions.

from θ. The Matlab or R code also plots the posterior distribution of the rate θ. A histogram of the samples looks something like the jagged line in Figure 3.2.

Exercises

Exercise 3.1.1 Carefully consider the posterior distribution for θ given $k = 5$ successes out of $n = 10$ trials. Based on a visual impression, what is your estimate of the probability that the rate θ is higher than 0.4 but smaller than 0.6? How did you arrive at your estimate?

Exercise 3.1.2 Consider again the posterior distribution for θ given $k = 5$ successes out of $n = 10$ trials. Based on a visual impression, what is your estimate of how much more likely it is that the rate θ is equal to 0.5 rather than 0.7? How did you arrive at your estimate?

Exercise 3.1.3 Alter the data to $k = 50$ and $n = 100$, and compare the posterior for the rate θ to the original with $k = 5$ and $n = 10$.

Exercise 3.1.4 For both the $k = 50$, $n = 100$ and $k = 5$, $n = 10$ cases just considered, re-run the analyses with many more samples (e.g., 10 times as many) by changing the `nsamples` variable in Matlab, or the `n.iter` variable

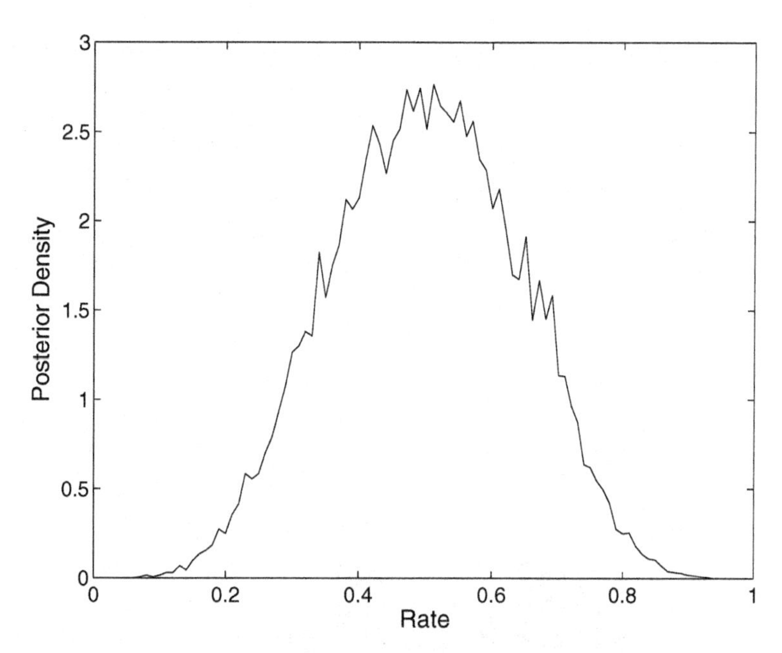

Fig. 3.2 Posterior distribution of rate θ for $k = 5$ successes out of $n = 10$ trials.

in R. This will take some time, but there is an important point to understand. What controls the width of the posterior distribution (i.e., the expression of uncertainty in the rate parameter θ)? What controls the quality of the approximation of the posterior (i.e., the smoothness of the histograms in the figures)?

Exercise 3.1.5 Alter the data to $k = 99$ and $n = 100$, and comment on the shape of the posterior for the rate θ.

Exercise 3.1.6 Alter the data to $k = 0$ and $n = 1$, and comment on what this demonstrates about the Bayesian approach.

3.2 Difference between two rates

Now suppose that we have two different processes, producing k_1 and k_2 successes out of n_1 and n_2 trials, respectively. First, we will make the assumption that the underlying rates are different, so they correspond to different latent variables θ_1 and θ_2. Our interest is in the values of these rates, as estimated from the data, and in the difference $\delta = \theta_1 - \theta_2$ between the rates.

The graphical model representation for this problem is shown in Figure 3.3. The new notation is that the deterministic variable δ is shown by a double-bordered node. A deterministic variable is one that is defined in terms of other variables, and inherits its distribution from them. Computationally, deterministic nodes are

Box 3.2 Interpreting distributions

Since the essence of Bayesian inference is using probability distributions to represent uncertainty, it is important to be able to interpret probability mass functions and probability density functions. Probability mass functions are for discrete variables, which take a finite number of values, while probability density functions are for continuous variables, which take infinitely many values.

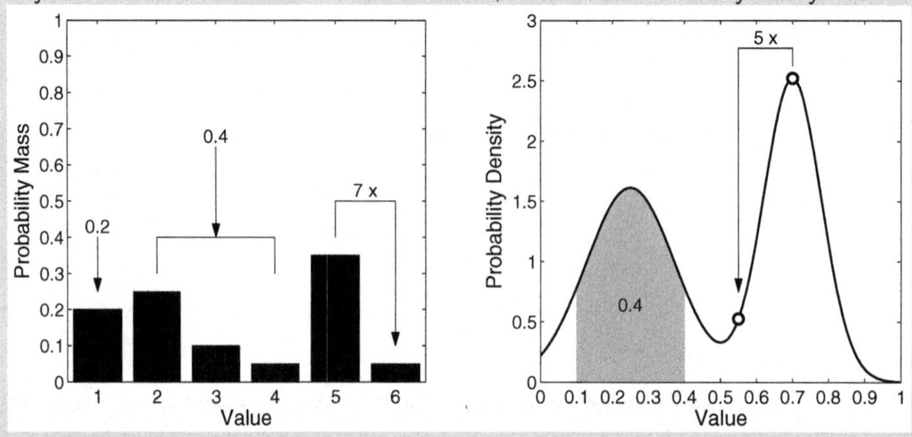

The panel on the left shows a probability mass function for a variable with 6 values. Each bar represents the probability of that value, so that, for example, the probability of the value 1 is 0.2. The probability of a range of values is the sum of their probabilities, so that the probability that the value is between 2 and 4 inclusive is 0.4. The ratio between the probabilities determines how much more likely one value is than another, so that the value 5 is 7 times more likely than the value 6. And, the sum of all of the probabilities (i.e., the height of the bars stacked on each other) is always 1. The panel on the right shows a probability density function for a variable that is between 0 and 1. The total area under the curve is always 1, which means the densities of individual points can (and often do) exceed 1. They cannot be interpreted as probabilities. But the probability of a range of values can be determined by the relevant area under the curve. In the right panel, the probability that the value is between 0.1 and 0.4 is 0.4. And ratios can still be interpreted in a relative way, so it is 5 times more likely the value is 0.7 than 0.55.

unnecessary—all inference could be done with the variables that define them—but they are often conceptually very useful to include, to communicate the meaning of a model.

The script `Rate_2.txt` implements the graphical model in WinBUGS:

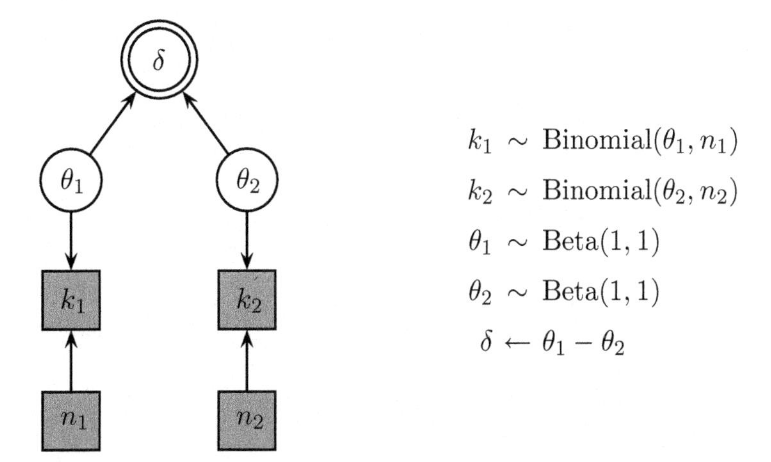

Fig. 3.3 Graphical model for inferring the difference, $\delta = \theta_1 - \theta_2$, in the rates of two binary processes.

```
# Difference Between Two Rates
model{
   # Observed Counts
   k1 ~ dbin(theta1,n1)
   k2 ~ dbin(theta2,n2)
   # Prior on Rates
   theta1 ~ dbeta(1,1)
   theta2 ~ dbeta(1,1)
   # Difference Between Rates
   delta <- theta1-theta2
}
```

The code `Rate_2.m` or `Rate_2.R` sets $k_1 = 5$, $k_2 = 7$, $n_1 = n_2 = 10$, and then calls WinBUGS to sample from the graphical model. WinBUGS returns to Matlab or R the posterior samples from θ_1, θ_2, and δ. If the main research question is how different the rates are, then δ is the most relevant variable, and its posterior distribution is shown in Figure 3.4.

There are many ways the full information in the posterior distribution of δ might usefully be summarized. The Matlab or R code produces a set of these from the posterior samples, including:

- The mean value, which approximates the expectation of the posterior. This summary tries to pick a single value close to the truth, with bigger deviations from the truth being punished more heavily. Statistically, it corresponds to the point estimate under quadratic loss.
- The value with maximum density in the posterior samples, approximating the posterior mode. This summary aims to pick the single most likely value. This is known as the maximum a posteriori (MAP) estimate, and is the same as the maximum likelihood estimate (MLE) for "flat" priors. Statistically, it corresponds to the point estimate under zero–one loss. Estimating the mode requires

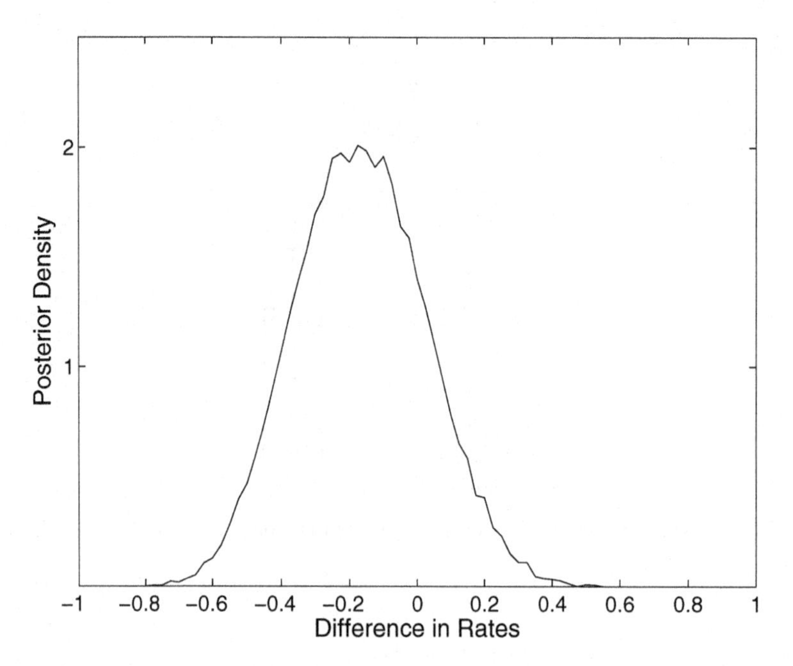

Fig. 3.4 Posterior distribution of the difference between two rates $\delta = \theta_1 - \theta_2$.

evaluating the likelihood function at each posterior sample, and so requires a bit more post-processing work in Matlab or R.

- The median value, which is the value that separates the highest 50% of the posterior distribution from the lowest 50%, and so finds the middle-most value. Statistically, it corresponds to the point estimate under linear loss.
- The 95% credible interval. This gives the upper and lower values between which 95% of samples fall. Thus, it approximates the bounds on the posterior distribution that contain 95% of the posterior density. The Matlab or R code can be modified to produce credible intervals for criteria other than 95%.

For the current problem, the mean of δ estimated from the returned samples is approximately -0.17, the mode is approximately -0.17, the median is approximately -0.17, and the 95% credible interval is approximately $[-0.52, 0.21]$.

Exercises

Exercise 3.2.1 Compare the data sets $k_1 = 8$, $n_1 = 10$, $k_2 = 7$, $n_2 = 10$ and $k_1 = 80$, $n_1 = 100$, $k_2 = 70$, $n_2 = 100$. Before you run the code, try to predict the effect that adding more trials has on the posterior distribution for δ.

Exercise 3.2.2 Try the data $k_1 = 0$, $n_1 = 1$ and $k_2 = 0$, $n_2 = 5$. Can you explain the shape of the posterior for δ?

Exercise 3.2.3 In what context might different possible summaries of the posterior distribution of δ (i.e., point estimates, or credible intervals) be reasonable, and when might it be important to show the full posterior distribution?

3.3 Inferring a common rate

We continue to consider two binary processes, producing k_1 and k_2 successes out of n_1 and n_2 trials, respectively, but now assume the underlying rate for both is the same. This means there is just one rate, θ.

The graphical model representation for this problem is shown in Figure 3.5.

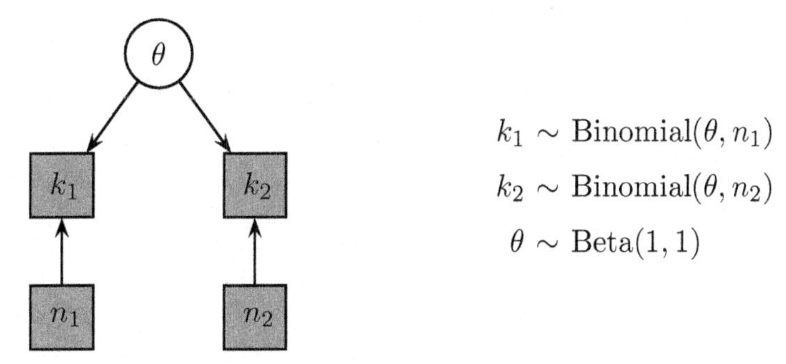

Fig. 3.5 Graphical model for inferring the common rate θ of two binary processes.

An equivalent graphical model, using plate notation, is shown in Figure 3.6. Plates are bounding rectangles that enclose independent replications of a graphical structure within a whole model. In this case, the plate encloses the two observed counts and numbers of trials. Because there is only one latent rate θ (i.e., the same probability drives both binary processes) it is not iterated inside the plate. One way to think of plates, which some people find helpful, is as "for loops" from programming languages (including WinBUGS itself).

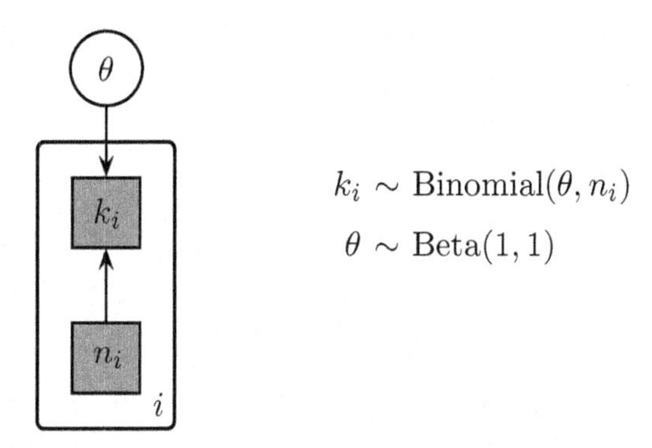

Fig. 3.6 Graphical model for inferring the common rate θ underlying a number of binary processes, using plate notation.

The script `Rate_3.txt` implements the graphical model in WinBUGS:

```
# Inferring a Common Rate
model{
   # Observed Counts
```

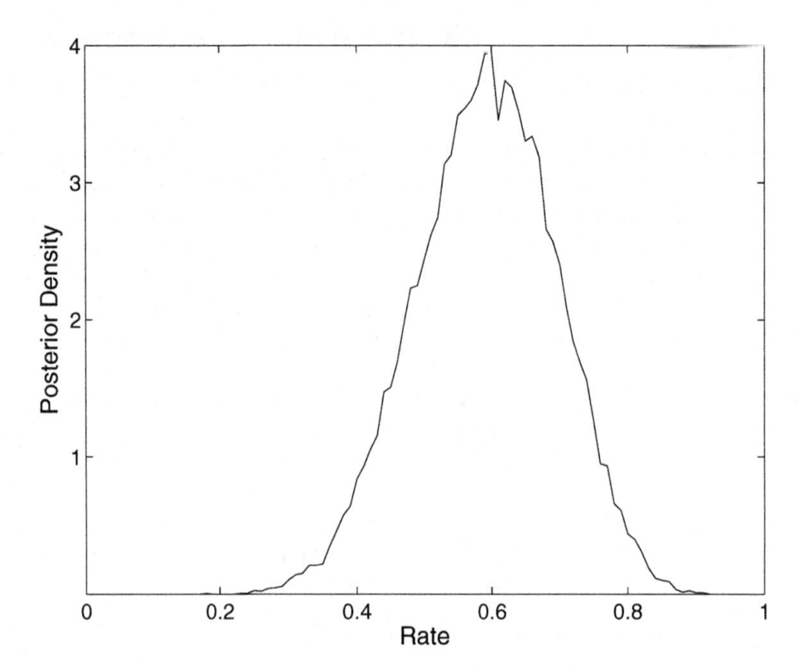

Fig. 3.7 Posterior distribution of the common rate θ of two binary processes.

```
    k1 ~ dbin(theta,n1)
    k2 ~ dbin(theta,n2)
    # Prior on Single Rate Theta
    theta ~ dbeta(1,1)
}
```

The code `Rate_3.m` or `Rate_3.R` sets k_1, k_2, n_1, and n_2, and then calls WinBUGS to sample from the graphical model.[1] The code also produces a plot of the posterior distribution for the common rate, as shown in Figure 3.7.

Exercises

Exercise 3.3.1 Try the data $k_1 = 14$, $n_1 = 20$, $k_2 = 16$, $n_2 = 20$. How could you report the inference about the common rate θ?

Exercise 3.3.2 Try the data $k_1 = 0$, $n_1 = 10$, $k_2 = 10$, $n_2 = 10$. What does this analysis infer the common rate θ to be? Do you believe the inference?

Exercise 3.3.3 Compare the data sets $k_1 = 7$, $n_1 = 10$, $k_2 = 3$, $n_2 = 10$ and $k_1 = 5$, $n_1 = 10$, $k_2 = 5$, $n_2 = 10$. Make sure, following on from the previous question, that you understand why the comparison works the way it does.

[1] Note that the R code specifies `debug=T`, and this means that WinBUGS needs to be closed (not minimized) before the sampling information can be returned to R. WinBUGS is ready as soon as the message "updates took x s" appears in the status bar.

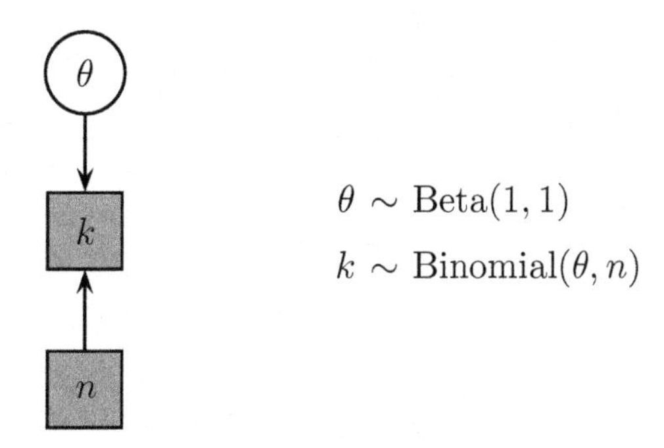

Fig. 3.8 Graphical model for inferring the rate θ of a binary process.

3.4 Prior and posterior prediction

One conceptual way to think about Bayesian analysis is that Bayes' rule provides a bridge between the unobserved parameters of models and the observed data. The most useful part of this bridge is that data allow us to update the uncertainty, represented by probability distributions, about parameters. But the bridge can handle two-way traffic, and so there is a richer set of possibilities for relating parameters to data. There are really four distributions available, and they are all important and useful.

- First, the *prior distribution* over parameters captures our initial assumptions or state of knowledge about the psychological variables they represent.
- Secondly, the *prior predictive distribution* tells us what data to expect, given our model and our current state of knowledge. The prior predictive is a distribution over data, and gives the relative probability of different observable outcomes before any data have been seen.
- Thirdly, the *posterior distribution* over parameters captures what we know about the psychological variables having updated the prior information with the evidence provided by data.
- Finally, the *posterior predictive distribution* tells us what data to expect, given the same model we started with, but with a current state of knowledge that has been updated by the observed data. Again, the posterior predictive is a distribution over data, and gives the relative probability of different observable outcomes after data have been seen.

As an example to illustrate these distributions, we return to the simple problem of inferring a single underlying rate. Figure 3.8 presents the graphical model, and is the same as Figure 3.1.

The script `Rate_4.txt` implements the graphical model in WinBUGS, and provides sampling not just for the posterior, but also for the prior, prior predictive, and posterior predictive:

```
# Prior and Posterior Prediction
model{
   # Observed Data
   k ~ dbin(theta,n)
   # Prior on Rate Theta
   theta ~ dbeta(1,1)
   # Posterior Predictive
   postpredk ~ dbin(theta,n)
   # Prior Predictive
   thetaprior ~ dbeta(1,1)
   priorpredk ~ dbin(thetaprior,n)
}
```

Posterior predictive sampling is achieved by the variable **postpredk** that samples predicted data using the same binomial as the actual observed data. To allow sampling from the prior, we use a dummy variable **thetaprior** that is identical to the one we actually do inference on, but is itself independent of the data, and so is never updated. Prior predictive sampling is achieved by the variable **priorpredk** that samples data using the same binomial, but relying on the prior rate.

The code Rate_4.m or Rate_4.R sets observed data with $k = 1$ successes out of $n = 15$ observations, and then calls WinBUGS to sample from the graphical model. The code also draws the four distributions, two in the parameter space (the prior and posterior for θ), and two in the data space (the prior predictive and posterior predictive for k). It should look something like Figure 3.9.

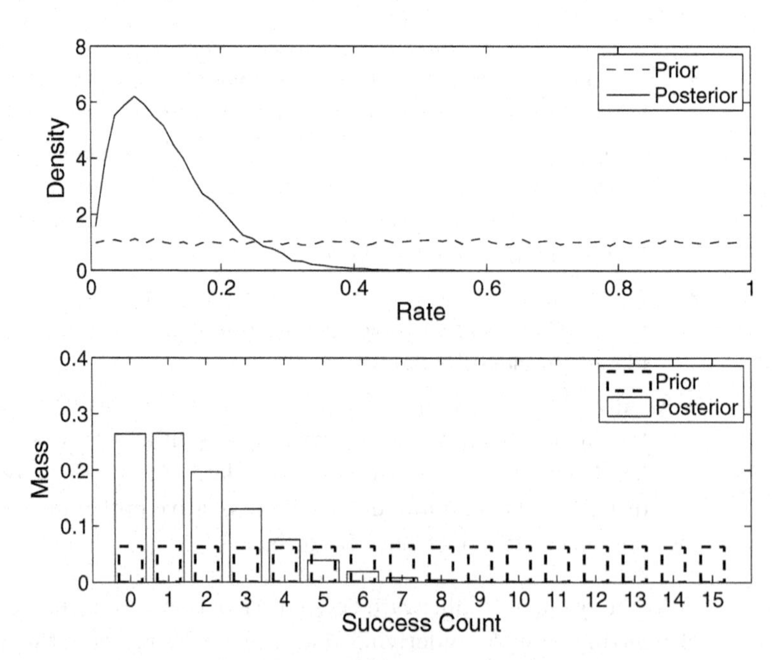

Fig. 3.9 Prior and posterior for the success rate θ (top panel), and prior and posterior predictive for counts of the number of successes (bottom panel), based on data giving $k = 1$ successes out of $n = 15$ trials.

Exercises

Exercise 3.4.1 Make sure you understand the prior, posterior, prior predictive, and posterior predictive distributions, and how they relate to each other (e.g., why is the top panel of Figure 3.9 a line plot, while the bottom panel is a bar graph?). Understanding these ideas is a key to understanding Bayesian analysis. Check your understanding by trying other data sets, varying both k and n.

Exercise 3.4.2 Try different priors on θ, by changing $\theta \sim \text{Beta}(1,1)$ to $\theta \sim \text{Beta}(10,10)$, $\theta \sim \text{Beta}(1,5)$, and $\theta \sim \text{Beta}(0.1,0.1)$. Use the figures produced to understand the assumptions these priors capture, and how they interact with the same data to produce posterior inferences and predictions.

Exercise 3.4.3 Predictive distributions are not restricted to exactly the same experiment as the observed data, and can be used in the context of any experiment where the inferred model parameters make predictions. In the current simple binomial setting, for example, predictive distributions could be found by an experiment that is different because it has $n' \neq n$ observations. Change the graphical model, and Matlab or R code, to implement this more general case.

Exercise 3.4.4 In October 2009, the Dutch newspaper *Trouw* reported on research conducted by H. Trompetter, a student from the Radboud University in the city of Nijmegen. For her undergraduate thesis, Trompetter had interviewed 121 older adults living in nursing homes. Out of these 121 older adults, 24 (about 20%) indicated that they had at some point been bullied by their fellow residents. Trompetter rejected the suggestion that her study may have been too small to draw reliable conclusions: "If I had talked to more people, the result would have changed by one or two percent at the most." Is Trompetter correct? Use the code `Rate_4.m` or `Rate_4.R`, by changing the `dataset` variable (Matlab) or changing the values for `k` and `n` (R), to find the prior and posterior predictive for the relevant rate parameter and bullying counts. Based on these distributions, do you agree with Trompetter's claims?

3.5 Posterior prediction

One important use of posterior predictive distributions is to examine the descriptive adequacy of a model. It can be viewed as a set of predictions about what data the model expects to see, based on the posterior distribution over parameters. If these predictions do not match the data already seen, the model is descriptively inadequate.

As an example to illustrate this idea of checking model adequacy, we return to the problem of inferring a common rate underlying two binary processes. Figure 3.10 presents the graphical model, and is the same as Figure 3.5.

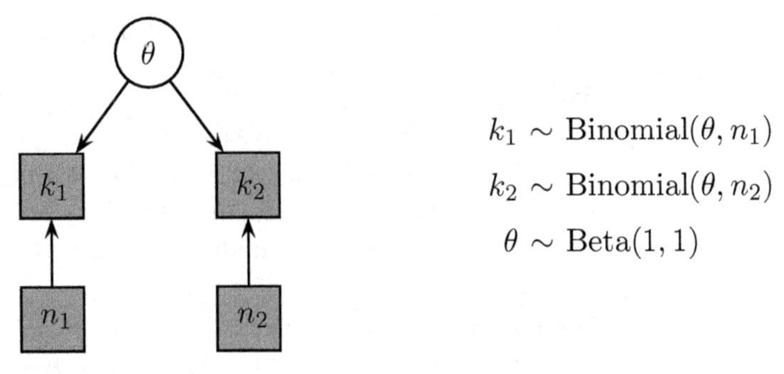

Fig. 3.10 Graphical model for inferring the common rate θ underlying two binary processes.

The script `Rate_5.txt` implements the graphical model in WinBUGS, and provides sampling for the posterior predictive distribution:

```
# Inferring a Common Rate, With Posterior Predictive
model{
   # Observed Counts
   k1 ~ dbin(theta,n1)
   k2 ~ dbin(theta,n2)
   # Prior on Single Rate Theta
   theta ~ dbeta(1,1)
   # Posterior Predictive
   postpredk1 ~ dbin(theta,n1)
   postpredk2 ~ dbin(theta,n2)
}
```

The code `Rate_5.m` or `Rate_5.R` sets observed data with $k_1 = 0$ successes out of $n_1 = 10$ observations, and $k_2 = 10$ successes out of $n_2 = 10$ observations, as considered in Exercise 3.3.2. The code draws the posterior distribution for the rate and the posterior predictive distribution, as shown in Figure 3.11.

The left panel shows the posterior distribution over the common rate θ for two binary processes, which gives density to values near 0.5. The right panel shows the posterior predictive distribution of the model, with respect to the two success counts. The size of each square is proportional to the predictive mass given to each

Box 3.3 The fundamental problem of inference

"The fundamental problem of inference and induction is to use past data to predict future data. Extensive observations on the motions of heavenly bodies enables their future positions to be calculated. Clinical studies on a drug allow a doctor to give a prognosis for a patient for whom the drug is prescribed. Sometimes the uncertain data are in the past, not the future. A historian will use what evidence he has to assess what might have happened where records are missing. A court of criminal law enquires about what had happened on the basis of later evidence." (Lindley, 2000, p. 304).

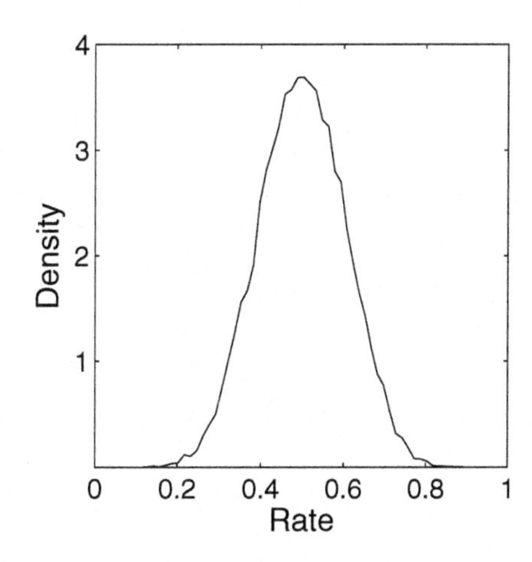

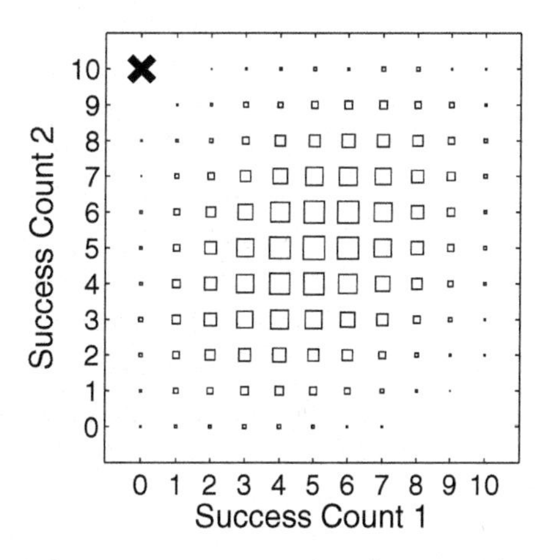

Fig. 3.11 The posterior distribution of the common rate θ for two binary processes (left panel), and the posterior predictive distribution (right panel), based on 0 and 10 successes out of 10 observations.

possible combination of success count observations. The actual data observed in this example, with 0 and 10 successes for the two counts, are shown by the cross.

Exercises

Exercise 3.5.1 Why is the posterior distribution in the left panel inherently one-dimensional, but the posterior predictive distribution in the right panel inherently two-dimensional?

Exercise 3.5.2 What do you conclude about the descriptive adequacy of the model, based on the relationship between the observed data and the posterior predictive distribution?

Exercise 3.5.3 What can you conclude about the parameter θ?

3.6 Joint distributions

So far, we have assumed that the number of successes k and number of total observations n is known, but that the underlying rate θ is unknown. This means that our parameter space has been one-dimensional. Everything learned from data is incorporated into a single probability distribution representing the relative probabilities of different values for the rate θ.

For many problems in cognitive science (and more generally), however, there will be more than one unknown variable of interest, and they will interact. A simple case of this general property is a binomial process in which both the rate θ *and* the total number n are unknown, and so the problem is to infer both simultaneously from counts of successes k.

Box 3.4 Today's posterior is tomorrow's prior

The idea that prior information about parameters can be transformed into posterior information, and hence prior predictive information about data can be transformed into posterior predictive information, can be continued indefinitely. As more information becomes available, usually as more data are collected, uncertainty about parameters and predictive distributions are naturally updated in the Bayesian approach.

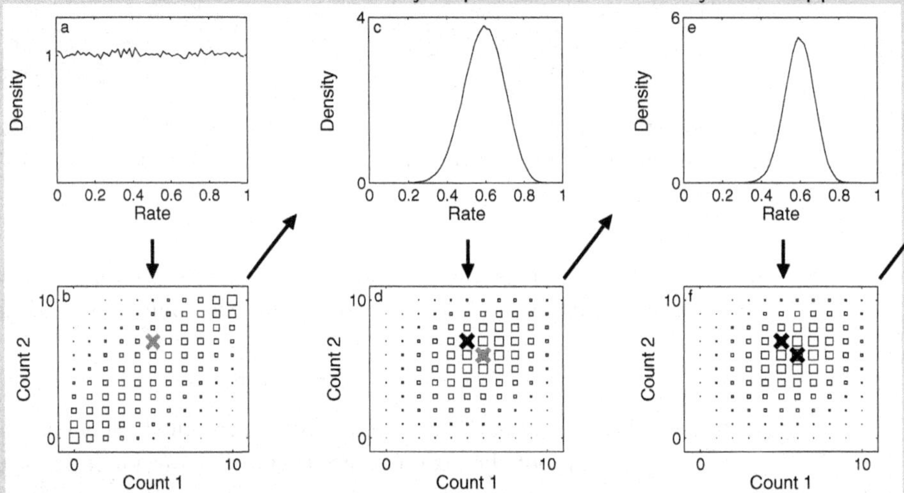

The figure shows the incorporation of a sequence of data for the common model in Figure 3.10. Panel "a" shows the uniform prior over the common rate. Panel "b" shows the prior predictive, for the two counts of successes out of 10 trials. The gray cross corresponds to the observed data, which has yet to be incorporated, but can be compared to the prior predictive distribution. Panel "c" shows the posterior on the rate that now incorporates the data, and panel "d" shows the resulting posterior predictive. The first data are now shown by the black cross in this posterior predictive, since they are incorporated, but a new second data set, in the form of the different gray cross, is about to arrive. These new data are incorporated into the posterior distribution over the rate in panel "e," which leads to the posterior prediction in panel "f." And so the process can continue. Notice how the distribution over the rate parameter in panel "c" is the posterior distribution with respect to the first data set, but acts as the prior for the second data set. This leads to Lindley's Bayesian motto "Today's posterior is tomorrow's prior."

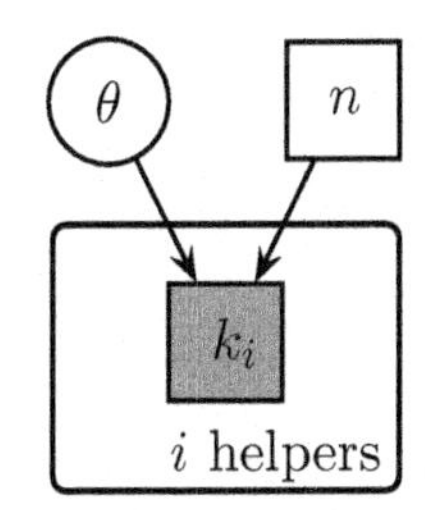

$$k_i \sim \text{Binomial}(\theta, n)$$
$$\theta \sim \text{Beta}(1,1)$$
$$n \sim \text{Categorical}(\underbrace{\frac{1}{n_{\max}}, \ldots, \frac{1}{n_{\max}}}_{m})$$

Fig. 3.12 Graphical model for the joint inference of n and θ from a set of m observed counts of successes $k_1, \ldots, k_m$.

To make the problem concrete, suppose there are five helpers distributing a bundle of surveys to houses. It is known that each bundle contained the same number of surveys, n, but the number itself is not known. The only available relevant information is that the maximum bundle is $n_{\max} = 500$, and so n must be between 1 and $n_{\max}$.

In this problem, it is also not known what the rate of return for the surveys is. But, it is assumed that each helper distributed to houses selected in a random enough way that it is reasonable to believe the return rates are the same. It is also assumed to be reasonable to set a uniform prior on this common rate $\theta \sim \text{Beta}(1,1)$.

Inferences can simultaneously be made about n and θ from the observed number of surveys returned for each of the helpers. Assuming the surveys themselves can be identified with their distributing helper when returned, the data will take the form of $m = 5$ counts, one for each helper, giving the number of returned surveys for each.

The graphical model for this problem is shown in Figure 3.12, and the script `Survey.txt` implements the graphical model in WinBUGS. Note the use of the categorical distribution, which gives probabilities to a finite set of nominal outcomes:

```
# Inferring Return Rate and Number of Surveys from Observed Returns
model{
  # Observed Returns
  for (i in 1:m){
     k[i] ~ dbin(theta,n)
  }
  # Priors on Rate Theta and Number n
  theta ~ dbeta(1,1)
  n ~ dcat(p[])
  for (i in 1:nmax){
    p[i] <- 1/nmax
  }
}
```

The code `Survey.m` or `Survey.R` uses the data $k = \{16, 18, 22, 25, 27\}$, and then calls WinBUGS to sample from the graphical model. Figure 3.13 shows the joint posterior distribution over n and θ as a scatter-plot, and the marginal distributions of each as histograms.

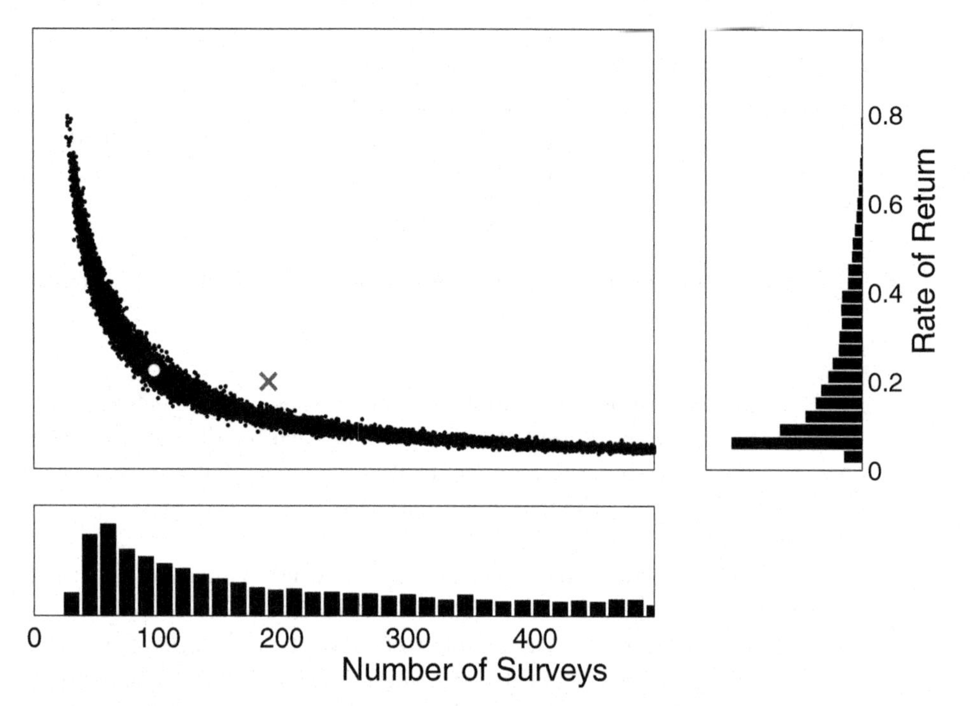

Fig. 3.13 Joint posterior distribution of the probability of return θ and the number of surveys n for $m = 5$ observed counts $k = \{16, 18, 22, 25, 27\}$. The histograms show the marginal densities. The cross shows the expected value of the joint posterior, and the circle shows the mode (i.e., maximum likelihood), both estimated from the posterior samples.

It is clear that the joint posterior distribution carries more information than the marginal posterior distributions. This is very important. It means that just looking at the marginal distributions will not give a complete account of the inferences made, and may provide a misleading account.

An intuitive graphical way to see that there is extra information in the joint posterior is to see if it is well approximated by the product of the marginal distributions. Imagine sampling a point from the histogram for n where there is non-negligible marginal density, such as at $n = 300$. Imagine also sampling points from the histogram for θ, where there is non-negligible marginal density, such as at $\theta = 0.4$. These choices correspond to a single point in the joint posterior density space. Now imagine repeating this process many times. It should be clear that the resulting scatter-plot would be different from the joint posterior scatter-plot in Figure 3.13. So, the joint distribution carries information not available from the marginal distributions.

For this example, it is intuitively obvious why the joint posterior distribution has the clear non-linear structure it does. One possible way in which 20 surveys might be returned is if there were only about 50 surveys, but 40% were returned. Another possibility is that there were 500 surveys, but only a 4% return rate. In general, the number and return rate can trade-off against each other, sweeping out the joint posterior distribution seen in Figure 3.13.

Exercises

Exercise 3.6.1 The basic moral of this example is that it is often worth thinking about joint posterior distributions over model parameters. In this case the marginal posterior distributions are probably misleading. Potentially even more misleading are common (and often perfectly appropriate) point estimates of the joint distribution. The cross in Figure 3.13 shows the expected value of the joint posterior, as estimated from the samples. Notice that it does not even lie in a region of the parameter space with any posterior mass. Does this make sense?

Exercise 3.6.2 The circle in Figure 3.13 shows an approximation to the mode (i.e., the sample with maximum likelihood) from the joint posterior samples. Does this make sense?

Exercise 3.6.3 Try the very slightly changed data $k = \{16, 18, 22, 25, 28\}$. How does this change the joint posterior, the marginal posteriors, the expectation, and the mode? If you were comfortable with the mode, are you still comfortable?

Exercise 3.6.4 If you look at the sequence of samples in the trace plot, some autocorrelation is evident. The samples "sweep" through high and low values in a systematic way, showing the dependency of a sample on those immediately preceding. This is a deviation from the ideal situation in which posterior samples are independent draws from the joint posterior. Try thinning the sampling, taking only every 100th sample, by setting `nthin=100` in Matlab or `n.thin=100` in R. To make the computational time reasonable, reduce the number of samples collected after thinning to just 500 (i.e., run 50,000 total samples, so that 500 are retained after thinning). How is the sequence of samples visually different with thinning?[2]

[2] A note for R2jags users: at the time of writing, R2jags mistakenly randomizes the values in the `sims.array` object whenever you run a single chain. Until this error is fixed it is safest to run multiple chains, at least when you are interested in examining autocorrelation. See also the last few posts here: `http://sourceforge.net/p/mcmc-jags/discussion/610037/thread/cc61b820/?limit=50#83b4`.

4 Inferences with Gaussians

4.1 Inferring a mean and standard deviation

One of the most common inference problems involves assuming data following a Gaussian (also known as a Normal, Central, or Maxwellian) distribution, and inferring the mean and standard deviation of this distribution from a sample of observed independent data.

The graphical model representation for this problem is shown in Figure 4.1. The data are the n observations $x_1, \ldots, x_n$. The mean of the Gaussian is μ and the standard deviation is σ. WinBUGS parameterizes the Gaussian distribution in terms of the mean and precision, not the mean and variance or the mean and standard deviation. These are all simply related, with the variance being σ^2 and the precision being $\lambda = 1/\sigma^2$.

Here the prior used for μ is intended to be only weakly informative. That is, it is a prior intended to convey little information about the mean, so that inference will be primarily dependent upon relevant data. It is a Gaussian centered on zero, but with very low precision (i.e., very large variance), and gives prior probability to a wide range of possible means for the data. When the goal is to estimate parameters, this sort of approach is relatively non-controversial.

Setting priors for standard deviations (or variances, or precisions) is trickier, and certainly more controversial. If there is any relevant information that helps put the data on scale, so that bounds can be set on reasonable possibilities for the standard deviation, then setting a uniform over that range is advocated by Gelman (2006). In this first example, we assume the data are all small enough that setting an upper bound of 10 on the standard deviation covers all the possibilities.

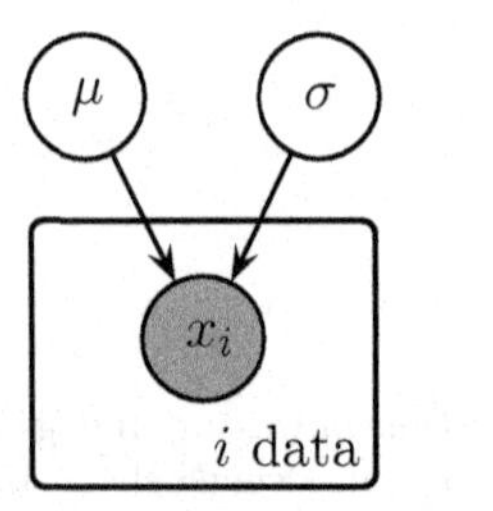

$$\mu \sim \text{Gaussian}(0, 0.001)$$
$$\sigma \sim \text{Uniform}(0, 10)$$
$$x_i \sim \text{Gaussian}(\mu, \tfrac{1}{\sigma^2})$$

Fig. 4.1 Graphical model for inferring the mean and standard deviation of data generated by a Gaussian distribution.

The script `Gaussian.txt` implements the graphical model in WinBUGS. Note the conversion of the standard deviation `sigma` into the precision parameter `lambda` used to sample from a Gaussian:

```
# Inferring the Mean and Standard Deviation of a Gaussian
model{
  # Data Come From A Gaussian
  for (i in 1:n){
    x[i] ~ dnorm(mu,lambda)
  }
  # Priors
  mu ~ dnorm(0,.001)
  sigma ~ dunif(0,10)
  lambda <- 1/pow(sigma,2)
}
```

The code `Gaussian.m` or `Gaussian.R` creates some artificial data, and applies the graphical model to make inferences from data. The code does not produce a graph, or any other output. But all of the information you need to analyze the results is in the returned variables `samples` and `stats`.

Exercises

Exercise 4.1.1 Try a few data sets, varying what you expect the mean and standard deviation to be, and how many data you observe.

Exercise 4.1.2 Plot the *joint* posterior of μ and σ. That is, plot the samples from μ against those of σ. Interpret the shape of the joint posterior.

Exercise 4.1.3 Suppose you knew the standard deviation of the Gaussian was 1.0, but still wanted to infer the mean from data. This is a realistic question: For example, knowing the standard deviation might amount to knowing the noise associated with measuring some psychological trait using a test instrument. The x_i values could then be repeated measures for the same person, and their mean the trait value you are trying to infer. Modify the WinBUGS script and Matlab or R code to do this. What does the revised graphical model look like?

Exercise 4.1.4 Suppose you knew the mean of the Gaussian was zero, but wanted to infer the standard deviation from data. This is also a realistic question: Suppose you know that the error associated with a measurement is unbiased, so its average or mean is zero, but you are unsure how much noise there is in the instrument. Inferring the standard deviation is then a sensible way to infer the noisiness of the instrument. Once again, modify the WinBUGS script and Matlab or R code to do this. Once again, what does the revised graphical model look like?

4.2 The seven scientists

This problem is from MacKay (2003, p. 309) where it is, among other things, treated to a Bayesian solution, but not quite using a graphical modeling approach, nor relying on computational sampling methods.

Seven scientists with wildly-differing experimental skills all make a measurement of the same quantity. They get the answers $x = \{-27.020, 3.570, 8.191, 9.898, 9.603, 9.945, 10.056\}$. Intuitively, it seems clear that the first two scientists are pretty inept measurers, and that the true value of the quantity is probably just a bit below 10. The main problem is to find the posterior distribution over the measured quantity, telling us what we can infer from the measurement. A secondary problem is to infer something about the measurement skills of the seven scientists.

The graphical model for one way of solving this problem is shown in Figure 4.2. The assumption is that all the scientists have measurements that follow a Gaussian distribution, but with different standard deviations. However, because they are all measuring the same quantity, each Gaussian has the same mean, and it is just the standard deviation that differs.

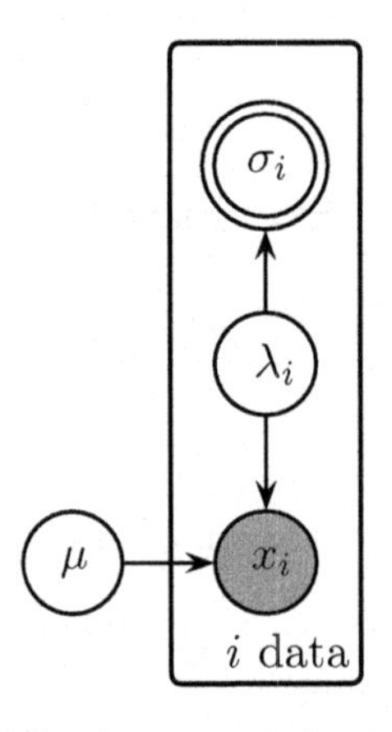

$$\mu \sim \text{Gaussian}(0, 0.001)$$
$$\lambda_i \sim \text{Gamma}(0.001, 0.001)$$
$$\sigma_i \leftarrow 1/\sqrt{\lambda_i}$$
$$x_i \sim \text{Gaussian}(\mu, \lambda_i)$$

Fig. 4.2 Graphical model for the seven scientists problem.

Notice that we have used a different approach to assign priors to the standard deviations. The previous example, as shown in Figure 4.1, used a uniform distribution. The current example, shown in Figure 4.2, uses a gamma distribution for the priors on the precisions. This is another standard approach, which has some attractive theoretical motivations, but is critically discussed by Gelman (2006).

The script `SevenScientists.txt` implements the graphical model in Figure 4.2 in WinBUGS:

```
# The Seven Scientists
model{
  # Data Come From Gaussians With Common Mean But Different Precisions
  for (i in 1:n){
    x[i] ~ dnorm(mu,lambda[i])
```

Box 4.1 Priors on precisions

The practice of assigning $\text{Gamma}(0.001, 0.001)$ priors on precision parameters is theoretically motivated by scale invariance arguments, meaning that priors are chosen so that changing the measurement scale of the data does not affect inference. The invariant prior on precision λ corresponds to a uniform distribution on $\log \sigma$, that is, $p(\sigma^2) \propto 1/\sigma^2$, or a $\text{Gamma}(a \rightarrow 0, b \rightarrow 0)$ distribution. This invariant prior distribution, however, is *improper* (i.e., the area under the curve is unbounded), which means it is not really a distribution, but the limit of a sequence of distributions (see Jaynes, 2003). WinBUGS requires the use of proper distributions, and the $\text{Gamma}(0.001, 0.001)$ prior is intended as a proper approximation to the theoretically motivated improper prior. This raises the issue of whether inference is sensitive to the essentially arbitrary value 0.001, and it is sometimes the case that using other small values such as 0.01 or 0.1 leads to more stable sampling in WinBUGS.

```
  }
  # Priors
  mu ~ dnorm(0,.001)
  for (i in 1:n){
    lambda[i] ~ dgamma(.001,.001)
    sigma[i] <- 1/sqrt(lambda[i])
  }
}
```

Notice that the graphical model implements the prior on the precisions, but also re-parameterizes to the standard deviation scale, which is often more easily interpretable.

The code `SevenScientists.m` or `SevenScientists.R` applies the seven scientist data to the graphical model.

Exercises

Exercise 4.2.1 Draw posterior samples using the Matlab or R code, and reach conclusions about the value of the measured quantity, and about the accuracies of the seven scientists.

Exercise 4.2.2 Change the graphical model in Figure 4.2 to use a uniform prior over the standard deviations, as was done in Figure 4.1. Experiment with the effect the upper bound of this uniform prior has on inference.

Box 4.2 Ill-posed problems

"If one fails to specify the prior information, a problem of inference is just as ill-posed as if one had failed to specify the data." (Jaynes, 2003, p. 373).

4.3 Repeated measurement of IQ

In this example, we consider how to estimate the IQ of a set of people, each of whom have done multiple IQ tests. The data are the measures x_{ij} for the $i = 1, \dots, n$ people and their $j = 1, \dots, m$ repeated test scores.

We assume that the differences in repeated test scores are distributed as Gaussian error terms with zero mean and unknown precision. The mean of the Gaussian of a person's test scores corresponds to their latent true IQ. This will be different for each person. The standard deviation of the Gaussians corresponds to the accuracy of the testing instruments in measuring the one underlying IQ value. We assume this is the same for every person, since it is conceived as a property of the tests themselves.

The graphical model for this problem is shown in Figure 4.3. Because we know quite a bit about the IQ scale, it makes sense to set priors for the mean and standard deviation using this knowledge. Our first attempt to set priors (these are revisited in the exercises) simply assume the actual IQ values are equally likely to be anywhere between 0 and 300, and standard deviations are anywhere between 0 and 100.

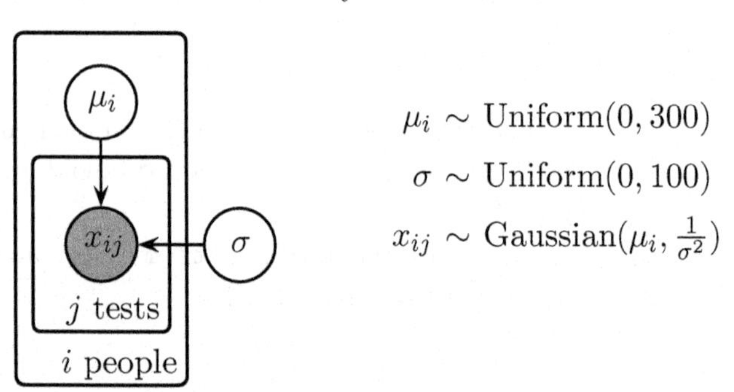

Fig. 4.3 Graphical model for inferring the IQ from repeated measures.

The script `IQ.txt` implements the graphical model in WinBUGS:

```
# Repeated Measures of IQ
model{
  # Data Come From Gaussians With Different Means But Common Precision
  for (i in 1:n){
    for (j in 1:m){
      x[i,j] ~ dnorm(mu[i],lambda)
    }
  }
```

```
  # Priors
  sigma ~ dunif(0,100)
  lambda <- 1/pow(sigma,2)
  for (i in 1:n){
    mu[i] ~ dunif(0,300)
  }
}
```

The code `IQ.m` or `IQ.R` creates a data set corresponding to there being three people, with test scores of $(90, 95, 100)$, $(105, 110, 115)$, and $(150, 155, 160)$, and applies the graphical model.

Exercises

Exercise 4.3.1 Use the posterior distribution for each person's μ_i to estimate their IQ. What can we say about the precision of the IQ test?

Exercise 4.3.2 Now, use a more realistic prior assumption for the μ_i means. Theoretically, IQ distributions should have a mean of 100, and a standard deviation of 15. This corresponds to having a prior of `mu[i]` $\sim$ `dnorm(100,.0044)`, instead of `mu[i]` $\sim$ `dunif(0,300)`, because $1/15^2 = 0.0044$. Make this change in the WinBUGS script, and re-run the inference. How do the estimates of IQ given by the means change? Why?

Exercise 4.3.3 Repeat both of the above stages (i.e., using both priors on μ_i) with a new, but closely related, data set that has scores of $(94, 95, 96)$, $(109, 110, 111)$, and $(154, 155, 156)$. How do the different prior assumptions affect IQ estimation for these data. Why does it not follow the same pattern as the previous data?

5 Some examples of data analysis

5.1 Pearson correlation

The Pearson product-moment correlation coefficient, usually denoted r, is a widely used measure of the relationship between two variables. It ranges from -1, indicating a perfect negative linear relationship, to $+1$, indicating a perfect positive relationship. A value of 0 indicates that there is no linear relationship. Usually the correlation r is reported as a single point estimate, perhaps together with a frequentist significance test.[1]

But, rather than just having a single number to measure the correlation, it would be nice to have a posterior distribution for r, saying how likely each possible level of correlation was. There are frequentist confidence interval methods that try to do this, as well as various analytic Bayesian results based on asymptotic approximations (e.g., Donner & Wells, 1986). An advantage of using a computational approach is the flexibility in the assumptions that can be made. It is possible to set up a graphical model that allows inferences about the correlation coefficient for any set of prior assumptions about the correlation.

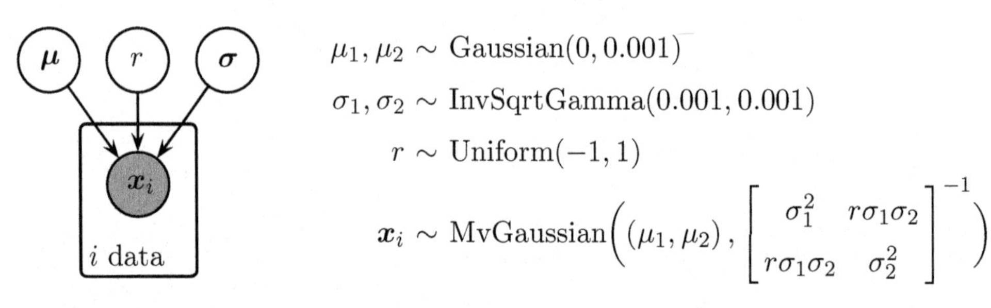

$$\mu_1, \mu_2 \sim \text{Gaussian}(0, 0.001)$$
$$\sigma_1, \sigma_2 \sim \text{InvSqrtGamma}(0.001, 0.001)$$
$$r \sim \text{Uniform}(-1, 1)$$
$$\boldsymbol{x}_i \sim \text{MvGaussian}\left((\mu_1, \mu_2), \begin{bmatrix} \sigma_1^2 & r\sigma_1\sigma_2 \\ r\sigma_1\sigma_2 & \sigma_2^2 \end{bmatrix}^{-1}\right)$$

Fig. 5.1 Graphical model for inferring a correlation coefficient.

One graphical model for doing this is shown in Figure 5.1. The observed data take the form $\boldsymbol{x}_i = (x_{i1}, x_{i2})$ for the ith observation, and, following the theory behind the correlation coefficient, are modeled as draws from a multivariate Gaussian distribution. The parameters of this distribution are the means $\boldsymbol{\mu} = (\mu_1, \mu_2)$ and standard deviations $\boldsymbol{\sigma} = (\sigma_1, \sigma_2)$ of the two variables, and the correlation coefficient r that links them.

[1] Frequentist or orthodox statistics is familiar to all cognitive scientists. Key frequentist concepts include the p-value, power, confidence intervals, and Type-I error rate. We believe that for scientific inference, the frequentist approach is inefficient at best and misleading at worst.

Box 5.1 Frequentist subjectivity

"Today one wonders how it is possible that orthodox logic continues to be taught in some places year after year and praised as 'objective', while Bayesians are charged with 'subjectivity'. Orthodoxians, preoccupied with fantasies about nonexistent data sets and, in principle, unobservable limiting frequencies—while ignoring relevant prior information—are in no position to charge anybody with 'subjectivity'." (Jaynes, 2003, p. 550).

In Figure 5.1, the standard deviations are assigned relatively uninformative inverse-square-root-gamma distributions. This is equivalent to placing gamma distributions on precisions, as was done in the seven scientists example in Section 4.2. The correlation coefficient itself is given a uniform prior over its possible range. All of these choices would be easily modified, with one obvious possible change being to give the prior for the correlation more density around 0.

The script `Correlation_1.txt` implements the graphical model in WinBUGS:

```
# Pearson Correlation
model{
  # Data
  for (i in 1:n){
    x[i,1:2] ~ dmnorm(mu[],TI[,])
  }
  # Priors
  mu[1] ~ dnorm(0,.001)
  mu[2] ~ dnorm(0,.001)
  lambda[1] ~ dgamma(.001,.001)
  lambda[2] ~ dgamma(.001,.001)
  r ~ dunif(-1,1)
  # Reparameterization
  sigma[1] <- 1/sqrt(lambda[1])
  sigma[2] <- 1/sqrt(lambda[2])
  T[1,1] <- 1/lambda[1]
  T[1,2] <- r*sigma[1]*sigma[2]
  T[2,1] <- r*sigma[1]*sigma[2]
  T[2,2] <- 1/lambda[2]
  TI[1:2,1:2] <- inverse(T[1:2,1:2])
}
```

The code `Correlation_1.m` or `Correlation_1.R` includes two data sets. Both involve fabricated data comparing response times in a semantic verification task (e.g., "Is a whale a fish?") on the x-axis with IQ measures on the y-axis, looking for a correlation between simple measures of decision-making and general intelligence.

For the first data set in the Matlab and R code, the results shown in Figure 5.2 are produced. The left panel shows a scatter-plot of the raw data. The right panel shows the posterior distribution of r, together with the standard frequentist point estimate.

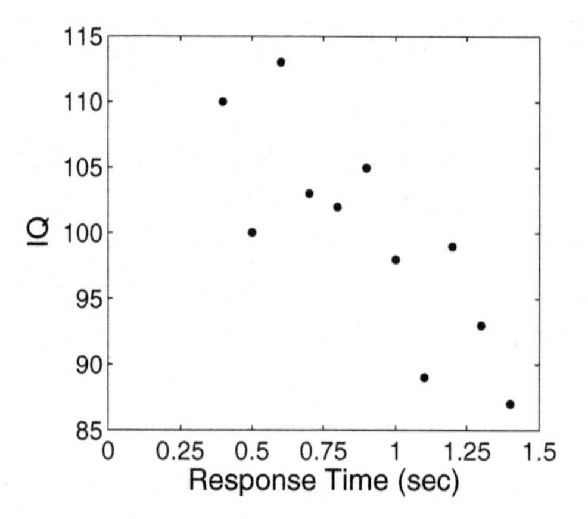

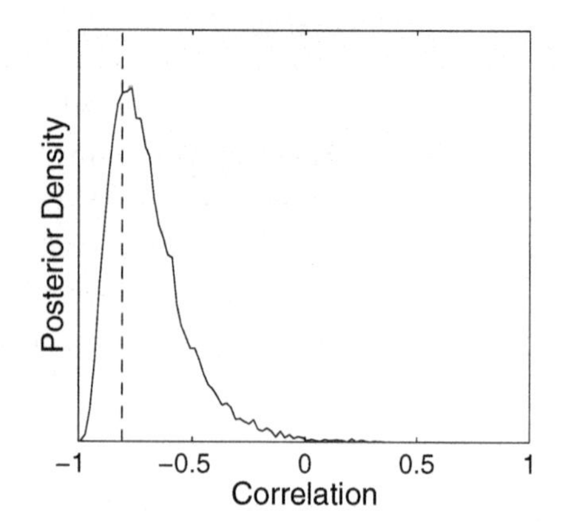

Fig. 5.2 Data (left panel) and posterior distribution for correlation coefficient (right panel). The broken line shows the frequentist point estimate.

Exercises

Exercise 5.1.1 The second data set in the Matlab and R code is just the first data set from Figure 5.2 repeated twice. Set `dataset=2` to consider these repeated data, and interpret the differences in the posterior distributions for r.

Exercise 5.1.2 Do you find the priors on μ_1 and μ_2 to be reasonable?

Exercise 5.1.3 The current graphical model assumes that the values from the two variables—the $\boldsymbol{x}_i = (x_{i1}, x_{i2})$—are observed with perfect accuracy. When might this be a problematic assumption? How could the current approach be extended to make more realistic assumptions?

5.2 Pearson correlation with uncertainty

We now tackle the problem asked by the last question in the previous section, and consider the correlations when there is uncertainty about the exact values of variables. It is likely that each individual response time is measured very accurately, since it is a physical quantity and good measurement tools exist. But the measurement of IQ seems likely to be less precise, since it is a psychological quantity, and measurement tools like IQ tests are less accurate. The uncertainty in measurement should be incorporated in an assessment of the correlation between the variables (e.g., Behseta, Berdyyeva, Olson, & Kass, 2009).

A simple approach for including this uncertainty is adopted by the graphical model in Figure 5.3. The observed data still take the form $\boldsymbol{x}_i = (x_{i1}, x_{i2})$ for the ith person's response time and IQ measure. But these observations are now sampled from a Gaussian distribution, centered on the unobserved true response time and IQ of that person, denoted $\boldsymbol{y}_i = (y_{i1}, y_{i2})$. These true values are then modeled as the

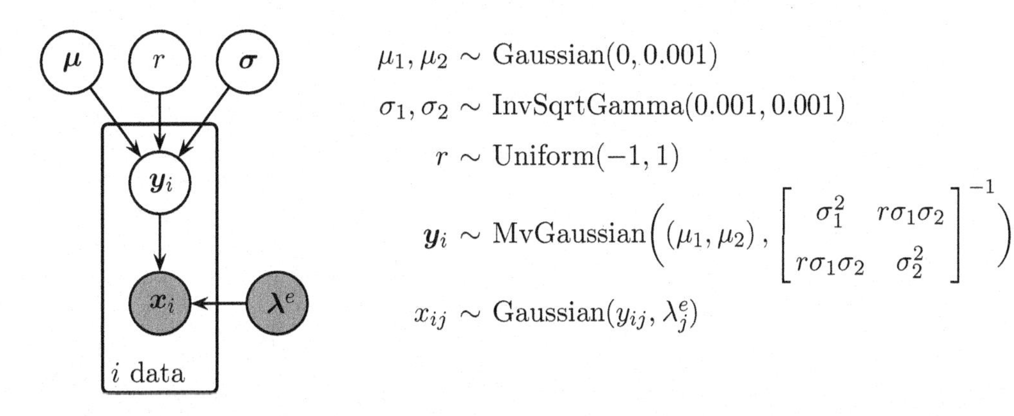

Fig. 5.3 Graphical model for inferring a correlation coefficient, when there is uncertainty inherent in the measurements.

$\boldsymbol{x}$ were in the previous model in Figure 5.1, as draws from a multivariate Gaussian distribution.

The precision of the measurements is captured by $\boldsymbol{\lambda}^e = (\lambda_1^e, \lambda_2^e)$ of the Gaussian draws for the observed data, $x_{ij} \sim \text{Gaussian}(y_{ij}, \lambda_j^e)$. The graphical model in Figure 5.3 assumes that these precisions are known.

The script `Correlation_2.txt` implements the graphical model shown in WinBUGS:

```
# Pearson Correlation With Uncertainty in Measurement
model{
  # Data
  for (i in 1:n){
    y[i,1:2] ~ dmnorm(mu[],TI[,])
    for (j in 1:2){
      x[i,j] ~ dnorm(y[i,j],lambdaerror[j])
    }
  }
  # Priors
  mu[1] ~ dnorm(0,.001)
  mu[2] ~ dnorm(0,.001)
  lambda[1] ~ dgamma(.001,.001)
  lambda[2] ~ dgamma(.001,.001)
  r ~ dunif(-1,1)
  # Reparameterization
  sigma[1] <- 1/sqrt(lambda[1])
  sigma[2] <- 1/sqrt(lambda[2])
  T[1,1] <- 1/lambda[1]
  T[1,2] <- r*sigma[1]*sigma[2]
  T[2,1] <- r*sigma[1]*sigma[2]
  T[2,2] <- 1/lambda[2]
  TI[1:2,1:2] <- inverse(T[1:2,1:2])
}
```

The code `Correlation_2.m` or `Correlation_2.R` uses the same data as in the previous section, but has different analyses because of the different assumptions about the uncertainty in measurement. In these new analyses, we assume that

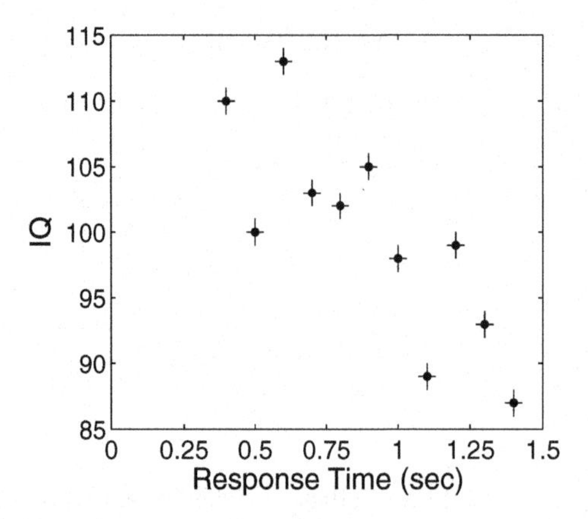

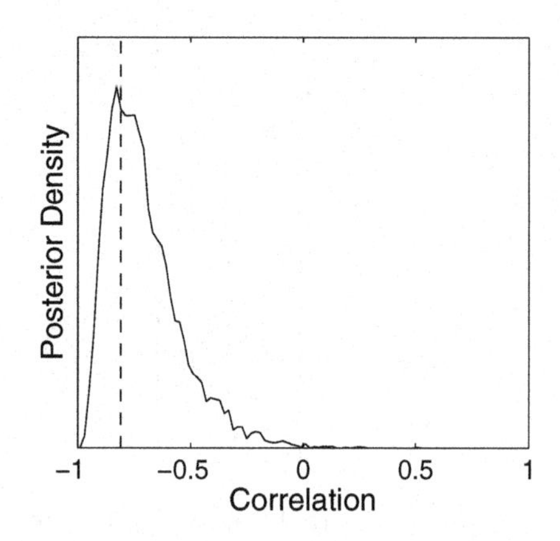

Fig. 5.4 Data (left panel), including error bars showing uncertainty in measurement, and posterior distribution for the correlation coefficient (right panel). The broken line shows the frequentist point estimate.

measurement uncertainty is originally expressed in terms of standard deviations, and then re-parameterized and supplied to the graphical model as precisions. The specific assumption is that $\sigma_1^e = .03$ for response times (which seem likely to be measured accurately) and $\sigma_2^e = 1$ for IQ (which seems near the smallest plausible value, so we assume that IQ is also measured accurately). The results of these assumptions using the model are shown in Figure 5.4. The left panel shows a scatter-plot of the raw data, together with error bars representing the uncertainty quantified by the assumed standard deviations σ_1^e and σ_2^e. The right panel shows the posterior distribution of r, together with the standard frequentist point estimate.

Exercises

Exercise 5.2.1 Compare the results obtained in Figure 5.4 with those obtained earlier using the same data, in Figure 5.2, for the model without any account of uncertainty in measurement.

Exercise 5.2.2 Generate results for the second data set, which changes $\sigma_2^e = 10$ for the IQ measurement. Compare these results with those obtained assuming $\sigma_2^e = 1$.

Exercise 5.2.3 The graphical model in Figure 5.3 assumes the uncertainty for each variable is known. How could this assumption be relaxed to the case where the uncertainty is unknown?

Exercise 5.2.4 The graphical model in Figure 5.3 assumes the uncertainty for each variable is the same for all observations. How could this assumption be relaxed to the case where, for example, extreme IQs are less accurately measured than IQs in the middle of the standard distribution?

5.3 The kappa coefficient of agreement

An important statistical inference problem in a range of physical, biological, behavioral, and social sciences is to decide how well one decision-making method agrees with another. An interesting special case considers only binary decisions, and views one of the decision-making methods as giving objectively true decisions to which the other aspires. This problem occurs often in medicine, when cheap or easily administered methods for diagnosis are evaluated in terms of how well they agree with a more expensive or complicated "gold standard" method.

For this problem, when both decision-making methods make n independent assessments, the data $\boldsymbol{y}$ take the form of four counts: a observations where both methods decide "one," b observations where the objective method decides "one" but the surrogate method decides "zero," c observations where the objective method decides "zero" but the surrogate method decides "one," and d observations where both methods decide "zero," with $n = a + b + c + d$.

A variety of orthodox statistical measures have been proposed for assessing agreement using these data (but see Basu, Banerjee, & Sen, 2000; Broemeling, 2009, for Bayesian approaches). Useful reviews are provided by Agresti (1992), Banerjee, Capozzoli, McSweeney, and Sinha (1999), Fleiss, Levin, and Paik (2003), Kraemer (1992), Kraemer, Periyakoil, and Noda (2004), and Shrout (1998). Of all the measures, however, it is reasonable to argue that the conclusion of Uebersax (1987) that "the kappa coefficient is generally regarded as the statistic of choice for measuring agreement" (p. 140) remains true.

Cohen's (1960) kappa statistic estimates the level of observed agreement

$$p_o = \frac{a + d}{n}$$

relative to the agreement that would be expected by chance alone (i.e., the overall probability for the first method to decide "one" times the overall probability for the second method to decide "one," and added to this the overall probability for the second method to decide "zero" times the overall probability for the first method to decide "zero")

$$p_e = \frac{(a + b)(a + c) + (b + d)(c + d)}{n^2},$$

and is given by

$$\kappa = \frac{p_o - p_e}{1 - p_e}.$$

Kappa lies on a scale of -1 to $+1$, with values below 0.4 often interpreted as "poor" agreement beyond chance, values between 0.4 and 0.75 interpreted as "fair to good" agreement beyond chance, and values above 0.75 interpreted as "excellent" agreement beyond chance (Landis & Koch, 1977). The key insight of kappa as a measure of agreement is its correction for chance agreement.

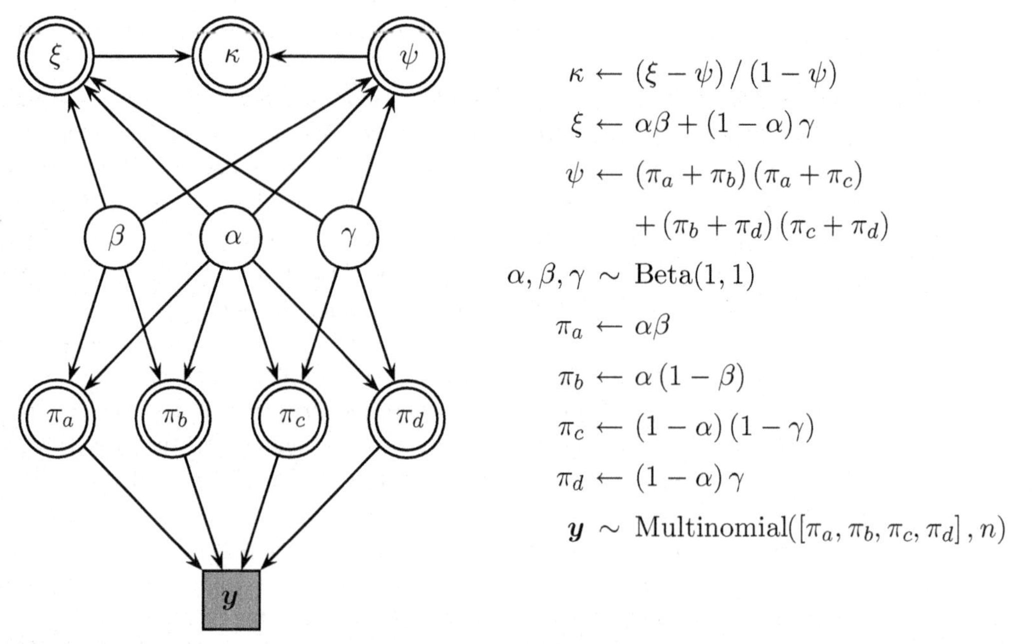

Fig. 5.5 Graphical model for inferring the kappa coefficient of agreement.

The graphical model for a Bayesian version of kappa is shown in Figure 5.5. The key latent variables are α, β, and γ. The rate α is the rate at which the gold standard method decides "one." This means $(1 - \alpha)$ is the rate at which the gold standard method decides "zero." The rate β is the rate at which the surrogate method decides "one" when the gold standard also decides "one." The rate γ is the rate at which the surrogate method decides "zero" when the gold standard decides "zero." The best way to interpret β and γ is that they are the rate of agreement of the surrogate method with the gold standard, for the "one" and "zero" decisions respectively.

Using the rates α, β, and γ, it is possible to calculate the probabilities that both methods will decide "one," $\pi_a = \alpha\beta$, that the gold standard will decide "one" but the surrogate will decide "zero," $\pi_b = \alpha\,(1 - \beta)$, the gold standard will decide "zero" but the surrogate will decide "one," $\pi_c = (1 - \alpha)\,(1 - \gamma)$, and that both methods will decide "zero," $\pi_d = (1 - \alpha)\,\gamma$.

These probabilities, in turn, describe how the observed data, $\boldsymbol{y}$, made up of the counts a, b, c, and d, are generated. They come from a multinomial distribution with n trials, where on each trial there is a π_a probability of generating an a count, π_b probability for a b count, and so on.

So, observing the data $\boldsymbol{y}$ allows inferences to be made about the key rates α, β, and γ. The remaining variables in the graphical model in Figure 5.5 just re-express these rates in the way needed to provide an analogue to the kappa measure of chance-corrected agreement. The ξ variable measures the rate of agreement, which is $\xi = \alpha\beta + (1-\alpha)\gamma$. The ψ variable measures the rate of agreement that would occur by chance, which is $\psi = (\pi_a + \pi_b)(\pi_a + \pi_c) + (\pi_b + \pi_d)(\pi_c + \pi_d)$, and could be expressed in terms of α, β, and γ. Finally κ is the chance-corrected measure of agreement on the -1 to $+1$ scale, given by $\kappa = (\xi - \psi)/(1-\psi)$.

The script `Kappa.txt` implements the graphical model in WinBUGS:

```
# Kappa Coefficient of Agreement
model{
  # Underlying Rates
  # Rate Objective Method Decides "one"
  alpha ~ dbeta(1,1)
  # Rate Surrogate Method Decides "one" When Objective Method Decides "one"
  beta ~ dbeta(1,1)
  # Rate Surrogate Method Decides "zero" When Objective Method Decides "zero"
  gamma ~ dbeta(1,1)
  # Probabilities For Each Count
  pi[1] <- alpha*beta
  pi[2] <- alpha*(1-beta)
  pi[3] <- (1-alpha)*(1-gamma)
  pi[4] <- (1-alpha)*gamma
  # Count Data
  y[1:4] ~ dmulti(pi[],n)
  # Derived Measures
  # Rate Surrogate Method Agrees With the Objective Method
  xi <- alpha*beta+(1-alpha)*gamma
  # Rate of Chance Agreement
  psi <- (pi[1]+pi[2])*(pi[1]+pi[3])+(pi[2]+pi[4])*(pi[3]+pi[4])
  # Chance-Corrected Agreement
  kappa <- (xi-psi)/(1-psi)
}
```

The code `Kappa.m` or `Kappa.R` includes several data sets, described in the exercises below, for WinBUGS to sample from the graphical model.

Exercises

Exercise 5.3.1 *Influenza Clinical Trial.* Poehling, Griffin, and Dittus (2002) reported data evaluating a rapid bedside test for influenza using a sample of 233 children hospitalized with fever or respiratory symptoms. Of the 18 children known to have influenza, the surrogate method identified 14 and missed 4. Of the 215 children known not to have influenza, the surrogate method correctly rejected 210 but falsely identified 5. These data correspond to $a = 14$, $b = 4$, $c = 5$, and $d = 210$. Examine the posterior distributions of the interesting variables, and reach a scientific conclusion. That is, pretend you are a consultant for the clinical trial. What would your two- or three-sentence "take home message" conclusion be to your customers?

Exercise 5.3.2 *Hearing Loss Assessment Trial.* Grant (1974) reported data from a screening of a pre-school population intended to assess the adequacy of a school nurse assessment in relation to expert assessment of hearing loss. Of those children assessed by the expert as having hearing loss, 20 were correctly identified by the nurse and 7 were missed. Of those assessed by the expert as not having hearing loss, 417 were correctly diagnosed by the nurse but 103 were incorrectly diagnosed as having hearing loss. These data correspond to $a = 20$, $b = 7$, $c = 103$, $d = 417$. Once again, examine the posterior distributions of the interesting variables, and reach a scientific conclusion. Once again, what would your two- or three-sentence "take home message" conclusion be to your customers?

Exercise 5.3.3 *Rare Disease.* Suppose you are testing a cheap instrument for detecting a rare medical condition. After 170 patients have been screened, the test results show that 157 did not have the condition, but 13 did. The expensive ground-truth assessment subsequently revealed that, in fact, none of the patients had the condition. These data correspond to $a = 0$, $b = 0$, $c = 13$, $d = 157$. Apply the kappa graphical model to these data, and reach a conclusion about the usefulness of the cheap instrument. What is special about this data set, and what does it demonstrate about the Bayesian approach?

5.4 Change detection in time series data

This case study involves near-infrared spectrographic data, in the form of oxygenated hemoglobin counts of frontal lobe activity during an attention task in Attention Deficit Hyperactivity Disorder (ADHD) adults. The interesting modeling problem is that a change is expected in the time series of counts because of the attention task. The statistical problem is to identify the change. To do this, we are going to make a number of strong assumptions. In particular, we will assume that the counts come from a Gaussian distribution that always has the same variance, but changes its mean at one specific point in time. The main interest is therefore in making an inference about this change point.

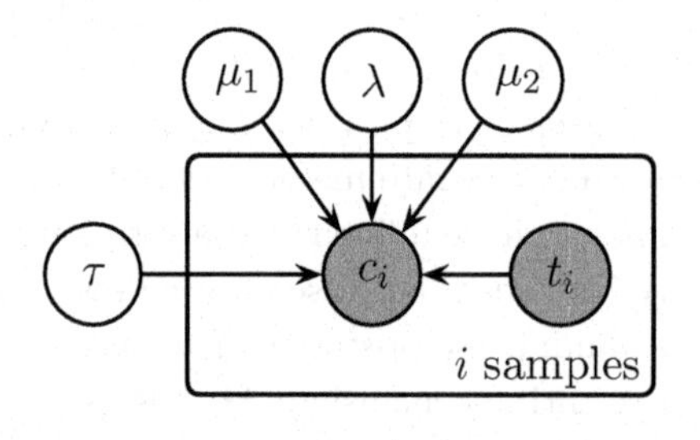

$$\mu_1, \mu_2 \sim \text{Gaussian}(0, 0.001)$$
$$\lambda \sim \text{Gamma}(0.001, 0.001)$$
$$\tau \sim \text{Uniform}(0, t_{\text{max}})$$
$$c_i \sim \begin{cases} \text{Gaussian}(\mu_1, \lambda) & \text{if } t_i < \tau \\ \text{Gaussian}(\mu_2, \lambda) & \text{if } t_i \geq \tau \end{cases}$$

Fig. 5.6 Graphical model for detecting a single change point in time series.

Figure 5.6 presents a graphical model for detecting the change point. The observed data are the counts c_i at time t_i for the ith sample. The unobserved variable τ is the time at which the change happens, which controls whether the counts have mean μ_1 or μ_2. A uniform prior over the full range of possible times is assumed for the change point, and generic weakly informative priors are given to the means and the precision.

The script ChangeDetection.txt implements this graphical model in WinBUGS:

```
# Change Detection
model{
  # Data Come From A Gaussian
  for (i in 1:n){
    c[i] ~ dnorm(mu[z1[i]],lambda)
  }
  # Group Means
  mu[1] ~ dnorm(0,.001)
  mu[2] ~ dnorm(0,.001)
  # Common Precision
  lambda  ~ dgamma(.001,.001)
  sigma  <- 1/sqrt(lambda)
  # Which Side is Time of Change Point?
  for (i in 1:n){
    z[i]  <- step(t[i]-tau)
    z1[i] <- z[i]+1
  }
  # Prior On Change Point
  tau ~ dunif(0,n)
}
```

Note the use of the step function. This function returns 1 if its argument is greater than or equal to zero, and 0 otherwise. The z1 variable, however, serves as an indicator variable for mu, and therefore it needs to take on values 1 and 2. This is the reason z is transformed to z1. Study this code and make sure you understand what the step function accomplishes in this example.

The code ChangeDetection.m or ChangeDetection.R applies the model to the near-infrared spectrographic data. Uniform sampling is assumed, so that $t = 1, \ldots, 1178$.

The code produces a simple analysis, finding the mean of the posteriors for τ, μ_1 and μ_2, and using these summary points to overlay the inferences over the raw data. The result looks something like Figure 5.7. The time series data themselves are shown by the jagged black lines. The expected value of the posterior mean for the pre- and post-change levels, given by the posterior means for μ_1 and μ_2, are shown by the horizontal lines. The expected change point, given by the posterior mean for τ, is just under 800 samples, and is used to separate the plotting of the pre-change level from the post-change level.

Exercises

Exercise 5.4.1 Draw the posterior distributions for the change point, the means, and the common standard deviation.

Exercise 5.4.2 Figure 5.7 shows the mean of the posterior distribution for the change point (this is the point in time where the two horizontal lines meet). Can you think of a situation in which such a plotting procedure can be misleading?

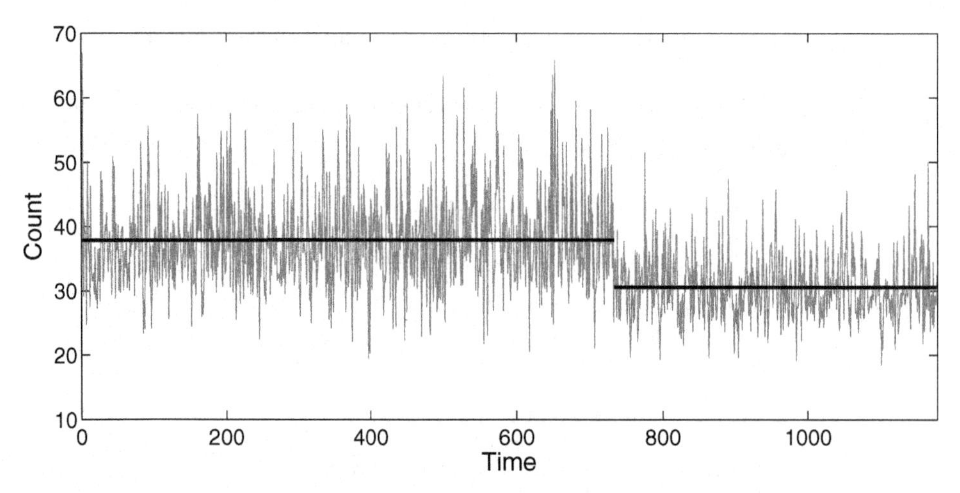

Fig. 5.7 Identification of a change point in time series data. The time series are shown by the jagged lines (note that these are observed data to be modeled; they are not chains from MCMC sampling), and the pre- and post-change levels around the expected change point are shown by the two overlaid horizontal lines.

Exercise 5.4.3 Imagine that you apply this model to a data set that has two change points instead of one. What could happen?

5.5 Censored data

Starting April 13 2005, Cha Sa-soon, a 68-year-old grandmother living in Jeonju, South Korea, repeatedly tried to pass the written exam for a driving license. In South Korea, this exam features 50 four-choice questions. In order to pass, one requires a score of at least 60 points out of a maximum of 100. Accordingly, we assume that each correct answer is worth two points, so that in order to pass, one needs to answer at least 30 questions correctly.

What has made Cha Sa-soon something of a national celebrity is that she failed to pass the test on 949 consecutive occasions, spending the equivalent of 4200 US dollars on application fees. In her last, 950th attempt, Cha Sa-soon scored the required minimum of 30 correct questions and finally passed her written exam. After her 775th failure, in February 2009, Mrs Cha told Reuters news agency, "I believe you can achieve your goal if you persistently pursue it. So don't give up your dream, like me. Be strong and do your best."

We know that on her final and 950th attempt, Cha Sa-soon answered 30 questions correctly. In addition, news agencies report that in her 949 unsuccessful attempts, the number of correct answers had ranged from 15 to 25. Armed with this knowledge, what can we say about θ, the latent probability that Cha Sa-soon can answer

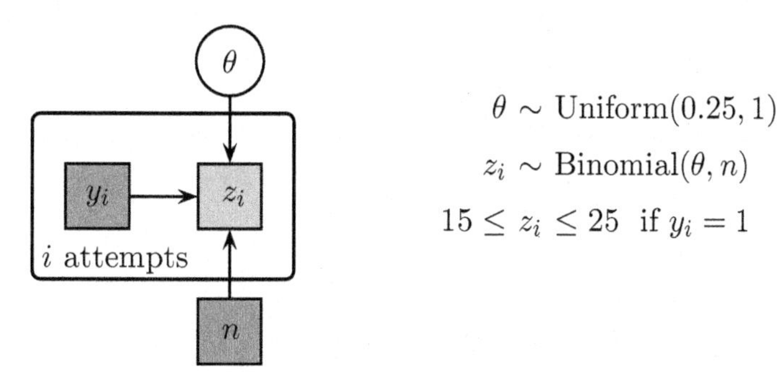

Fig. 5.8 Graphical model for inferring a rate from observed and censored data.

any one question correctly? Note that we assume each question is equally difficult, and that Cha Sa-soon does not learn from her earlier attempts.

The Cha Sa-soon data are special because we do not know the precise scores for the failed attempts. We only know that these scores range from 15 to 25. In statistical terms, these data are said to be censored, both from below and from above. We follow an approach inspired by Gelman and Hill (2007, p. 405) to apply WinBUGS to the problem of dealing with censored data.

Figure 5.8 presents a graphical model for dealing with the censored data. The variable z_i represents both the first 949 unobserved, and the final observed attempt. This means z_i is observed once, but not observed the other times. This sort of variable is known as *partially observed*, and is denoted in the graphical model by a lighter shading, between the dark shading of fully observed nodes, and the lack of shading for fully unobserved or latent nodes.

The variable y_i is a simple binary indicator variable, denoting whether or not the ith attempt is observed. The bounds $z^{\text{lo}} = 15$ and $z^{\text{hi}} = 25$ give the known censored interval for the unobserved attempts. Finally, $n = 50$ is the number of questions in the test. This means that $z_i \sim \text{Binomial}(\theta, n)_{\mathcal{I}(z^{\text{lo}}, z^{\text{hi}})}$ when y_i indicates a censored attempt, but that z_i is not censored for the final known score $z_{950} = 30$. The probability of a correct answer to a question, θ, is given a uniform prior between 0.25 and 1, corresponding to the assumption that chance accuracy of 1 in 4 is the lowest possible probability.

The script `ChaSaSoon.txt` implements this graphical model in WinBUGS:

```
# ChaSaSoon Censored Data
model{
  for (i in 1:nattempts){
    # If the Data Were Unobserved y[i]=1, Otherwise y[i]=0
    z.low[i]  <- 15*equals(y[i],1)+0*equals(y[i],0)
    z.high[i] <- 25*equals(y[i],1)+n*equals(y[i],0)
    z[i] ~ dbin(theta,n)I(z.low[i],z.high[i])
  }
  # Uniform Prior on Rate Theta
  theta ~ dbeta(1,1)I(.25,1)
}
```

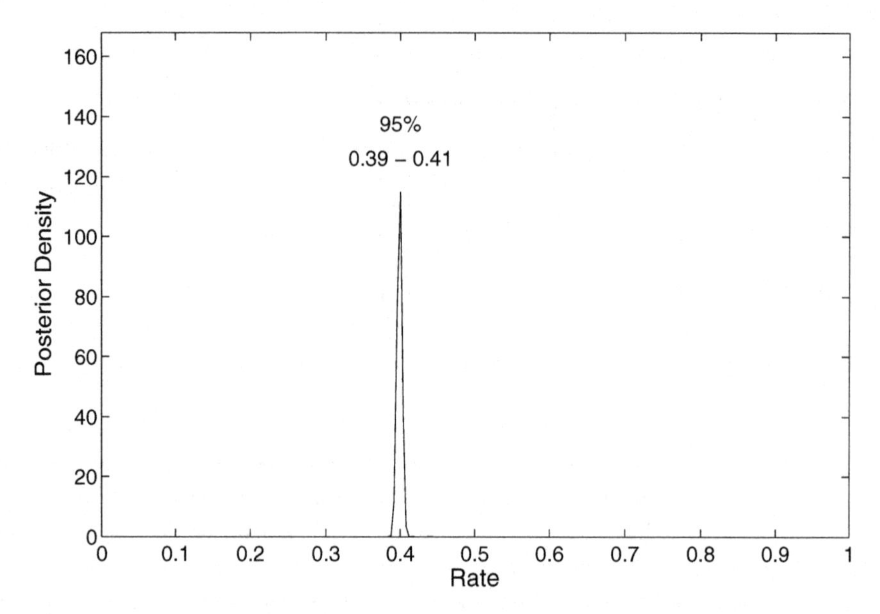

Fig. 5.9 Posterior density for Cha Sa-soon's rate of answering questions correctly.

Note the use of the `equals` command, which returns 1 when its arguments match, and 0 when they mismatch. Thus, when `y[i]=1`, for censored data, `z.low[i]` is set to 15 and `z.hi[i]` is set to 25. When `y[i]=0`, `z.low[i]` is set to 0 and `z.hi[i]` is set to n. These `z.low[i]` and `z.hi[i]` values are then applied to censor the binomial distribution that generates the test scores, using the WinBUGS `I` ("interval") command. In this way, the use of `equals` implements what might be considered the "case" or "if-then-else" logic of the model.

The code `ChaSaSoon.m` or `ChaSaSoon.R` applies the model to the data from Cha Sa-soon.[2] The posterior density for θ is shown in Figure 5.9, and can be seen to be relatively peaked. Despite the fact that we do not know the actual scores for 949 of the 950 results, we are still able to infer a lot about θ.

Exercises

Exercise 5.5.1 Do you think Cha Sa-soon could have passed the test by just guessing?

Exercise 5.5.2 What happens when you increase the interval in which you know the data are located, from 15–25 to something else?

Exercise 5.5.3 What happens when you decrease the number of failed attempts?

Exercise 5.5.4 What happens when you increase Cha Sa-soon's final score from 30?

[2] On some computers, WinBUGS will persistently return the mysterious error message "value of binomial `z[950]` must be greater than lower bound." If you know how to fix this error, we would love to hear from you. Otherwise, we can only suggest you run the code on a different computer.

Exercise 5.5.5 Do you think the assumption that all of the scores follow a binomial distribution with a single rate of success is a good model for these data?

5.6 Recapturing planes

An interesting inference problem that occurs in a number of fields is to estimate the size of a population, when a census is impossible, but repeated surveying is possible. For example, the goal might be to estimate the number of animals in a large woodland area that cannot be searched exhaustively. Or, the goal might be to decide how many students are on a campus, but it is not possible to count them all. Or, the goal might be to find out how many words in a given language a person knows, but it is not feasible to ask the person to list them all.

A clever sampling approach to this problem is given by capture-and-recapture methods. The basic idea is to capture (i.e., identify, tag, or otherwise remember) a sample at one time point, and then collect another sample. The number of items in the second sample that were also in the first then provides relevant information as to the population size. High recapture counts suggest that the population is small, and low recapture counts suggest that the population is large.

Probably the simplest possible version of this approach can be formalized with t as the unknown population size, x as the size of the first sample (i.e., number of units captured), and n as the size of the second sample from which a subset of k units were also present in the first sample (i.e., number of units recaptured). That is, first x animals are tagged or people remembered or words produced, then k out of n are seen again when a second sample is taken.

The statistical model to relate the counts and make inferences about the population size t is based on the hypergeometric distribution. The probability of seeing k items recaptured in a sample of size n, from the x originally captured in a population of size t, is

$$\Pr(K = k) = \binom{x}{k}\binom{t-x}{n-k}\bigg/\binom{t}{n}.$$

Intuitively, the second sample involves taking n items from a population of t, and has k out of x recaptures, and $n - k$ other items out of the other $t - x$ in the population. Another way to formalize this is to say that the number of recaptures k is a sample from a hypergeometric distribution

$$k \sim \text{Hypergeometric}(n, x, t).$$

To make these ideas concrete, consider the challenge of estimating how many aircraft a small airline company has in its fleet. One day at an airport, you see 10 of the airline company's planes parked at adjacent gates, and record their unique identifying tail numbers. A few days later, at a different airport, you see 5 of the same company's planes. Looking at the tail number of those planes, you observe

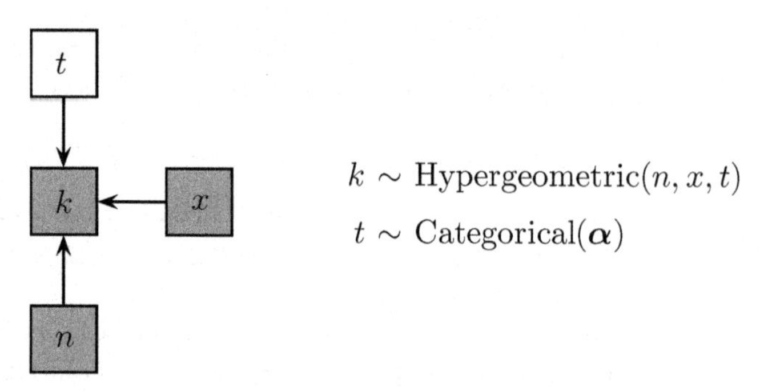

Fig. 5.10 Graphical model for inferring a population from capture-and-recapture data.

that 4 of the 5 were part of your original list. This is a capture-and-recapture problem with $x = 10$, $k = 4$, and $n = 5$.

The Bayesian approach to this problem involves assigning a prior to t, and using the hypergeometric distribution as the appropriate likelihood function. Conceptually, this means $k \sim \text{Hypergeometric}(n, x, t)$, as in the graphical model in Figure 5.10. The vector $\boldsymbol{\alpha}$ allows for any sort of prior mass to be given to all the possible counts for the population total. Since $x + (n - k)$ items are known to exist, one reasonable choice of prior might be to make every possibility from $x+(n-k)$ to t^{max} equally likely, where t^{max} is a sensible upper bound on the possible population. Suppose, for example, in the airplane problem that you know that the maximum number the company could possibly have is 50 planes, so that $t^{\text{max}} = 50$.

While it is simple conceptually, there is a difficulty in implementing the graphical model in Figure 5.10. The problem is that WinBUGS does not provide the hypergeometric distribution. It is, however, possible to implement distributions that are not provided, but for which the likelihood function can be expressed in WinBUGS. This can be done using either the so-called "ones trick" or the "zeros trick."[3] These tricks rely on simple properties of the Poisson and Bernoulli distributions. By implementing the likelihood function of the new distribution within the Poisson or Bernoulli distribution, and forcing values of 1 or 0 to be sampled, it can be shown that the samples actually generated will come from the desired distribution.

The script `Planes.txt` implements the graphical model in Figure 5.10 in WinBUGS, using the zeros trick. Note how the terms in the log-likelihood expression for the hypergeometric distribution are built up to define `phi`, and a constant `C` is used to ensure the Poisson distribution is used with a positive value:

```
# Planes
model{
  # Hypergeometric Likelihood Via Zeros Trick
  logterm1 <- logfact(x)-logfact(k)-logfact(x-k)
  logterm2 <- logfact(t-x)-logfact(n-k)-logfact((t-x)-(n-k))
  logterm3 <- logfact(t)-logfact(n)-logfact(t-n)
  C <- 1000
```

[3] Using the zeros trick or ones trick in JAGS involves putting the assignment of `zeros` or `ones` inside the data definition block, rather than inside the model definition block.

Box 5.2 The zeros trick, ones trick, and WBDev

The zeros trick and ones trick are extremely useful, and relatively easy to implement in many cases, but a little difficult to understand conceptually. The key insight is that the negative log-likelihood of a sample of 0 from $\text{Poisson}(\phi)$ is ϕ, and similarly for a sample of 1 from $\text{Bernoulli}(\theta)$ it is θ. So, by setting $\log \phi$ or θ appropriately, and forcing 1 or 0 to be observed, sampling effectively proceeds from the distribution defined by ϕ or θ.

More complicated extensions to the distributions and functions available in WinBUGS require using the WinBUGS Development Interface (WBDev: Lunn, 2003). This is an add-on program that allows the user to hand-code functions and distributions in Component Pascal. Wetzels, Lee, and Wagenmakers (2010) provide a tutorial on WBDev that includes simple worked examples of defining new distributions and functions. More detailed cognitive science applications are provided by Wetzels, Vandekerckhove, et al. (2010) implementing the Expectancy-Valence model of decision-making as a function in WBDev, and Vandekerckhove et al. (2011) implementing the drift-diffusion model as a distribution in WBDev. Both of these applications would be impractical without WBDev.

```
  phi <- -(logterm1+logterm2-logterm3)+C
  zeros <- 0
  zeros ~ dpois(phi)
  # Prior on Population Size
  for (i in 1:tmax){
    tptmp[i] <- step(i-(x+n-k))
    tp[i] <- tptmp[i]/sum(tptmp[1:tmax])
  }
  t ~ dcat(tp[])
}
```

The code `Planes.m` or `Planes.R` applies the model to the data $x = 10$, $k = 4$, and $n = 5$, using uniform prior mass for all possible sizes between $x + (n - k) = 11$ and $t^{\max} = 50$. The posterior distribution for t is shown in Figure 5.11. The inference is that it is mostly likely there are not many more than 11 planes, which makes intuitive sense, since 4 out of 5 in the second sample were from the original set of 10.

Exercises

Exercise 5.6.1 Try changing the number of planes seen again in the second sample from $k = 4$ to $k = 0$. What inference do you draw about the population size now?

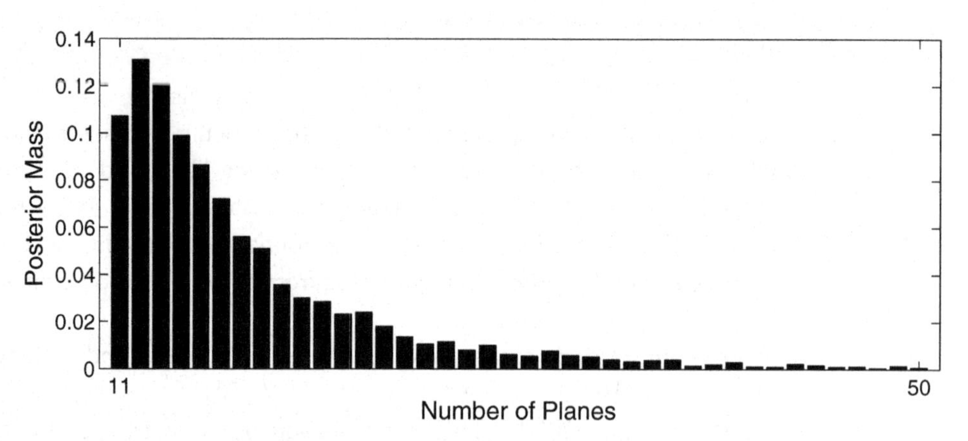

Fig. 5.11 Posterior mass for the number of planes, known to be 50 or fewer, based on a capture-recapture experiment with $x = 10$ planes in the first sample, and $k = 4$ out of $n = 5$ seen again in the second sample.

Exercise 5.6.2 How much impact does the upper bound $t^{\mathrm{max}} = 50$ have on the final conclusions when $k = 4$ and when $k = 0$? Develop your answer by trying both the $k = 4$ and $k = 0$ cases with $t^{\mathrm{max}} = 100$.

Exercise 5.6.3 Suppose, having obtained the posterior mass in Figure 5.11, the same fleet of planes was subjected to a new sighting at a different airport at a later day. What would be an appropriate prior for t?

6 Latent-mixture models

6.1 Exam scores

Suppose a group of 15 people sit an exam made up of 40 true-or-false questions, and they get 21, 17, 21, 18, 22, 31, 31, 34, 34, 35, 35, 36, 39, 36, and 35 right. These scores suggest that the first 5 people were just guessing, but the last 10 had some level of knowledge.

One way to make statistical inferences along these lines is to assume there are two different groups of people. These groups have different probabilities of success, with the guessing group having a probability of 0.5, and the knowledge group having a probability greater than 0.5. Whether each person belongs to the first or the second group is a latent or unobserved variable that can take just two values. Using this approach, the goal is to infer to which group each person belongs, and also the rate of success for the knowledge group.

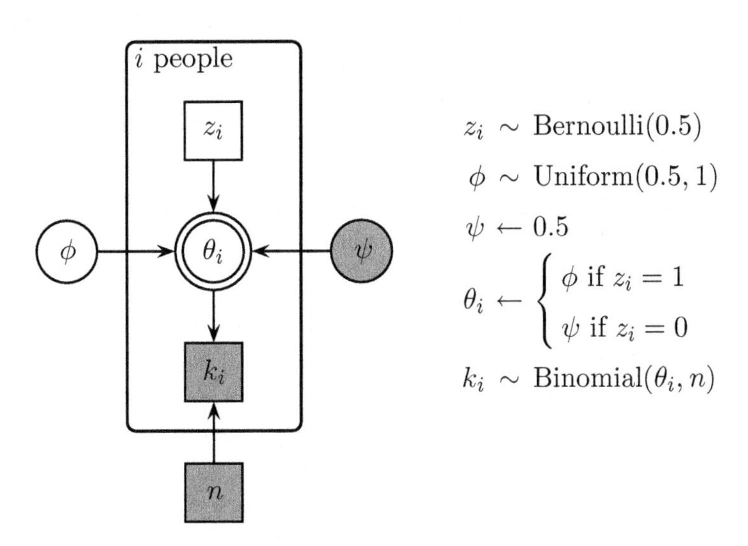

Fig. 6.1 Graphical model for inferring membership of two latent groups, with different rates of success in answering exam questions.

A graphical model for doing this is shown in Figure 6.1. The number of correct answers for the ith person is k_i, and is out of $n = 40$. The probability of success on each question for the ith person is the rate θ_i. This rate is either ψ, if the person is in the guessing group, or ϕ if the person is in the knowledge group. Which group the ith person belongs to is determined by a binary indicator variable z_i, with $z_i = 0$ if the ith person is in the guessing group, and $z_i = 1$ if the ith person is in the knowledge group.

We assume each of these indicator variables is equally likely to be 0 or 1 a priori, so they have the prior $z_i \sim \text{Bernoulli}(1/2)$. For the guessing group, we assume that the rate is $\psi = 1/2$. For the knowledge group, we use a prior where all rate possibilities greater than $1/2$ are equally likely, so that $\phi \sim \text{Uniform}(0.5, 1)$.

This type of model is known as a *latent-mixture* model, because the data are assumed to be generated by two different processes that combine or mix, and important properties of that mixture are unobserved or latent. In this case, the two components that mix are the guessing and knowledge processes, and the group membership of each person is latent.

The script `Exams_1.txt` implements the graphical model in WinBUGS:

```
# Exam Scores
model{
  # Each Person Belongs To One Of Two Latent Groups
  for (i in 1:p){
    z[i] ~ dbern(0.5)
  }
  # First Group Guesses
  psi <- 0.5
  # Second Group Has Some Unknown Greater Rate Of Success
  phi ~ dbeta(1,1)I(0.5,1)
  # Data Follow Binomial With Rate Given By Each Person's Group Assignment
  for (i in 1:p){
    theta[i] <- equals(z[i],0)*psi+equals(z[i],1)*phi
    k[i] ~ dbin(theta[i],n)
  }
}
```

The code `Exams_1.m` or `Exams_1.R` makes inferences about group membership, and the success rate of the knowledge group, using the model.

Exercises

Exercise 6.1.1 Draw some conclusions about the problem from the posterior distribution. Who belongs to what group, and how confident are you?

Exercise 6.1.2 The initial allocations of people to the two groups in this code is random, and so will be different every time you run it. Check that this does not affect the final results from sampling.

Exercise 6.1.3 Include an extra person in the exam, with a score of 28 out of 40. What does their posterior for z tell you? Now add four extra people, all with the score 28 out of 40. Explain the change these extra people make to the inference.

Exercise 6.1.4 What happens if you change the prior on the success rate of the second group to be uniform over the whole range from 0 to 1, and so allow for worse-than-guessing performance?

Exercise 6.1.5 What happens if you change the initial expectation that everybody is equally likely to belong to either group, and have an expectation that people generally are not guessing, with (say), $z_i \sim \text{Bernoulli}(0.9)$?

6.2 Exam scores with individual differences

The previous example shows how sampling can model data as coming from a mixture of sources, and infer properties of these latent groups. But the specific model has at least one big weakness, which is that it assumes all the people in the knowledge group have exactly the same rate of success on the questions.

One straightforward way to allow for individual differences in the knowledge group is to extend the model hierarchically. This involves drawing the success rate for each of the people in the knowledge group from an over-arching distribution. One convenient (but not perfect) choice for this "individual differences" distribution is a Gaussian. It is a natural statistical model for individual variation, at least in the absence of any richer theory. But it has the problem of allowing for success rates below zero and above one. An inelegant but practical and effective way to deal with this is simply to restrict the sampled success rates to the valid range.

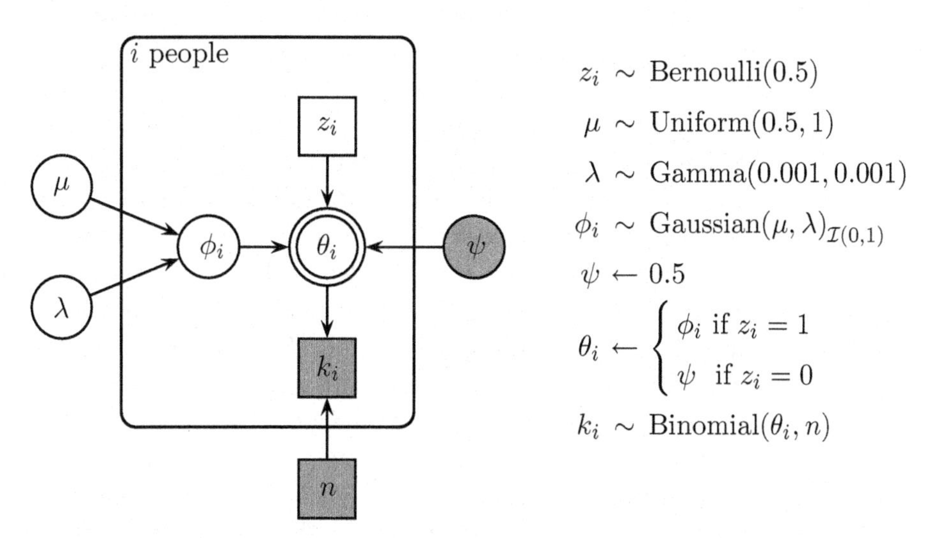

Fig. 6.2 Graphical model for inferring membership of two latent groups, with different rates of success in answering exam questions, allowing for individual differences in the knowledge group.

A graphical model that implements this idea is shown in Figure 6.2. It extends the original model by having a knowledge group success rate ϕ_i for the ith person. These success rates are drawn from a Gaussian distribution with mean μ and precision

Box 6.1 Assessing and improving convergence

In a perfect world, a single MCMC chain would immediately begin drawing samples from the posterior distribution, and the only computational issue would be how many are needed to form a sufficiently precise approximation. This ideal state of affairs is often not what happens, and latent-mixture models are notorious for needing convergence checks. So, this is a good place to list some checks (see also Gelman, 1996; Gelman & Hill, 2007).

The basic principle is that, when the sampling process has converged, chains with substantially different starting values should be indistinguishable from each other. One implication of this requirement is that chains should vary around a constant mean, so a slow drift up or down signals a problem. And, if the sampling process has converged, each individual chain should look like a "fat hairy caterpillar," because this visual appearance is generated when successive values are relatively independent. As a formal test for convergence, the $\hat{R}$ statistic (Gelman & Rubin, 1992) is widely used. It is basically a measure of between-chain to within-chain variance, and so values close to 1 indicate convergence. As a rule of thumb, values higher than 1.1 are (deeply) suspect. If you were not paying much attention to the `rhat` values WinBUGS is returning to Matlab and R in previous modeling exercises, now is a good time to start checking them.

There are three basic remedies for a lack of convergence, easily implemented in WinBUGS for any model. The first is simply to collect many more samples, or more chains of samples, and wait (and hope) for convergence. The second is to increase the number of *burn-in* samples, which are initial samples in a chain that are discarded. This will be effective if separate chains are sensitive to their starting points, and take some time to converge. A worked example of this is presented in Section 11.2. The third is to *thin* the samples, by retaining only one out of every n. This will be effective if a chain is autocorrelated, with lack of independence between samples. A worked example of this is presented in Section 3.6. There are other, more advanced, methods for improving convergence in WinBUGS, involving changing the model itself. Worked examples of the *parameter expansion* method are presented in Sections 11.3 and 14.2.

λ. The mean μ is given a uniform prior between 0.5 and 1.0, consistent with the original assumption that people in the knowledge group have a greater-than-chance success rate.

Box 6.2 Scripts for graphical models

The scripts that implement graphical models in WinBUGS are declarative, rather than procedural. This means the order of the commands does not matter. All that a script does is define the observed and unobserved variables in a graphical model, saying how they are distributed, and how they relate to each other. This is inherently a structure, rather than a process, and so order is not important. In practice this means, for example, that a separate loop is not needed in a script like `Exam_2.txt` to define `k[i]`, `z[i]`, and `phi[i]`. Exactly the same graphical model would be defined if they were all placed inside one `for (i in 1:p)` loop. Sometimes, however, it is conceptually clearer to use separate loops to implement different parts of a graphical model.

The script `Exams_2.txt` implements the graphical model in WinBUGS:

```
# Exam Scores With Individual Differences
model{
  # Rates Given By Each Person's Group Assignment
  for (i in 1:p){
    theta[i] <- equals(z[i],0)*psi+equals(z[i],1)*phi[i]
    k[i] ~ dbin(theta[i],n)
  }
  # Each Person Belongs To One Of Two Latent Groups
  for (i in 1:p){
    z[i] ~ dbern(0.5)
  }
  # The Second Group Allows Individual Differences
  for (i in 1:p){
    phi[i] ~ dnorm(mu,lambda)I(0,1)
  }
  # First Group Guesses
  psi <- 0.5
  # Second Group Mean, Precision (And Standard Deviation)
  mu ~ dbeta(1,1)I(.5,1) # >0.5 Average Success Rate
  lambda ~ dgamma(.001,.001)
  sigma <- 1/sqrt(lambda)
  # Posterior Predictive For Second Group
  predphi ~ dnorm(mu,lambda)I(0,1)
}
```

Notice that the code includes a variable `predphi` that draws success rates from the inferred Gaussian distribution of the knowledge group.

The code `Exams_2.m` or `Exams_2.R` makes inferences about group membership, the success rate of each person in the knowledge group, and the mean and standard deviation of the over-arching Gaussian for the knowledge group.

Exercises

Exercise 6.2.1 Compare the results of the hierarchical model with the original model that did not allow for individual differences.

Exercise 6.2.2 Interpret the posterior distribution of the variable `predphi`. How does this distribution relate to the posterior distribution for `mu`?

Exercise 6.2.3 In what sense could the latent assignment of people to groups in this case study be considered a form of model selection?

6.3 Twenty questions

Suppose a group of 10 people attend a lecture, and are asked a set of 20 questions afterwards, with every answer being either correct or incorrect. The pattern of data is shown in Table 6.1. From this pattern of correct and incorrect answers we want to infer two things. The first is how well each person attended to the lecture. The second is how hard each of the questions was.

Table 6.1 Correct and incorrect answers for 10 people on 20 questions.

	Question																			
	A	B	C	D	E	F	G	H	I	J	K	L	M	N	O	P	Q	R	S	T
Person 1	1	1	1	1	0	0	1	1	0	1	0	0	1	0	0	1	0	1	0	0
Person 2	0	1	1	0	0	0	0	0	0	0	0	0	0	0	0	0	0	0	0	0
Person 3	0	0	1	0	0	0	1	1	0	0	0	0	1	0	0	0	0	0	0	0
Person 4	0	0	0	0	0	0	1	0	1	1	0	0	0	0	0	0	0	0	0	0
Person 5	1	0	1	1	0	1	1	1	0	1	0	0	1	0	0	0	0	1	0	0
Person 6	1	1	0	1	0	0	0	1	0	1	0	1	1	0	0	1	0	1	0	0
Person 7	0	0	0	0	0	0	0	0	0	0	0	1	0	0	0	0	0	0	0	0
Person 8	0	0	0	0	0	0	0	0	0	0	0	0	0	0	0	0	0	0	0	0
Person 9	0	1	1	0	0	0	0	1	0	1	0	0	1	0	0	0	0	1	0	1
Person 10	1	0	0	0	0	0	1	0	0	1	0	0	1	0	0	0	0	0	0	0

One way to make these inferences is to specify a model of how a person's attentiveness and a question's difficulty combine to give an overall probability that the question will be answered correctly. A very simple model involves assuming that each person listens to some proportion of the lecture, and that each question has some probability of being answered correctly if the person was listening at the right point in the lecture.

A graphical model that implements this idea is shown in Figure 6.3. Under the model, if the ith person's probability of listening is p_i, and the jth question's probability of being answered correctly if the relevant information is heard is q_j,

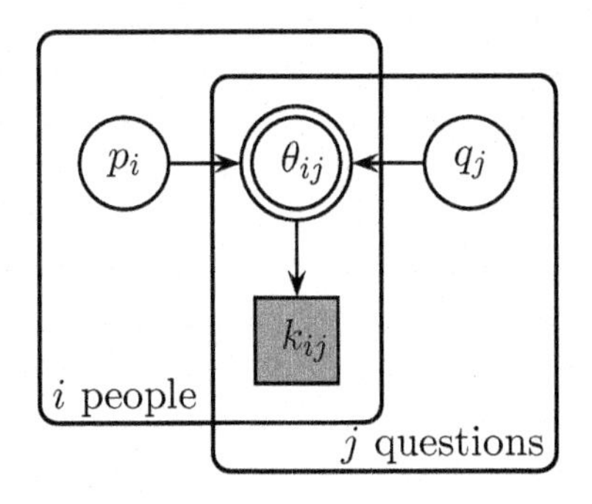

$$p_i, q_j \sim \text{Beta}(1, 1)$$
$$\theta_{ij} \leftarrow p_i q_j$$
$$k_{ij} \sim \text{Bernoulli}(\theta_{ij})$$

Fig. 6.3 Graphical model for inferring the rate people listened to a lecture, and the difficulty of the questions.

then the probability the ith person will answer the jth question correctly is just $\theta_{ij} = p_i q_j$. The observed pattern of correct and incorrect answers, where $k_{ij} = 1$ if the ith person answered the jth question correctly, and $k_{ij} = 0$ if they did not, then is a draw from a Bernoulli distribution with probability θ_{ij}.

The script `TwentyQuestions.txt` implements the graphical model in WinBUGS:

```
# Twenty Questions
model{
  # Correctness Of Each Answer Is Bernoulli Trial
  for (i in 1:np){
    for (j in 1:nq){
      k[i,j] ~ dbern(theta[i,j])
    }
  }
  # Probability Correct Is Product Of Question By Person Rates
  for (i in 1:np){
    for (j in 1:nq){
      theta[i,j] <- p[i]*q[j]
    }
  }
  # Priors For People and Questions
  for (i in 1:np){
    p[i] ~ dbeta(1,1)
  }
  for (j in 1:nq){
    q[j] ~ dbeta(1,1)
  }
}
```

The code `TwentyQuestions.m` or `TwentyQuestions.R` makes inferences about the data in Table 6.1 using the model.

Exercises

Exercise 6.3.1 Draw some conclusions about how well the various people listened, and about the difficulties of the various questions. Do the marginal posterior distributions you are basing your inference on seem intuitively reasonable?

Exercise 6.3.2 Now suppose that three of the answers were not recorded, for whatever reason. Our new data set, with missing data, now takes the form shown in Table 6.2. Bayesian inference will automatically make predictions about these missing values (i.e., "fill in the blanks") by using the same probabilistic model that generated the observed data. Missing data are entered as `nan` ("not a number") in Matlab, and `NA` ("not available") in R or WinBUGS. Including the variable `k` as one to monitor when sampling will then provide posterior values for the missing values. That is, it provides information about the relative likelihood of the missing values being each of the possible alternatives, using the statistical model and the available data. Look through the Matlab or R code to see how all of this is implemented in the second data set. Run the code, and interpret the posterior distributions for the three missing values. Are they reasonable inferences?

Table 6.2 Correct, incorrect, and missing answers for 10 people on 20 questions.

	Question																			
	A	B	C	D	E	F	G	H	I	J	K	L	M	N	O	P	Q	R	S	T
Person 1	1	1	1	1	0	0	1	1	0	1	0	0	?	0	0	1	0	1	0	0
Person 2	0	1	1	0	0	0	0	0	0	0	0	0	0	0	0	0	0	0	0	0
Person 3	0	0	1	0	0	0	1	1	0	0	0	0	1	0	0	0	0	0	0	0
Person 4	0	0	0	0	0	0	1	0	1	1	0	0	0	0	0	0	0	0	0	0
Person 5	1	0	1	1	0	1	1	1	0	1	0	0	1	0	0	0	0	1	0	0
Person 6	1	1	0	1	0	0	0	1	0	1	0	1	1	0	0	1	0	1	0	0
Person 7	0	0	0	0	0	0	0	0	0	0	0	1	0	0	0	0	0	0	0	0
Person 8	0	0	0	0	?	0	0	0	0	0	0	0	0	0	0	0	0	0	0	0
Person 9	0	1	1	0	0	0	0	1	0	1	0	0	1	0	0	0	0	1	0	1
Person 10	1	0	0	0	0	0	1	0	0	1	0	0	1	0	0	0	0	?	0	0

Exercise 6.3.3 The definition of the accuracy for a person on a question in terms of the product $\theta_{ij} = p_i q_j$ is very simple to understand, but other models of the interaction between person ability and question difficulty are used in psychometric models. For example, the Rasch model (e.g., Andrich, 1988) uses $\theta_{ij} = \exp(p_i - q_j) / (1 + \exp(p_i - q_j))$. Change the graphical model to implement the Rasch model.

6.4 The two-country quiz

Suppose a group of people take a historical quiz, and each answer for each person is scored as correct or incorrect. Some of the people are Thai, and some are Moldovan.

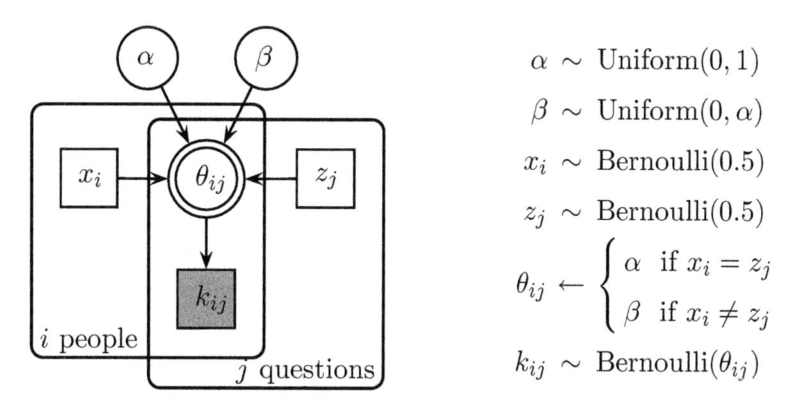

Fig. 6.4 Graphical model for inferring the country of origin for people and questions.

Some of the questions are about Thai history, and it is more likely the answer would be known by a Thai person than a Moldovan. The rest of the questions are about Moldovan history, and it is more likely the answer would be known by a Moldovan than a Thai.

We do not know who is Thai or Moldovan, and we do not know the content of the questions. All we have are the data shown in Table 6.3. Spend some time just looking at the data, and try to infer which people are from the same country, and which questions relate to their country.

Table 6.3 Correct and incorrect answers for 8 people on 8 questions.

	Question							
	A	B	C	D	E	F	G	H
Person 1	1	0	0	1	1	0	0	1
Person 2	1	0	0	1	1	0	0	1
Person 3	0	1	1	0	0	1	0	0
Person 4	0	1	1	0	0	1	1	0
Person 5	1	0	0	1	1	0	0	1
Person 6	0	0	0	1	1	0	0	1
Person 7	0	1	0	0	0	1	1	0
Person 8	0	1	1	1	0	1	1	0

A good way to make these inferences formally is to assume there are two types of answers. For those where the nationality of the person matches the origin of the question, the answer will be correct with high probability. For those where a person is being asked about the other country, the answer will have a very low probability of being correct.

A graphical model that implements this idea is shown in Figure 6.4. The rate α is the (expected to be high) probability of a person from a country correctly answering a question about their country's history. The rate β is the (expected to be low) probability of a person correctly answering a question about the other country's history. To capture the knowledge about the rates, the priors constrain $\alpha \geq \beta$, by defining `alpha` $\sim$ `dunif(0,1)` and `beta` $\sim$ `dunif(0,alpha)`. At first glance, this might seem inappropriate, since it specifies a prior for one parameter in terms of another (unknown, and being inferred) parameter. Conceptually, it is clearer to think of this syntax as a (perhaps clumsy) way to specify a *joint* prior over α and β in which the $\alpha \geq \beta$. Graphically, the parameter space over (α, β) is a unit square, and the prior being specified is the half of the square on one side of the diagonal line $\alpha = \beta$.

In the remainder of the graphical model, the binary indicator variable x_i assigns the ith person to one or other country, and z_j similarly assigns the jth question to one or other country. The probability the ith person will answer the jth question correctly is θ_{ij}, which is simply α if the country assignments match, and β if they do not. Finally, the actual data k_{ij} indicating whether or not the answer was correct follow a Bernoulli distribution with rate θ_{ij}.

The script `TwoCountryQuiz.txt` implements the graphical model in WinBUGS:

```
# The Two Country Quiz
model{
  # Probability of Answering Correctly
  alpha ~ dunif(0,1)     # Match
  beta ~ dunif(0,alpha) # Mismatch
  # Group Membership For People and Questions
  for (i in 1:nx){
    x[i] ~ dbern(0.5)
    x1[i] <- x[i]+1
  }
  for (j in 1:nz){
    z[j] ~ dbern(0.5)
    z1[j] <- z[j]+1
  }
  # Probability Correct For Each Person-Question Combination By Groups
  for (i in 1:nx){
    for (j in 1:nz){
      theta[i,j,1,1] <- alpha
      theta[i,j,1,2] <- beta
      theta[i,j,2,1] <- beta
      theta[i,j,2,2] <- alpha
    }
  }
  # Data Are Bernoulli By Rate
  for (i in 1:nx){
    for (j in 1:nz){
      k[i,j] ~ dbern(theta[i,j,x1[i],z1[j]])
    }
  }
}
```

The code `TwoCountryQuiz.m` or `TwoCountryQuiz.R` makes inferences about the data in Table 6.3 using the model.

Exercises

Exercise 6.4.1 Interpret the posterior distributions for `x[i]`, `z[j]`, `alpha`, and `beta`. Do the formal inferences agree with your original intuitions?

Exercise 6.4.2 The priors on the probabilities of answering correctly capture knowledge about what it means to match and mismatch, by imposing an order constraint $\alpha \geq \beta$. Change the code so that this information is not included, by using priors `alpha~dbeta(1,1)` and `beta~dbeta(1,1)`. Run a few chains against the same data, until you get an inappropriate, and perhaps counter-intuitive, result. The problem that is being encountered is known as model indeterminacy or label-switching. Describe the problem, and discuss why it comes about.

Exercise 6.4.3 Now suppose that three extra people enter the room late, and begin to take the quiz. One of them (Late Person 1) has answered the first four questions, the next (Late Person 2) has only answered the first question, and the final new person (Late Person 3) is still sharpening their pencil, and has not started the quiz. This situation can be represented as an updated data set, now with missing data, as in Table 6.4. Interpret the inferences the model makes about the nationality of the late people, and whether or not they will get the unfinished questions correct.

Table 6.4 Correct, incorrect, and missing answers for 8 people and 3 late people on 8 questions.

	Question							
	A	B	C	D	E	F	G	H
Person 1	1	0	0	1	1	0	0	1
Person 2	1	0	0	1	1	0	0	1
Person 3	0	1	1	0	0	1	0	0
Person 4	0	1	1	0	0	1	1	0
Person 5	1	0	0	1	1	0	0	1
Person 6	0	0	0	1	1	0	0	1
Person 7	0	1	0	0	0	1	1	0
Person 8	0	1	1	1	0	1	1	0
Late Person 1	1	0	0	1	?	?	?	?
Late Person 2	0	?	?	?	?	?	?	?
Late Person 3	?	?	?	?	?	?	?	?

Exercise 6.4.4 Finally, suppose that you are now given the correctness scores for a set of 10 new people, whose data were not previously available, but who form part of the same group of people we are studying. The updated data set is shown in Table 6.5. Interpret the inferences the model makes about the nationality of the new people. Revisit the inferences about the late people, and whether or not they will get the unfinished questions correct. Does the

inference drawn by the model for the third late person match your intuition? There is a problem here. How could it be fixed?

Table 6.5 Correct, incorrect, and missing answers for 8 people, 3 late people, and 10 new people on 8 questions.

	Question							
	A	B	C	D	E	F	G	H
New Person 1	1	0	0	1	1	0	0	1
New Person 2	1	0	0	1	1	0	0	1
New Person 3	1	0	0	1	1	0	0	1
New Person 4	1	0	0	1	1	0	0	1
New Person 5	1	0	0	1	1	0	0	1
New Person 6	1	0	0	1	1	0	0	1
New Person 7	1	0	0	1	1	0	0	1
New Person 8	1	0	0	1	1	0	0	1
New Person 9	1	0	0	1	1	0	0	1
New Person 10	1	0	0	1	1	0	0	1
Person 1	1	0	0	1	1	0	0	1
Person 2	1	0	0	1	1	0	0	1
Person 3	0	1	1	0	0	1	0	0
Person 4	0	1	1	0	0	1	1	0
Person 5	1	0	0	1	1	0	0	1
Person 6	0	0	0	1	1	0	0	1
Person 7	0	1	0	0	0	1	1	0
Person 8	0	1	1	1	0	1	1	0
Late Person 1	1	0	0	1	?	?	?	?
Late Person 2	0	?	?	?	?	?	?	?
Late Person 3	?	?	?	?	?	?	?	?

6.5 Assessment of malingering

Armed with the knowledge from the previous sections, we now consider the practical challenge of detecting if people cheat on a test. For example, people who have been in a car accident may seek financial compensation from insurance companies by feigning cognitive impairment such as pronounced memory loss. When these people are confronted with a memory test that is intended to measure the extent of their impairment, they may deliberately under-perform. This behavior is called malingering, and it may be accompanied by performance much worse than that displayed by real amnesiacs. Sometimes, for example, malingerers may perform substantially below chance.

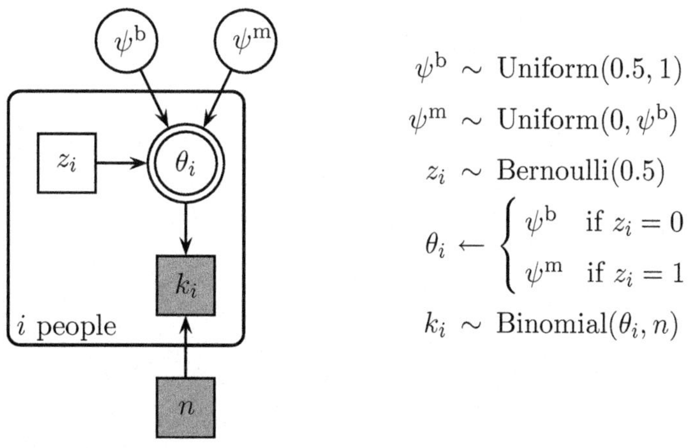

Fig. 6.5 Graphical model for the detection of malingering.

Malingering is not, however, always easy to detect, but is naturally addressed by latent-mixture modeling. Using this approach, it is possible to infer which of two categories—those who malinger, and those who are truthful or *bona fide*—each person belongs to, and quantify the confidence in each of these classifications.

We consider an experimental study on malingering, in which each of $p = 22$ participants completed a memory test (Ortega, Wagenmakers, Lee, Markowitsch, & Piefke, 2012). One group of participants was told to do their best. These are the *bona fide* participants. The other group of participants was told to under-perform by deliberately simulating amnesia. These are the malingerers. Out of a total of $n = 45$ test items, the participants get 45, 45, 44, 45, 44, 45, 45, 45, 45, 45, 30, 20, 6, 44, 44, 27, 25, 17, 14, 27, 35, and 30 correct. Because this was an experimental study, we know that the first 10 participants were *bona fide* and the next 12 were instructed to malinger.

The first analysis is straightforward, and uses the graphical model shown in Figure 6.5. We assume that all *bona fide* participants have the same ability, and so have the same rate ψ_{b} of answering each question correctly. For the malingerers, the rate of answering questions correctly is given by ψ_{m}, and $\psi_{\text{b}} > \psi_{\text{m}}$.

The script `Malingering_1.txt` implements the graphical model in WinBUGS:

```
# Malingering
model{
  # Each Person Belongs to One of Two Latent Groups
  for (i in 1:p){
    z[i] ~ dbern(0.5)
    z1[i] <- z[i]+1
  }
  # Bona Fide Group has Unknown Success Rate Above Chance
  psi[1] ~ dunif(0.5,1)
  # Malingering Group has Unknown Success Rate Below Bona Fide
  psi[2] ~ dunif(0,psi[1])
  # Data are Binomial with Group Rate for Each Person
  for (i in 1:p){
```

```
    theta[i] <- psi[z1[i]]
    k[i] ~ dbin(theta[i],n)
  }
}
```

Notice the restriction in the dunif definition of psi[2], which prevents the indeterminacy or label-switching problem by ensuring that $\psi_b > \psi_m$.

The code Malingering_1.m or Malingering_1.R applies the model to the data.

Exercise

Exercise 6.5.1 What are your conclusions about group membership? Did all of the participants follow the instructions?

6.6 Individual differences in malingering

As before, it may seem restrictive to assume that all members of a group have the same chance of answering correctly. So, now we assume that the ith participant in each group has a unique rate of answering questions correctly, θ_i, which is constrained by group-level distributions. In Section 6.2, we used group-level Gaussians. The problem with that approach is that values can lie outside the range 0 to 1. These values were just censored in Section 6.2, but this is not quite technically correct, and is certainly not elegant.[1]

One of several alternatives is to assume that instead of being Gaussian, the group-level distribution is $\text{Beta}(\alpha, \beta)$. Because the Beta distribution is defined on the interval from 0 to 1 it respects the natural boundaries of rates. So we now have a model in which each individual binomial rate parameter is constrained by a group-level beta distribution. This complete model is known as the beta-binomial (e.g., Merkle, Smithson, & Verkuilen, 2011; J. B. Smith & Batchelder, 2010).

It is useful to transform the α and β parameters from the beta distribution to a group mean $\mu = \alpha/(\alpha + \beta)$ and a measure $\lambda = \alpha + \beta$ that can be conceived of as a precision, in the sense that as it increases the variability of the distribution decreases. It is then straightforward to assign uniform priors to both μ_b, the group-level mean for the *bona fide* participants, and μ_m, the group-level mean for the malingerers. This assignment does not, however, reflect our knowledge that $\mu_b > \mu_m$. To capture this knowledge, we could define dunif(0,mubon), as done in the previous model.

However, for this model we apply a different approach. We first define μ_m as the additive combination of μ_b and a difference parameter, so that $\text{logit}(\mu_m) = \text{logit}(\mu_b) - \mu_d$. Note that this is an additive combination on the logit scale,

[1] WinBUGS conceptually conflates censoring and truncation in the I(,) notation, which is the cause of the technical problem. JAGS has the advantage of dealing with these two related concepts coherently.

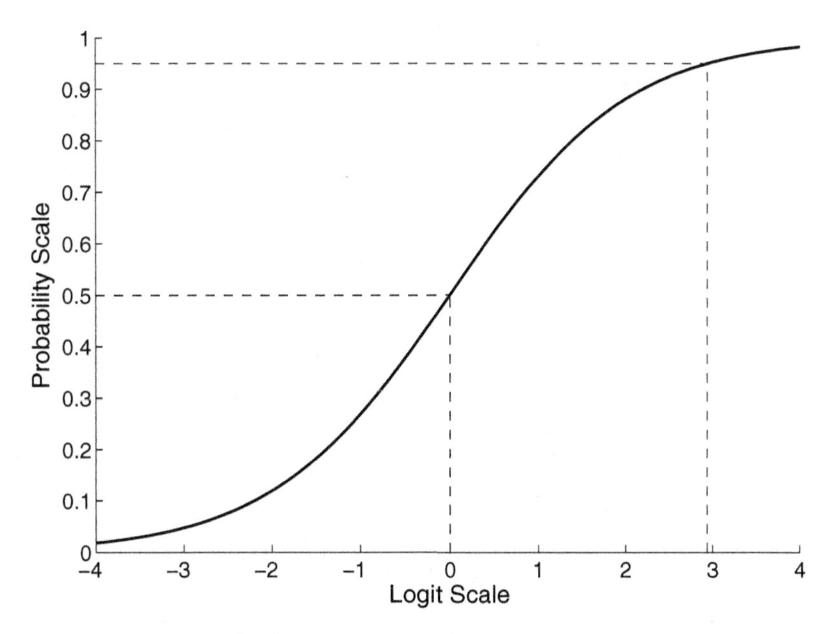

Fig. 6.6 The logit transformation. Probabilities range from 0 to 1 and are mapped to the entire set of real numbers using the logit transform.

as is customary in beta-binomial models. The logit transformation is defined as $\text{logit}(\theta) \equiv \ln(\theta/(1-\theta))$ and it transforms values on the rate scale, ranging from 0 to 1, to values on the logit scale, ranging from $-\infty$ to ∞. The logit transformation is shown in Figure 6.6, including two specific examples with the logit value 0 corresponding to probability 0.5, and the logit probability 2.94 corresponding to probability 0.95.

The prior for $\mu_{\text{d}} \sim \text{Gaussian}(0, 0.5)_{\mathcal{I}(0,\infty)}$ is a positive-only Gaussian distribution. This ensures that the group mean of the *bona fide* participants is always larger than that of the malingerers. Finally, note that the base rate of malingering ϕ, which was previously fixed to 0.5, is now assigned a relatively wide beta prior distribution that is centered around 0.5. This means the model uses the data to infer group membership and at the same time learn about the base rate.

A graphical model that implements the above ideas is shown in Figure 6.7. The script `Malingering_2.txt` implements the graphical model in WinBUGS:

```
# Malingering, with Individual Differences
model{
  # Each Person Belongs to One of Two Latent Groups
  for (i in 1:p){
    z[i] ~ dbern(phi) # phi is the Base Rate
    z1[i] <- z[i]+1
  }
  # Relatively Uninformative Prior on Base Rate
  phi ~ dbeta(5,5)
  # Data are Binomial with Rate Given by
  # Each Person's Group Assignment
```

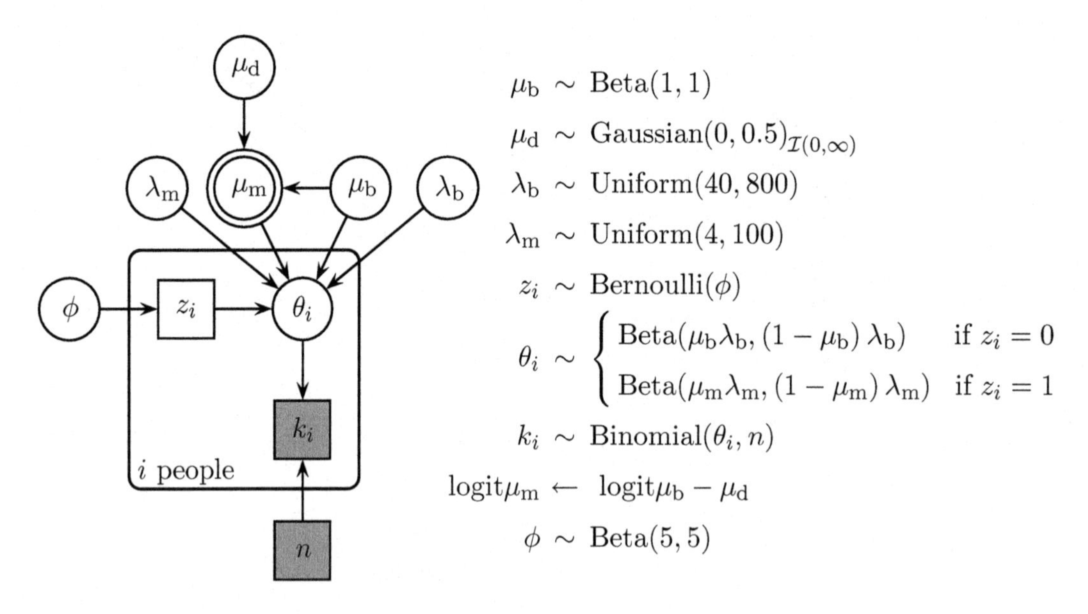

Fig. 6.7 Graphical model for inferring membership of two latent groups, consisting of malingerers and *bona fide* participants.

```
  for (i in 1:p){
    k[i] ~ dbin(theta[i,z1[i]],n)
    theta[i,1] ~ dbeta(alpha[1],beta[1])
    theta[i,2] ~ dbeta(alpha[2],beta[2])
  }
  # Transformation to Group Mean and Precision
  alpha[1] <- mubon * lambdabon
  beta[1] <- lambdabon * (1-mubon)
  # Additivity on Logit Scale
  logit(mumal) <- logit(mubon) - mudiff
  alpha[2] <- mumal * lambdamal
  beta[2]  <- lambdamal * (1-mumal)
  # Priors
  mubon ~ dbeta(1,1)
  mudiff ~ dnorm(0,0.5)I(0,) # Constrained to be Positive
  lambdabon ~ dunif(40,800)
  lambdamal ~ dunif(4,100)
}
```

The code `Malingering_2.m` or `Malingering_2.R` allows you to draw conclusions about group membership and the success rate of the two groups.

Exercises

Exercise 6.6.1 Is the inferred rate of malingering consistent with what is known about the instructions given to participants?

Exercise 6.6.2 Assume you know that the base rate of malingering is 10%. Change the WinBUGS script to reflect this knowledge. Do you expect any differences?

Exercise 6.6.3 Assume you know for certain that participants 1, 2, and 3 are *bona fide*. Change the code to reflect this knowledge.

Exercise 6.6.4 Suppose you add a new participant. What number of questions answered correctly by this participant would lead to the greatest uncertainty about their group membership?

Exercise 6.6.5 Try to solve the label-switching problem by using the `dunif(0,mubon)` approach instead of the logit transform.

Exercise 6.6.6 Why are the priors for λ_b and λ_m different?

6.7 Alzheimer's recall test cheating

In this section, we apply the same latent-mixture model shown in Figure 6.7 to different memory test data. Simple recognition and recall tasks are an important part of screening for Alzheimer's Disease and Related Disorders (ADRD), and are sometimes administered over the telephone. This practice raises the possibility of people cheating by, for example, writing down the words they are being asked to remember.

The data we use come from an informal experiment, in which 118 people were either asked to complete the test normally, or instructed to cheat. The particular test used was a complicated sequence of immediate and delayed free recall tasks, which we simplify to give a simple score correct out of 40 for each person. By design, there are 61 *bona fide* people who are known to have done the task as intended, and 57 people who are known to have cheated.

This graphical model is shown in Figure 6.8, and is essentially the same as for the previous example on malingering in Figure 6.7. It changes the names of variables from malingering to cheating as appropriate, uses different priors on the precisions of the group distributions, and makes the mean of accuracy rate for the cheaters *higher* than that of the *bona fide* people, since the impact of cheating is to recall more words than would otherwise be the case.

The script `Cheating.txt` implements the analysis in WinBUGS:

```
# Cheating Latent-Mixture Model
model{
  # Each Person Belongs to One of Two Latent Groups
  for (i in 1:p){
     z[i] ~ dbern(phi) # phi is the Base Rate
     z1[i] <- z[i]+1
  }
  # Relatively Uninformative Prior on Base Rate
  phi ~ dbeta(5,5)
  # Data are Binomial with Rate Given by
  # Each Person's Group Assignment
  for (i in 1:p){
```

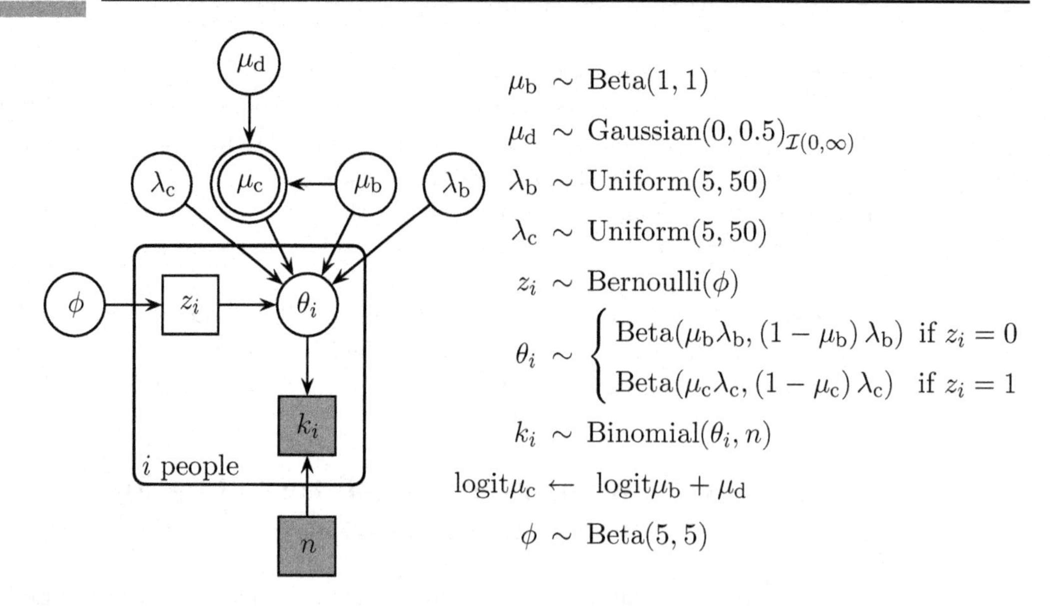

Fig. 6.8 Graphical model for inferring membership of two latent groups, consisting of cheaters and *bona fide* people in a memory test.

```
      k[i] ~ dbin(theta[i,z1[i]],n)
      thetatmp[i,1] ~ dbeta(alpha[1],beta[1])
      theta[i,1] <- max(.01,min(.99,thetatmp[i,1]))
      thetatmp[i,2] ~ dbeta(alpha[2],beta[2])
      theta[i,2] <- max(.01,min(.99,thetatmp[i,2]))
    }
    # Transformation to Group Mean and Precision
    alpha[1] <- mubon * lambdabon
    beta[1] <- lambdabon * (1-mubon)
    # Additivity on Logit Scale
    logit(muche) <- logit(mubon) + mudiff # Note the "+"
    alpha[2] <- muche * lambdache
    beta[2] <- lambdache * (1-muche)
    # Priors
    mubon ~ dbeta(1,1)
    mudiff ~ dnorm(0,0.5)I(0,) # Constrained to be Positive
    lambdabon ~ dunif(5,40)
    lambdache ~ dunif(5,40)
    # Correct Count
    for (i in 1:p){
      pct[i] <- equals(z[i],truth[i])
    }
    pc <- sum(pct[1:p])
  }
```

Note that the script includes a variable pc that keeps track of the accuracy of each classification made in sampling by comparing each person's latent assignment to the known truth from the experimental design.

The code `Cheating.m` or `Cheating.R` applies the graphical model to the data. We focus our analysis of the results firstly on the classification accuracy of the

Box 6.3 Undefined real result

In WinBUGS, error messages are called traps, and some traps are more serious than others. One of the more serious regularly occurring traps is "undefined real result." This trap indicates numerical overflow or underflow caused by a sample with very low likelihood. This can happen when you have not specified your model well enough. In particular, the prior distribution for a standard deviation may be too wide, thus allowing extreme values that are highly unlikely in light of the data (and, most often, also unlikely in light of prior knowledge). Initial values may also be to blame, and this is why it can be better to specify those yourself instead of having WinBUGS pick them automatically. Although the "undefined real result" trap can be a nuisance, in the end it may actually help you improve your model.

model. The top panel of Figure 6.9 summarizes the data, showing the distribution of correctly recalled words in both the *bona fide* and cheater groups. It is clear that cheaters generally recall more words, but that there is overlap between the groups.

One way to provide a benchmark classification accuracy is to consider the best possible cut-off. This is a total correct score below which a person is classified as *bona fide*, and at or above which they are classified as a cheater. The line in the bottom panel in Figure 6.9 shows the classification accuracy for all possible cut-offs, which peaks at 86.4% accuracy using the cut-off of 35. The gray distribution at the left of the panel is the posterior distribution of the `pc` variable, showing the range of accuracy achieved by the latent-mixture model.

Using a generative model to solve classification problems is unlikely to work as well as the best discriminative methods from machine learning and statistics. This is not because of failings of the Bayesian approach, but because the models we develop are imperfect accounts of how data are generated. If the focus is purely on prediction, other statistical approaches, including especially ones that combine the best aspects of generative and discriminative modeling, may be superior (e.g., Lasserre, Bishop, & Minka, 2006).

The advantage of the generative model is in providing details about the underlying processes assumed to produce the data, particularly by quantifying uncertainty. A good example of this important feature is shown in Figure 6.10, which shows the relationship between the total correct raw data score, and the posterior uncertainty about classification as a cheater, for each person. The broken lines connecting 35 people and a classification probability of 0.5 shows that the model infers people with scores above 35 as more likely than not to be cheaters. But it also shows how certain the model is about each classification, which provides more information (and more probabilistically coherent information) than many machine-learning methods.

This information about uncertainty is useful, for example, if there are costs or utilities associated with different classification decisions. Suppose that raising a

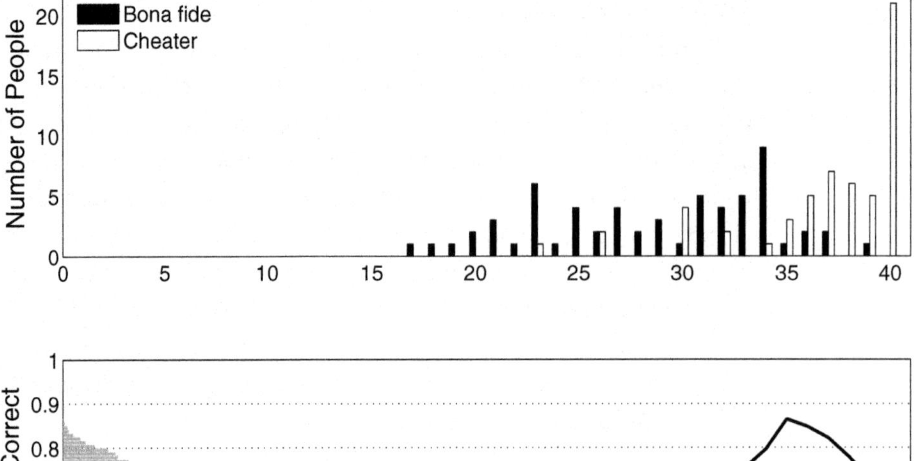

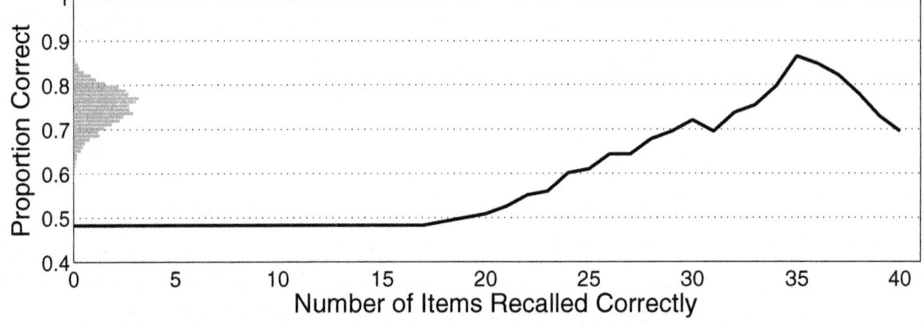

Fig. 6.9 The distribution of total correct recall scores for the Alzheimer's data, and classification performance. The top panel shows the distribution of scores for the *bona fide* and cheater groups. The bottom panel shows, with the line, the accuracy achieved using various cut-offs to separate the groups, and, with the distribution, the accuracy achieved by the latent-mixture model.

false-alarm and suspecting someone of cheating on the screening test costs \$25, perhaps through a wasted follow-up-in-person test, but that missing someone who cheated on the screening test costs \$100, perhaps through providing insurance that should have been withheld. With these utilities, the decision should be to classify people as *bona fide* only if it is four times more likely than them being a cheater. In other words, we need 80% certainty they are *bona fide*. The posterior distribution of the latent assignment variable $\mathbf{z}$ provides exactly this information. Under this set of utilities, as shown by the broken lines connecting 30 people and a classification probability of 0.2 in Figure 6.10, only people with a total correct score below 30 (not 35) should be treated as *bona fide*.

Exercises

Exercise 6.7.1 Suppose the utilities are very different, so that a false alarm costs \$100, because of the risk of litigation in a false accusation, but misses are relatively harmless, costing \$10 in wasted administrative costs. What decisions should be made about *bona fide* and cheating people now?

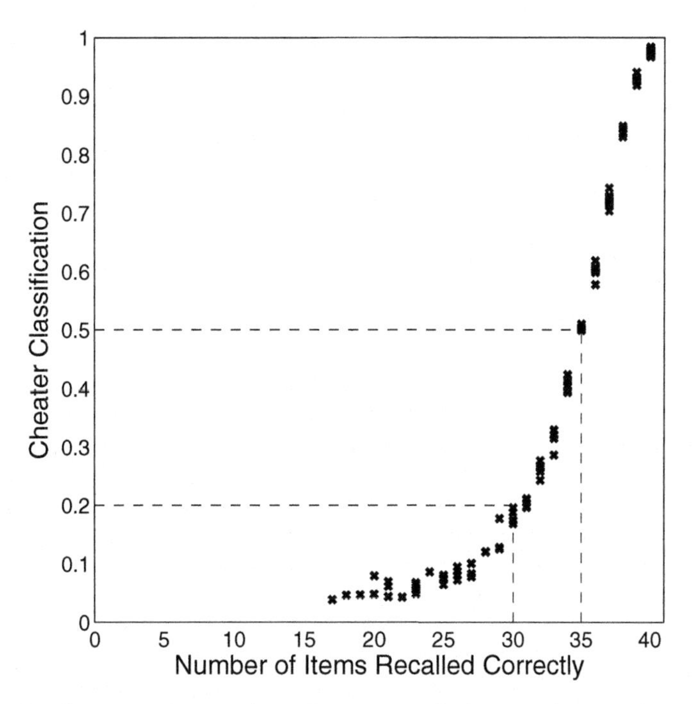

Fig. 6.10 The relationship between the number of items recalled correctly, and the posterior classification as belonging to the cheater group. Each cross corresponds to a person.

Exercise 6.7.2 What other potential information, besides the uncertainty about classification, does the model provide? Give at least one concrete example.

PART III

MODEL SELECTION

We consider it a good principle to explain the phenomena by the simplest hypotheses that can be established, provided this does not contradict the data in an important way.

Ptolemy, 85–165 AD

7 Bayesian model comparison

In the previous chapters we concerned ourselves with parameter estimation, often staying within the context of a single model. In much of cognitive science, however, researchers entertain more than just a single model. Different models often represent competing theories or hypotheses, and the focus of interest is on which substantive theory or hypothesis is more plausible, more useful, and better supported by the data. In order to address these questions we need to move beyond parameter estimation and turn to Bayesian methods for *model comparison*.

7.1 Marginal likelihood

To understand the Bayesian solution to the problem of selecting between competing models, we return to the very first equation of this book: Bayes' rule. We now indicate explicitly that the parameter θ depends on a specific model $\mathcal{M}_1$ that is entertained:

$$\begin{aligned} \text{posterior} = p(\theta \mid D, \mathcal{M}_1) &= \frac{p(D \mid \theta, \mathcal{M}_1)\, p(\theta \mid \mathcal{M}_1)}{p(D \mid \mathcal{M}_1)} \\ &= \frac{\text{likelihood} \times \text{prior}}{\text{marginal likelihood}}. \end{aligned} \tag{7.1}$$

The marginal likelihood $p(D \mid \mathcal{M}_1)$ is a single number that is sometimes called the *evidence*. It indicates the probability of the observed data D in light of the model specification $\mathcal{M}_1$. One interpretation is that the marginal likelihood measures the average quality of the predictions that a model has made for the observed data. The better the predictions, the greater the evidence.

As a simple example, suppose you construct a model $\mathcal{M}_x$ with a single parameter ξ. Furthermore, suppose that this parameter can take on only three values, $\xi_1 = -1$, $\xi_2 = 0$, and $\xi_3 = 1$. You assign these parameter values the following prior probability masses: $p(\xi_1) = 0.6$, $p(\xi_2) = 0.3$, and $p(\xi_3) = 0.1$. These assignments reflect the belief or knowledge that low values of ξ are more likely than high values of ξ. Next, you obtain data D, and you compute the likelihood for all parameter values. For example, assume that $p(D \mid \xi_1) = 0.001$, $p(D \mid \xi_2) = 0.002$, and $p(D \mid \xi_3) = 0.003$.[1]

[1] The likelihood $p(D \mid \xi_\star)$ quantifies the degree to which the observed data are expected, given a particular parameter value $\xi_\star$. Hence, you can think of the likelihood as a measure of goodness-of-fit.

Then, the marginal likelihood of your model $\mathcal{M}_x$ is given by

$$\begin{aligned} p(D \mid \mathcal{M}_x) &= p(\xi_1)\, p(D \mid \xi_1) + p(\xi_2)\, p(D \mid \xi_2) + p(\xi_3)\, p(D \mid \xi_3) \\ &= 0.6 \times 0.001 + 0.3 \times 0.002 + 0.1 \times 0.003 \\ &= 0.0015. \end{aligned}$$

The marginal likelihood is computed by averaging the likelihood across the parameter space, with prior probabilities acting as averaging weights. Thus, in order to determine how well a model predicted the data, we need to take into account *all* predictions that the model made and weight these by their prior probability. We can restate this mathematically by saying that the marginal likelihood is obtained by averaging out the model parameters in accordance with the *law of total probability*. For a parameter ξ that can take on k discrete values, the marginal likelihood is given by

$$p(D \mid \mathcal{M}_1) = \sum_{i=1}^{k} p(D \mid \xi_i, \mathcal{M}_1)\, p(\xi_i \mid \mathcal{M}_1). \tag{7.2}$$

For a continuously varying parameter θ—such as a binomial rate parameter that can take on any value between 0 and 1—the sum needs to be replaced by an integral, so that

$$p(D \mid \mathcal{M}_1) = \int p(D \mid \theta, \mathcal{M}_1)\, p(\theta \mid \mathcal{M}_1)\, \mathrm{d}\theta. \tag{7.3}$$

Despite the difference in notation between Equations 7.2 and 7.3, the computation is conceptually the same. The likelihood is evaluated for every possible parameter value, weighted by its prior plausibility, and added to the total.

The foregoing shows that in order to obtain firm evidence, a model needs to make a high proportion of good predictions. This is precisely the problem with models that are overly complex. These models are able to make many predictions, but a high proportion of these predictions will turn out to be false. Complex models need to divide their prior predictive probability across all of their predictions, and, in the limit, a model that predicts almost everything has its prior predictive probability spread out thinly: so thinly, in fact, that the occurrence of any particular event cannot substantially increase that model's credibility. This is the Bayesian justification for the adage "a model that predicts everything predicts nothing." As was illustrated above, the marginal likelihood for a model $\mathcal{M}_1$ is calculated by averaging the likelihood $p(D \mid \theta, \mathcal{M}_1)$ over the prior $p(\theta \mid \mathcal{M}_1)$.

The basic principle is that a model is complex when it makes many predictions. In practice, this can come about in a number of ways. The most obvious factor is that the inclusion of more parameters in a model allows it to make more predictions.

More subtly, models also become more complex as prior distributions over parameters become broad. When prior distributions become very broad, relatively low prior probability is assigned to those parts of the parameter space where the likelihood is high (i.e., where the predictions are good). It also means that relatively high prior probability is assigned to the remaining parts of the parameter

Box 7.1 Ockham's razor

Ockham's razor is also known as the principle of parsimony, and it embodies a preference for assumptions, theories, and hypotheses that are as simple as possible without being false. The metaphorical razor cuts away all theorizing that is needlessly complex. The razor is named after the English logician and Franciscan friar Father William of Ockham (c.1288–c.1348), who stated "*Numquam ponenda est pluralitas sine necessitate*" (Plurality must never be posited without necessity), and "*Frustra fit per plura quod potest fieri per pauciora*" (It is futile to do with more what can be done with less). However, Ockham's razor appears to be an example of Stigler's law of eponymy, which says that no scientific discovery is named after its original discoverer. Indeed, the principle of parsimony already features in work by Aristotle and Ptolemy. The latter even stated "We consider it a good principle to explain the phenomena by the simplest hypotheses possible." Hence, it may be historically correct to speak not of Ockham's razor, but of Ptolemy's principle of parsimony. Regardless of nomenclature, what is important is that the marginal likelihood acts as an automatic Ockham's razor (Jefferys & Berger, 1992; Myung & Pitt, 1997): models are punished for making predictions that are needlessly flexible with respect to the observed data.

space, those parts where the likelihood is almost zero (i.e., where the predictions are false). These effects combine to lower the average or marginal likelihood. Thus, a rate model that has a prior $\theta \sim \text{Uniform}(0.5, 1)$ is simpler than a rate model with the prior $\theta \sim \text{Uniform}(0, 1)$.

A final important factor that influences model complexity is the functional form of the model parameters. Consider for instance two laws of psychophysics that each relate the objective intensity I of a stimulus (e.g., a sound, a flash of light) to its subjective experience $\Psi(I)$. The first, Fechner's law, states that $\Psi(I) = k \ln (I + \beta)$, so that experienced intensity is a negatively accelerating function of stimulus intensity. The second, Stevens' law, states that $\Psi(I) = kI^{\beta}$, so that experienced intensity can be a negatively or a positively accelerating function of stimulus intensity. Fechner's law and Stevens' law each have two parameters, k and β, but nonetheless Stevens' law is more complex, because it can capture more data patterns and is therefore more difficult to falsify than Fechner's law (Townsend, 1975; Myung & Pitt, 1997).

The marginal likelihood, by assessing the average quality of a model's predictions for the data at hand, automatically takes all of these considerations into account.

Exercises

Exercise 7.1.1 Suppose you construct a second model for the same data D. This model, $\mathcal{M}_y$, has a parameter ζ that can take on two values, ζ_1 and ζ_2. You assign prior probability mass $p(\zeta_1) = 0.3$ and $p(\zeta_2) = 0.7$. For these two values, the likelihoods are 0.002 and 0.003, respectively. Compute the marginal likelihood for $\mathcal{M}_y$.

Exercise 7.1.2 What is the relative support of the data D for $\mathcal{M}_y$ versus $\mathcal{M}_x$?

Exercise 7.1.3 Suppose you construct a third model for the same data D. This model, $\mathcal{M}_z$, has a parameter μ that can take on 5 values, $\mu_1, \mu_2, \ldots, \mu_5$ with equal prior probability. The likelihoods are $p(D \mid \mu_1) = 0.001$, $p(D \mid \mu_2) = 0.001$, $p(D \mid \mu_3) = 0.001$, $p(D \mid \mu_4) = 0.001$, and $p(D \mid \mu_5) = 0.006$. Note that the likelihood for μ_5 is twice as high as the best possible likelihood for $\mathcal{M}_y$ and $\mathcal{M}_x$. Calculate the marginal likelihood for $\mathcal{M}_z$. Do you prefer it over $\mathcal{M}_y$ and $\mathcal{M}_x$? What is the lesson here?

Exercise 7.1.4 Consider Bart and Lisa, who each get 100 euros to bet on the winner of the world cup soccer tournament. Bart decides to divide his money evenly over 10 candidate teams, including those from Brazil and Germany. Lisa divides her money over just two teams, betting 60 euros on the team from Brazil and 40 euros on the team from Germany. Now if either Brazil or Germany turn out to win the 2010 world cup, Lisa wins more money than Bart. Explain in what way this scenario is analogous to the computation of marginal likelihood.

Exercise 7.1.5 Holmes and Watson[2] are involved in a rather simple game of darts, in which, with each dart, the player tries to score as many points as possible. The maximum score per dart is 60, and the minimum score is 0 (when the dart lands outside the board). After 5 darts, Holmes scored {38, 10, 0, 0, 0} and Watson scored {20,20,20,18,16}. How do you determine who is the better player? Consider another game of darts, but now one of the players gets to throw 50 times instead of 5. Explain how this scenario shows the importance of averaging instead of maximizing in order to penalize complexity.

7.2 The Bayes factor

Marginal likelihood is a measure of absolute evidence, in the sense that it is an index of a single model's overall predictive performance. In model selection, however, one is specifically interested in *relative* evidence, that is, the comparison of predictive performance for one model versus another. This comparison is accomplished simply by dividing the marginal likelihoods, yielding a quantity known as the Bayes factor (Jeffreys, 1961; Kass & Raftery, 1995):

$$BF_{12} = \frac{p(D \mid \mathcal{M}_1)}{p(D \mid \mathcal{M}_2)}. \tag{7.4}$$

Here, BF_{12} indicates the extent to which the data support $\mathcal{M}_1$ over $\mathcal{M}_2$, and as such it represents "the standard Bayesian solution to the hypothesis testing and

[2] We thank Wolf Vanpaemel for suggesting this example.

model selection problems" (Lewis & Raftery, 1997, p. 648). For example, when $BF_{12} = 5$ the observed data are 5 times more likely to have occurred under $\mathcal{M}_1$ than under $\mathcal{M}_2$, and when $BF_{12} = 0.2$ the observed data are 5 times more likely to have occurred under $\mathcal{M}_2$ than under $\mathcal{M}_1$.[3]

Even though the Bayes factor has an unambiguous and continuous scale, calibrated by betting, it is sometimes useful to summarize the Bayes factor in terms of discrete categories of evidential strength. Jeffreys (1961, Appendix B) proposed the classification scheme shown in Table 7.1.[4] This set of labels facilitates scientific communication but should only be considered an approximate descriptive articulation of different standards of evidence.[5]

Table 7.1 Evidence categories for the Bayes factor BF_{12} (Jeffreys, 1961).

Bayes factor BF_{12}			Interpretation
	>	100	Extreme evidence for $\mathcal{M}_1$
30	–	100	Very strong evidence for $\mathcal{M}_1$
10	–	30	Strong evidence for $\mathcal{M}_1$
3	–	10	Moderate evidence for $\mathcal{M}_1$
1	–	3	Anecdotal evidence for $\mathcal{M}_1$
	1		No evidence
1/3	–	1	Anecdotal evidence for $\mathcal{M}_2$
1/10	–	1/3	Moderate evidence for $\mathcal{M}_2$
1/30	–	1/10	Strong evidence for $\mathcal{M}_2$
1/100	–	1/30	Very strong evidence for $\mathcal{M}_2$
	<	1/100	Extreme evidence for $\mathcal{M}_2$

To illustrate, consider again our binomial example of 9 correct responses out of 10 questions, and the test between two models for performance: guessing (i.e., $\mathcal{M}_1 : \theta = 0.5$) versus not guessing (i.e., $\mathcal{M}_2 : \theta \neq 0.5$). In order to calculate the Bayes factor we need to be explicit about what we mean with "not guessing", which corresponds to defining a prior for θ. Here we use the uniform distribution for θ as a prior, such that $p(\theta \mid \mathcal{M}_2) \sim \text{Uniform}(0, 1) = \text{Beta}(1, 1)$.[6] After having specified both $\mathcal{M}_1$ and $\mathcal{M}_2$ we can proceed to calculate the separate marginal likelihoods, and then divide these to obtain the Bayes factor.

The marginal likelihood for $\mathcal{M}_1$ is calculated simply by plugging in the value $\theta = 0.5$ in the binomial equation: $p(D \mid \mathcal{M}_1) = \binom{10}{9}\left(\frac{1}{2}\right)^{10}$. The marginal likelihood

[3] Note that $BF_{12} = 1/BF_{21}$.

[4] We replaced the labels "worth no more than a bare mention" with "anecdotal", "decisive" with "extreme", and "substantial" with "moderate".

[5] The fact that the labels are only approximate is aptly illustrated by Jeffreys himself when he describes a Bayes factor of 5.33 as "odds that would interest a gambler, but would be hardly worth more than a passing mention in a scientific paper" (Jeffreys, 1961, pp. 256-257).

[6] This distribution includes the point $\theta = 0.5$, which may seem odd. However, when we compute the marginal likelihood we integrate over the prior distribution, and the inclusion of any single point is inconsequential, as $\int_a^a f(x)\, dx = 0$.

Box 7.2 **Bayes versus Fisher**

"The Bayesian method is comparative. It compares the probabilities of the observed event on the null hypothesis and on the alternatives to it. In this respect it is quite different from Fisher's approach which is absolute in the sense that it involves only a single consideration, the null hypothesis. All our uncertainty judgements should be comparative: there are no absolutes here. A striking illustration of this arises in legal trials. When a piece of evidence E is produced in a court investigating the guilt G or innocence I of the defendant, it is not enough merely to consider the probability of E assuming G; one must also contemplate the probability of E supposing I. In fact, the relevant quantity is the ratio of the two probabilities. Generally if evidence is produced to support some thesis, one must also consider the reasonableness of the evidence were the thesis false. Whenever courses of action are contemplated, it is not the merits or demerits of any course that matter, but only the comparison of these qualities with those of other courses." (Lindley, 1993, p. 25)

for model $\mathcal{M}_2$ is more difficult to calculate. As we have seen above, the marginal likelihood is obtained by averaging the likelihood over the prior parameter space, according to Equation 7.3. When we assume $p(\theta \mid \mathcal{M}_2) \sim \text{Beta}(1,1)$, then Equation 7.3 simplifies to $p(D \mid \mathcal{M}_2) = 1/(n+1)$. Thus, in our binomial example, $BF_{12} = \binom{10}{9}\left(\frac{1}{2}\right)^{10}(n+1) \approx 0.107$. This means that the data are $1/0.107 \approx 9.3$ times more likely under M_2 than they are under M_1.

Exercises

Exercise 7.2.1 Suppose you entertain a set of three models, x, y, and z. Assume you know $BF_{xy} = 4$ and $BF_{xz} = 3$. What is BF_{zy}?

Exercise 7.2.2 Suppose the $BF_{ab} = 1,000,000$, such that the data are one million times more likely to have occurred under $\mathcal{M}_a$ than under $\mathcal{M}_b$. Give two arguments for why you may still believe that $\mathcal{M}_a$ provides an inadequate or incorrect account of the data.

7.3 Posterior model probabilities

The Bayes factor compares the predictive performance of one model versus another, for the data at hand. A complete assessment of relative model preference, however, also requires us to consider how plausible the models are a priori. For example, let $\mathcal{M}_1$ be the hypothesis "neutrinos can travel faster than the speed of light," and let $\mathcal{M}_2$ be the hypothesis "neutrinos cannot travel faster than the speed of light."

The first hypothesis has been described as rather unlikely. Drew Baden, chairman of the physics department at the University of Maryland, compared its plausibility to that of finding a flying carpet. In such cases, even a very large Bayes factor in favor of $\mathcal{M}_1$ may be insufficient to make us believe that $\mathcal{M}_1$ is more likely than $\mathcal{M}_2$.

Hence, the assessment of the relative plausibility of two models after having seen the data requires that we combine information from the models' predictive performance for the data under consideration with the models' a priori plausibility. Expressed more formally,

$$\frac{p(\mathcal{M}_1 \mid D)}{p(\mathcal{M}_2 \mid D)} = \frac{p(D \mid \mathcal{M}_1)\, p(\mathcal{M}_1)}{p(D \mid \mathcal{M}_2)\, p(\mathcal{M}_2)}. \tag{7.5}$$

Or, in words,

$$\text{posterior odds} = \text{Bayes factor} \times \text{prior odds}. \tag{7.6}$$

This equation also yields another interpretation of the Bayes factor, namely as the change from prior odds $p(\mathcal{M}_1)/p(\mathcal{M}_2)$ to posterior odds $p(\mathcal{M}_1 \mid D)/p(\mathcal{M}_2 \mid D)$ that is brought about by the data. When the prior odds are 1, such that $\mathcal{M}_1$ and $\mathcal{M}_2$ are equally likely a priori, the Bayes factors can be converted to posterior probabilities $p(\mathcal{M}_1 \mid D) = BF_{12}/(BF_{12} + 1)$. This means that, for example, $BF_{12} = 2$ translates to $p(\mathcal{M}_1 \mid D) = 2/3$.

Exercises

Exercise 7.3.1 In this book, you have now encountered two qualitatively different kinds of priors. Briefly describe what they are.

Exercise 7.3.2 Consider one model, $\mathcal{M}_1$, that predicts a post-surgery survival rate by gender, age, weight, and history of smoking. A second model, $\mathcal{M}_2$, includes two additional predictors, namely body-mass index and fitness. We compute posterior model probabilities and find that $p(\mathcal{M}_1 \mid D) = .6$ and consequently $p(\mathcal{M}_2 \mid D) = .4$. For a patient Bob, $\mathcal{M}_1$ predicts a survival rate of 90%, and $\mathcal{M}_2$ predicts a survival rate of 80%. What is your prediction for Bob's probability of survival?

7.4 Advantages of the Bayesian approach

Bayesian hypothesis tests (i.e., Bayes factors and posterior model probabilities) implement an automatic Ockham's razor, describe the relative support or preference for a set of two (or more) candidate models, and can be used for model-averaged predictions. Here we highlight two additional advantages of Bayesian hypothesis tests that are of key importance to cognitive science.

First, Bayes factors can be used to obtain evidence in favor of the null hypothesis. Because theories and models often predict the absence of an effect, it is important

Box 7.3 Extraordinary claims require extraordinary evidence

This is quite possibly the single most underestimated maxim in current-day cognitive science. It was stated most eloquently by Scottish philosopher David Hume (1711–1776): "... no testimony is sufficient to establish a miracle, unless the testimony be of such a kind, that its falsehood would be more miraculous, than the fact, which it endeavors to establish; and even in that case there is a mutual destruction of arguments, and the superior only gives us an assurance suitable to that degree of force, which remains, after deducting the inferior." The first real Bayesian, Pierre-Simon Laplace (1749–1827), formulated the same sentiment more concisely: "The weight of evidence for an extraordinary claim must be proportioned to its strangeness." American astronomer Carl Sagan (1934–1996) coined the exact phrase "extraordinary claims require extraordinary evidence." The maxim is incorporated in the Bayesian computation where prior odds are combined with the Bayes factor to yield posterior odds. In many studies in cognitive science, attention is focused almost exclusively on the evidence that the observed data provide for or against a hypothesis. However, even strong evidence may fail to make an implausible claim acceptable. The fact that prior plausibility is difficult to quantify "objectively" is a poor excuse for ignoring it altogether. Nevertheless, most Bayesian statisticians are content when they provide only the Bayes factor; each researcher is then free to multiply that Bayes factor by his or her own prior odds in order to arrive at the posterior estimate of relative model plausibility.

to be able to quantify evidence in support of such predictions (e.g., Gallistel, 2009; Rouder, Speckman, Sun, Morey, & Iverson, 2009). In the field of visual word recognition, for instance, the entry-opening theory (Forster, Mohan, & Hector, 2003) predicts that masked priming is absent for items that do not have a lexical representation. Another example from that literature concerns the work by Bowers, Vigliocco, and Haan (1998), who hypothesized that priming depends on abstract letter identities—hence, priming should be equally effective for words that look the same in lower case and upper case (e.g., kiss/KISS) or different (e.g., edge/EDGE). A final example comes from the field of recognition memory, where Dennis and Humphreys' Bind Cue Decide model of episodic MEMory (BCDMEM) predicts the absence of a list-length effect and the absence of a list-strength effect (Dennis & Humphreys, 2001). In contrast to p-value hypothesis testing, Bayesian statistics assigns no special status to the null hypothesis and this means that Bayes factors can be used to quantify evidence for the null hypothesis just as for any other hypothesis.

A second advantage of Bayes factors is that they allow one to monitor the evidence as the data come in (Berger & Berry, 1988). In Bayesian hypothesis testing, "the

Box 7.4 **Problems with p-values**

This is a Bayesian book, and its focus is not on the deficiencies of p-values. But we will say that p-values are often misinterpreted, that they cannot quantify evidence in favor of the null hypothesis, that they depend on the (possibly unknown) intention with which the researcher collected the data, and that they focus only on what is expected under the null hypothesis, thus ignoring altogether what is expected under the alternative hypothesis. For scholarly details, see Berger and Wolpert (1988); Dennis et al. (2008); Dienes (2011); Edwards et al. (1963); Lindley (1993); Sellke et al. (2001); Wagenmakers (2007); Wagenmakers et al. (2008). For slogans, we like the following:

"The most important conclusion is that, for testing 'precise' hypotheses, p values should not be used directly, because they are too easily misinterpreted. The standard approach in teaching—of stressing the formal definition of a p value while warning against its misinterpretation—has simply been an abysmal failure." (Sellke et al., 2001, p. 71)

"Bayesian procedures can strengthen a null hypothesis, not only weaken it, whereas classical theory is curiously asymmetric. If the null hypothesis is classically rejected, the alternative hypothesis is willingly embraced, but if the null hypothesis is not rejected, it remains in a kind of limbo of suspended disbelief." (Edwards et al., 1963, p. 235)

"What the use of P implies, therefore, is that a hypothesis that may be true may be rejected because it has not predicted observable results that have not occurred. This seems a remarkable procedure." (Jeffreys, 1961, p. 385)

rules governing when data collection stops are irrelevant to data interpretation. It is entirely appropriate to collect data until a point has been proven or disproven, or until the data collector runs out of time, money, or patience" (Edwards et al., 1963, p. 193). This means that researchers are free to continue data collection in case the evidence (i.e., the Bayes factor) is inconclusive. Likewise, they are free to terminate data collection as soon as the interim evidence is sufficiently compelling. This freedom of "optional stopping" is denied to the researchers who use p-values to test their hypotheses (Wagenmakers, 2007).

7.5 Challenges for the Bayesian approach

Bayesian hypothesis testing comes with two main challenges, one conceptual and one computational. The conceptual challenge arises because the Bayesian hypothesis test is sensitive to the prior distributions for the model parameters (e.g., Bartlett, 1957; Liu & Aitkin, 2008; Vanpaemel, 2010). This occurs because the marginal likelihood is an average taken with respect to the prior. For example, consider a test for the mean μ of a Gaussian distribution with known variance. The null hypothesis $\mathcal{H}_0$ states that μ is zero. Specification of the alternative hypothesis $\mathcal{H}_1$ requires that we assign μ a prior distribution; in other words, we need to quantify $p(\mu \mid \mathcal{H}_1)$, our uncertainty about μ under $\mathcal{H}_1$. One tempting option is to use an "uninformative" prior for μ that does not express much preference for one value of μ over the other. For example, one could use a Gaussian distribution with mean zero and variance 10,000. This approach with uninformative priors often works well for parameter estimation. From a marginal likelihood perspective, however, the use of a low-precision prior effectively creates a model that predicts almost any observed result. When one hedges one's bets to such an extreme degree, the Bayes factor is likely to show a preference for $\mathcal{H}_0$, even when the data appear inconsistent with it.

The problem is not that the Bayesian hypothesis test is sensitive to the prior distribution. This feature merely reflects the workings of the automatic Ockham's razor that is an asset, not a liability, of the Bayesian hypothesis test. The prior distributions are part of the model specification, and low-precision priors correspond to complex models.[7] Instead, the problem is that researchers sometimes only have a vague idea about the vagueness of their prior knowledge, or the relevant available prior information. When the vagueness of the prior is more or less arbitrary, so are the results from the Bayesian hypothesis test.

Several procedures have been proposed to ensure that the results from the Bayesian hypothesis test do not simply reflect the arbitrary precision of the prior distributions. First, one can invest more effort in the *subjective* specification of prior distributions (e.g., Dienes, 2011). This means that the researcher attempts to translate substantive knowledge about the problem at hand into prior probability distributions. Such knowledge may be obtained by eliciting prior beliefs from experts, or by consulting the literature for earlier work on similar problems. Unfortunately, the substantive knowledge that is encoded in the prior distributions does not generalize well to other problems, and consequently each new problem requires its own prior elicitation process. Most researchers have neither the expertise nor the energy to carry out the careful problem-specific elicitation steps that define the subjective approach. In addition, some modeling problems are so large and complex that they defy a careful subjective specification of the prior distribution.

[7] As an aside, model selection methods that are insensitive to prior distributions also have difficulties dealing with order-restricted inference, such as when the complex model has parameters θ_1 and θ_2 free to vary, and the simpler model has the order-restriction $\theta_1 > \theta_2$ (e.g., Hoijtink, Klugkist, & Boelen, 2008).

Box 7.5 Confusion about priors

A persistent confusion when the results from a Bayesian hypothesis test are discussed involves the supposedly arbitrary and profound impact of "the prior". For example, one researcher may conduct an experiment and find that the presence of a big box fails to make people more creative (even though, with a box present, people are perhaps encouraged to "think outside the box"). To quantify the evidence that the data provide in favor of $\mathcal{H}_0$ this researcher may present a Bayes factor, say, $BF_{01} = 15.5$. Invariably, another researcher will object that this Bayesian result depends on the prior plausibility that was assigned to $\mathcal{H}_1$. Obviously, or so the argument goes, when you are skeptical about $\mathcal{H}_1$ you will assign $\mathcal{H}_1$ a low prior plausibility, and hence the end result of the Bayesian test simply reaffirms your initial bias. This argument is false, and reveals a misunderstanding of what it is that the Bayes factor measures. As is clear from Equations 7.4 and 7.5, the Bayes factor does *not* involve the prior probabilities on the models. One researcher may believe that the big-box hypothesis is silly, and another may believe it is entirely plausible, but these different prior opinions about the model's plausibility do not affect the Bayes factor. There is another kind of prior, however, that does influence the Bayes factor. This is not the prior on the models, but the prior distribution for the relevant parameters. This prior distribution reflects our uncertainty about the size of the effect in case $\mathcal{H}_1$ is true. To set this prior one can use a default specification, a subjective specification, and carry out a sensitivity analysis. It is important to understand the difference between the prior on the models and the prior distribution on the models' parameters.

Second, one can try to use formal rules and desiderata to specify prior distributions that yield reasonable results across a wide range of different research contexts (Kass & Wasserman, 1996). Such priors are called *objective* because they do not depend on information specific to the research topic under investigation. For example, one can use a unit-information prior, that is, a prior that contains as much information as a single observation.[8] Similar objective prior distributions have been developed by Jeffreys (1961), Zellner and Siow (1980), Liang, Paulo, Molina, Clyde, and Berger (2008), and others. The results from such objective hypothesis tests may not be definitive, but these tests can nevertheless serve as a good reference analysis that can later be refined, if necessary, by the inclusion of problem-specific information.

[8] This assumption also underlies the popular Bayesian information criterion (BIC: Schwarz, 1978). Masson (2011) provides a tutorial on how to use the BIC for statistical problems such as ANOVA.

Box 7.6 — Prior sensitivity

"When different reasonable priors yield substantially different answers, can it be right to state that there is a single answer? Would it not be better to admit that there is scientific uncertainty, with the conclusion depending on prior beliefs?" (Berger, 1985, p. 125)

Third, one can use sophisticated procedures such as the local Bayes factor (A. F. M. Smith & Spiegelhalter, 1980), the intrinsic Bayes factor (Berger & Mortera, 1999; Berger & Pericchi, 1996), the fractional Bayes factor (O'Hagan, 1995), and the partial Bayes factor (O'Hagan, 1995). Gill (2002, Chapter 7) provides a summary of this class of methods. The idea of the partial Bayes factor is to sacrifice a small part of the data to obtain a posterior that is relatively insensitive to the various priors one might entertain. The Bayes factor is then calculated by integrating the likelihood over this posterior instead of over the original prior. Procedures such as these are still undergoing further development and deserve more study.

Fourth, the dependence of the results on the width of the prior can be studied explicitly, by means of a sensitivity analysis. In such an analysis one varies the width of the prior distribution (across a reasonable range) and studies the corresponding fluctuations in the Bayes factor. Whenever these fluctuations cause meaningful, qualitative differences in conclusions one should acknowledge that the interpretation of the data is strongly dependent on prior beliefs, and that additional data may need to be collected before inference is robust across plausible prior beliefs.

The computational challenge for Bayesian hypothesis testing is that the marginal likelihood and the Bayes factor are often quite difficult to calculate. Earlier, we saw that with a uniform prior on the binomial rate parameter θ—$p(\theta \mid \mathcal{M}_1) \sim \text{Beta}(1,1)$—the marginal likelihood simplifies from $\int p(D \mid \theta, \mathcal{M}_1) p(\theta \mid \mathcal{M}_1)\, \mathrm{d}\theta$ to $1/(1+n)$.[9] However, in all but a few simple models, such simplifications are impossible. In order to be able to compute the marginal likelihood or the Bayes factor for more complex models, a range of computational methods have been developed. A recent summary lists as many as 15 different methods (Gamerman & Lopes, 2006, Chapter 7).

For example, one method computes the marginal likelihood by means of the *candidates' formula* (Besag, 1989) or the *basic marginal likelihood identity* (Chib, 1995; Chib & Jeliazkov, 2001). One simply exchanges the roles of posterior and marginal likelihood in Equation 7.1 to obtain

$$p(D \mid \mathcal{M}_1) = \frac{p(D \mid \theta, \mathcal{M}_1)\, p(\theta \mid \mathcal{M}_1)}{p(\theta \mid D, \mathcal{M}_1)}, \tag{7.7}$$

[9] This becomes intuitively clear from an inspection of the prior predictive distribution in the lower panel of Figure 3.9.

which holds for any one value of θ. When the posterior is available analytically, one only needs to plug in a single value of θ and obtain the marginal likelihood immediately. This method can however also be applied when the posterior is only available through MCMC output, either from the Gibbs sampler (Chib, 1995) or the Metropolis–Hastings algorithm (Chib & Jeliazkov, 2001).

Another method that computes the marginal likelihood is to sample repeatedly parameter values from the prior, calculate the associated likelihoods, and then take the likelihood average. When the posterior is highly peaked compared to the prior—as will happen with many data or with a medium-sized parameter space—it becomes necessary to employ more efficient sampling methods, with a concomitant increase in computational complexity.

Finally, it is also possible to compute the Bayes factor directly, without first calculating the constituent marginal likelihoods. The basic idea is to generalize the MCMC sampling routines for parameter estimation to incorporate a "model indicator" variable. In the case of two competing models, the model indicator variable z, say, can take on two values. For example, it can take $z = 1$ when the sampler is in model M_1, and $z = 2$ when the sampler is in model M_2. The Bayes factor is then estimated by the relative frequency with which $z = 1$ versus $z = 2$. This MCMC approach to model selection is called transdimensional MCMC (e.g., Sisson, 2005), an approach that encompasses both reversible jump MCMC (P. J. Green, 1995) and the product space technique (Carlin & Chib, 1995; Lodewyckx et al., 2011; Scheibehenne, Rieskamp, & Wagenmakers, 2013).[10]

Almost all of these computational methods suffer from the fact that they become less efficient and more difficult to implement as the underlying models become more complex. We now turn to an alternative method, whose implementation is extremely straightforward. The method's main limitation is that it applies only to *nested* models, a limitation that also holds for p-values.

7.6 The Savage–Dickey method

In the simplest classical hypothesis testing framework, one contemplates two models. One is the null hypothesis that fixes one of its parameters to a pre-specified value of substantive interest, say $\mathcal{H}_0 : \phi = \phi_0$; the other model is the alternative hypothesis, in which that parameter is free to vary, say $\mathcal{H}_1 : \phi \neq \phi_0$. Hence, the null hypothesis is nested under the alternative hypothesis, that is, $\mathcal{H}_0$ can be obtained from H_1 by setting ϕ equal to ϕ_0. Note that in the classical framework, $\mathcal{H}_0$ is generally a sharp null hypothesis, or a "point null". That is, the null hypothesis states that ϕ is exactly equal to ϕ_0.

[10] We recommend the Lodewyckx et al. article and accompanying software to anyone who wants to learn how to apply transdimensional MCMC techniques in WinBUGS or JAGS. In this book we focus on the Savage–Dickey technique because it is easier to understand and implement.

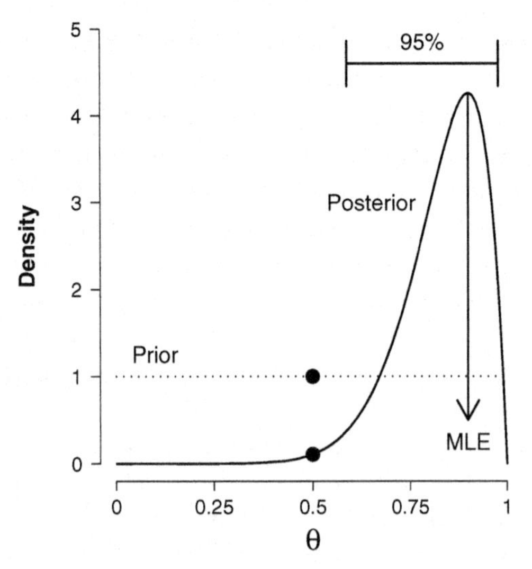

Fig. 7.1 Prior and posterior distributions for binomial rate parameter θ, after observing 9 correct responses and 1 incorrect response. The mode of the posterior distribution for θ is 0.9, equal to the maximum likelihood estimate, and the 95% credible interval extends from 0.59 to 0.98. The height of the distributions at $\theta = 0.5$ is indicated by a black dot; the ratio of these heights quantifies the evidence for $\mathcal{H}_0 : \theta = 0.5$ versus $\mathcal{H}_1 : \theta \sim \text{Beta}(1, 1)$.

For example, in our binomial example you answered 9 out of 10 questions correctly. Were you guessing or not? The Bayesian and the frequentist framework define $\mathcal{H}_0 : \theta = 0.5$ as the null hypothesis for chance performance. The alternative hypothesis under which $\mathcal{H}_0$ is nested could be defined as $\mathcal{H}_1 : \theta \neq .5$, or, more specifically, as $\mathcal{H}_1 : \theta \sim \text{Beta}(1, 1)$, which states that θ is free to vary from 0 to 1, and that it has a uniform prior distribution as shown in Figure 7.1.

For the binomial example, the Bayes factor for $\mathcal{H}_0$ versus $\mathcal{H}_1$ could be obtained by analytically integrating out the model parameter θ. However, the Bayes factor may likewise be obtained by only considering $\mathcal{H}_1$, and dividing the height of the posterior for θ by the height of the prior for θ, at the point of interest. This surprising result was first published by Dickey and Lientz (1970), who attributed it to Leonard J. "Jimmie" Savage. The result is now generally known as the *Savage–Dickey density ratio* (e.g., Dickey, 1971); for extensions and generalizations, see Chen (2005), Verdinelli and Wasserman (1995), and Wetzels, Grasman, and Wagenmakers (2010). Mathematically, the Savage–Dickey density ratio says that

$$BF_{01} = \frac{p(D \mid \mathcal{H}_0)}{p(D \mid \mathcal{H}_1)} = \frac{p(\theta = 0.5 \mid D, \mathcal{H}_1)}{p(\theta = 0.5 \mid \mathcal{H}_1)}. \tag{7.8}$$

A straightforward mathematical proof is presented in O'Hagan and Forster (2004, pp. 174–177).

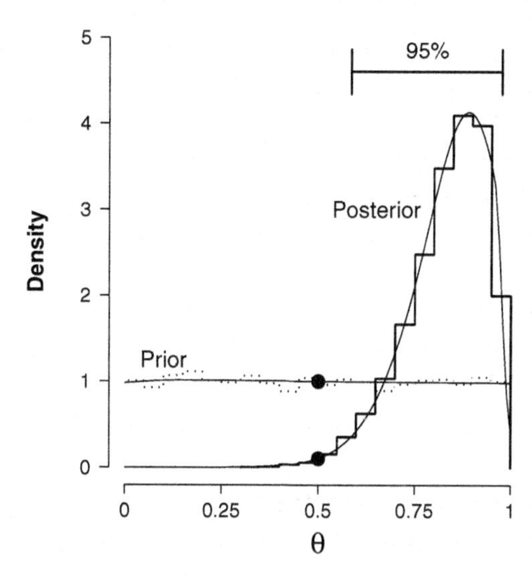

Fig. 7.2 MCMC-based prior and posterior distributions for the binomial rate parameter θ, after observing 9 correct responses and 1 incorrect response. The thin solid lines indicate the fit of a non-parametric density estimator. Based on this density estimator, the mode of the posterior distribution for θ is approximately 0.89, and the 95% credible interval extends from 0.59 to 0.98, closely matching the analytical results from Figure 7.1.

In Figure 7.1, the two thick dots located at $\theta = .5$ provide the required information. It is evident from the figure that after observing 9 out of 10 correct responses, the height of the density at $\theta = 0.5$ has decreased, so that one would expect these data to cast doubt on the null hypothesis and support the alternative hypothesis. Specifically, the height of the prior distribution at $\theta = 0.5$ equals 1, and the height of the posterior distribution at $\theta = 0.5$ equals 0.107. From Equation 7.8 the corresponding Bayes factor is $BF_{01} = 0.107/1 = 0.107$, and this corresponds exactly to the Bayes factor that was calculated by integrating out θ.

It is clear that the same procedure can be followed when the height of the posterior is not available in closed form, but instead has to be estimated from the histogram of MCMC samples. Figure 7.2 shows the estimates for the prior and the posterior densities as obtained from MCMC output (Stone, Hansen, Kooperberg, & Truong, 1997). The estimated height of the prior and posterior distributions at $\theta = 0.5$ equal 1.00 and 0.107, respectively.

In most nested model comparisons, $\mathcal{H}_0$ and $\mathcal{H}_1$ have several free parameters in common. These parameters are usually not of direct interest, and they are not the focus of the hypothesis test. Hence, the common parameters are known as *nuisance parameters*. For example, one might want to test whether or not the mean of a Gaussian distribution is zero—$\mathcal{H}_0 : \mu = \mu_0$ versus $\mathcal{H}_1 : \mu \neq \mu_0$—whereas the variance σ^2 is common to both models and not of immediate interest.

In general then, the framework of nested models features a parameter vector $\theta = (\phi, \psi)$, where ϕ denotes the parameter of substantive interest that is subject to test, and ψ denotes the set of nuisance parameters. The null hypothesis $\mathcal{H}_0$ posits that ϕ is constrained to some special value, so that $\phi = \phi_0$. The alternative hypothesis $\mathcal{H}_1$ assumes that ϕ is free to vary. Now consider $\mathcal{H}_1$, and let $\phi \to \phi_0$. This effectively means that $\mathcal{H}_1$ reduces to $\mathcal{H}_0$, and it is therefore reasonable to assume that $p(\psi \mid \phi \to \phi_0, \mathcal{H}_1) = p(\psi \mid \mathcal{H}_0)$. In other words, when $\phi \to \phi_0$ the prior for the nuisance parameters under $\mathcal{H}_1$ should equal the prior for the nuisance parameters under $\mathcal{H}_0$. When this condition holds, the nuisance parameters can be ignored, so that again

$$BF_{01} = \frac{p(D \mid \mathcal{H}_0)}{p(D \mid \mathcal{H}_1)} = \frac{p(\phi = \phi_0 \mid D, \mathcal{H}_1)}{p(\phi = \phi_0 \mid \mathcal{H}_1)}, \tag{7.9}$$

which equals the ratio of the heights for the posterior and the prior distribution for ϕ at ϕ_0. Thus, the Savage–Dickey density ratio holds under relatively general conditions. The next chapters use concrete examples to illustrate how cognitive scientists can use the Savage–Dickey density ratio test to their advantage.

Exercises

Exercise 7.6.1 The Bayes factor is relatively sensitive to the width of the prior distributions for the model parameters. Use Equation 7.9 to argue why this is the case.

Exercise 7.6.2 The Bayes factor is relatively sensitive to the width of the prior distributions, but only for the parameters that differ between the models under consideration. Use Equation 7.9 to argue why this is the case.

Exercise 7.6.3 What is the main advantage of the Savage–Dickey procedure?

7.7 Disclaimer and summary

The material covered in this chapter is controversial. Several professional Bayesian statisticians advise against the use of Bayes factors and instead recommend methods based on an assessment of the posterior distribution. And indeed, Bayes factors should be used with care, as their interpretation stands or falls with the plausibility of the models under comparison. For example, is it ever plausible to assume the complete absence of an effect? In other words, can the null hypothesis $\mathcal{H}_0$ ever be exactly true? If one does not believe that such a point null hypothesis can ever be true, even as an approximation, than the entire comparison between $\mathcal{H}_0$ and $\mathcal{H}_1$ may become meaningless (but see Berger & Delampady, 1987). We believe that, at least for experimental studies, point null hypotheses may often be true exactly. Homeopathy does not cure cancer, people cannot look into the future, and we doubt that people are any more creative when they stand next to a big box (but see Leung et al., 2012).

The most prominent argument against Bayes factors is that they depend largely on the specification of the prior distributions. This argument is true, but it presupposes that the specification of a model is somehow separate from the specification of the prior distributions. The Bayes factor perspective on model selection is that, before models can be compared, they need to be specified completely, and this includes the number of parameters, their prior distributions, and their functional form. All of these properties jointly determine an essential characteristic of a model, which is its ability to generate predictions. The evidence that the data provide for and against a model can only be properly assessed when one is able to discount the ability of that model to fit all kinds of other data as well.

In sum, the Bayes factor should be used with care, preferably in combination with other methods and a sensitivity analysis. That said, the Bayes factor has a number of undeniable advantages, some of which we have already discussed and others that will become evident in later chapters. One general advantage is that the Bayes factor directly addresses the key question that cognitive scientists care about: "To what extent do my data support $\mathcal{H}_1$ over $\mathcal{H}_0$?" Another general advantage is that the Bayes factor follows from the basic tenets of probability theory. Not only does this impart to the Bayes factor all kinds of pleasant properties—including consistency, transitivity, and immunity against optional stopping—it also means that alternative methods necessarily violate these basic tenets. Consequently, although alternative methods may work well for some situations, it is always possible to find situations in which these alternative methods fail and the Bayes factor succeeds.

Model selection and hypothesis testing are difficult topics, and it may take you a while to grasp the concepts involved. The next chapters provide concrete examples that show Bayesian model selection in action. From the current chapter, it is important that you understand the following key points:

- Complex models are models that make many predictions. This may happen because they have many parameters, because they have prior parameter distributions that are relatively non-precise and spread out over a wide range, or because they have parameters that have a complicated functional form. Complex models are difficult to falsify.
- The Bayes factor penalizes models for needless complexity and therefore it implements Ptolemy's principle of parsimony.
- The Bayes factor provides a comparative measure of evidence as it pits the predictive adequacy of one model against that of another.
- The Bayes factor requires a careful selection of prior distributions, as these form an integral part of the model specification.
- Extraordinary claims require extraordinary evidence.

Exercise

Exercise 7.7.1 Browse the empirical literature of your subfield of study. Do you find the null hypotheses plausible? That is, could they ever be exactly true?

8 Comparing Gaussian means

WITH RUUD WETZELS

Popular theories are difficult to overthrow. Consider the following hypothetical sequence of events. First, Dr John proposes a Seasonal Memory Model (SMM). The model is intuitively attractive and quickly gains in popularity. Dr Smith, however, remains unconvinced and decides to put one of SMM's predictions to the test. Specifically, SMM predicts that a glucose-driven increase in recall performance is more pronounced in summer than in winter. Dr Smith conducts the relevant experiment using a within-subjects design and finds the opposite result: as shown in the fictitious data in Table 8.1, the increase in recall performance is *smaller* in summer than in winter, although this difference is not significant. With $n = 41$ and a t value of 0.79, the corresponding two-sided p-value equals 0.44.

Table 8.1 Glucose-driven increase in recall performance in summer and winter.

Season	N	Mean	SD
Winter	41	0.11	0.15
Summer	41	0.07	0.23

Clearly, Dr Smith's data do not support SMM's prediction that the glucose-driven increase in performance is larger in summer than in winter. Instead, the data seem to suggest that the null hypothesis is plausible, and that no difference between summer and winter is evident. Dr Smith submits his findings to the *Journal of Experimental Psychology: Learning, Memory, and the Seasons.* Three months later, Dr Smith receives the reviews. Inevitably, one of the reviews is from Dr John, and it includes the following comment:

From a null result, we cannot conclude that no difference exists, merely that we cannot reject the null hypothesis. Although some have argued that with enough data we can argue for the null hypothesis, most agree that this is only a reasonable thing to do in the face of a sizeable number of data that have been collected over many experiments that control for all concerns. These conditions are not met here. Thus, the empirical contribution here does not enable readers to conclude very much, and so is quite weak.

Formally, Dr John's first statement is completely correct, since p-values cannot be used to quantify the support in favor of the null hypothesis. A p-value of 0.44 could indicate that the data support $\mathcal{H}_0$, but it could also indicate that the data are too few in number to result in a rejection of $\mathcal{H}_0$. In this chapter we show how this ambiguity can be overcome, and how Dr Smith and other researchers can use the Bayes factor to quantify evidence in favor of $\mathcal{H}_0$. As explained in Chapter 7, the Bayes factor measures the change from prior model odds to posterior model odds brought about by the data. This means that, in contrast to the p-value, the Bayes factor is able to quantify evidence both in favor of $\mathcal{H}_0$ and in favor of $\mathcal{H}_1$.

The sections below highlight some properties of the Bayes factor in the context of the popular t-test (Rouder et al., 2009). We show how to specify $\mathcal{H}_0$ and $\mathcal{H}_1$, and then use the Savage–Dickey density ratio to calculate the Bayes factor.[1] This then allows us to address the key point of contention between Dr John and Dr Smith. To what extent, if at all, do the observed data contradict the prediction from the Seasonal Memory Model?

8.1 One-sample comparison

When we use the one-sample t-test, we assume that the data follow a Gaussian distribution with unknown mean μ and unknown variance σ^2. This is a natural assumption for a within-subjects experimental design, like that undertaken by Dr Smith. The data consist of one sample of standardized difference scores (i.e., "winter scores – summer scores"). The null hypothesis states that the mean of the difference scores is equal to zero, that is, $\mathcal{H}_0 : \mu = 0$. The alternative hypothesis states that the mean is not equal to zero, that is, $\mathcal{H}_1 : \mu \neq 0$.

We follow Rouder et al. (2009) and use a $\text{Cauchy}(0, 1)$ prior for effect size δ. The advantage of defining a prior on effect size, instead of on the mean, is that it is very general. The same prior can be used across many experiments, dependent variables, and measurement scales. The Cauchy distribution used for this prior is a t-distribution with 1 degree of freedom, and resembles a Gaussian distribution with fatter tails. The choice for the Cauchy is theoretically motivated, and details are provided by Jeffreys (1961), Liang et al. (2008), and Zellner and Siow (1980). For the standard deviation we use a half-Cauchy distribution, so that $\sigma \sim \text{Cauchy}(0, 1)_{\mathcal{I}(0,\infty)}$. This is a $\text{Cauchy}(0, 1)$ distribution that is defined only for positive numbers (Gelman & Hill, 2007).

The graphical model for the one-sample comparison of means is shown in Figure 8.1. In the graphical model, x represents the observed data that follow a Gaussian distribution with mean μ and a variance σ^2. The effect size δ is defined as $\delta = \mu/\sigma$, and so μ is given by $\mu = \delta\sigma$. The null hypothesis puts all prior mass for δ on a single point, that is, $\mathcal{H}_0 : \delta = 0$, whereas the alternative hypothesis assumes that δ has a Cauchy distribution, with $\mathcal{H}_1 : \delta \sim \text{Cauchy}(0, 1)$.

The script `OneSample.txt` implements the graphical model in WinBUGS:

[1] More information can be found on the website of Ruud Wetzels, `www.ruudwetzels.com`, and the website of Jeff Rouder, `pcl.missouri.edu`.

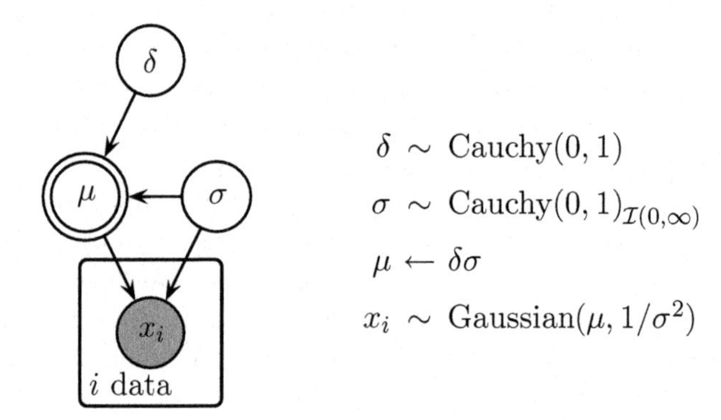

Fig. 8.1 Graphical model for the one-sample within-subjects comparison of means.

```
# One-Sample Comparison of Means
model{
  # Data
  for (i in 1:ndata){
    x[i] ~ dnorm(mu,lambda)
  }
  mu <- delta*sigma
  lambda <- pow(sigma,-2)
  # delta and sigma Come From (Half) Cauchy Distributions
  lambdadelta ~ dchisqr(1)
  delta ~ dnorm(0,lambdadelta)
  lambdasigma ~ dchisqr(1)
  sigmatmp ~ dnorm(0,lambdasigma)
  sigma <- abs(sigmatmp)
  # Sampling from Prior Distribution for Delta
  deltaprior ~ dnorm(0,lambdadeltaprior)
  lambdadeltaprior ~ dchisqr(1)
}
```

Note that the Cauchy distribution is not directly available in WinBUGS. This can be addressed by assigning $\delta \sim \text{Gaussian}(0, \lambda_\delta)$, where the precision λ_δ has a chi-square distribution with one degree of freedom, $\lambda_\delta \sim \chi^2(1)$. This two-step assignment procedure corresponds to $\delta \sim \text{Cauchy}(0, 1)$. Note also that the WinBUGS script generates prior as well as posterior samples of the effect size δ.

The code `OneSample.m` or `OneSample.R` applies the model to the data in Table 8.1, plots the prior and the posterior distributions for δ, and applies the Savage–Dickey density ratio test to the posterior samples of δ to compute the Bayes factor for $\mathcal{H}_0 : \delta = 0$ versus $\mathcal{H}_1 : \delta \sim \text{Cauchy}(0, 1)$.

Figure 8.2 shows the results. The posterior distribution is peaked near zero, with a little more density given to positive, rather than negative, effect sizes. The critical point $\delta = 0$ is about 5 times more likely in the posterior distribution than it is in the prior distribution. This means that the Bayes factor is about 5:1 in favor of the null hypothesis $\mathcal{H}_0$.

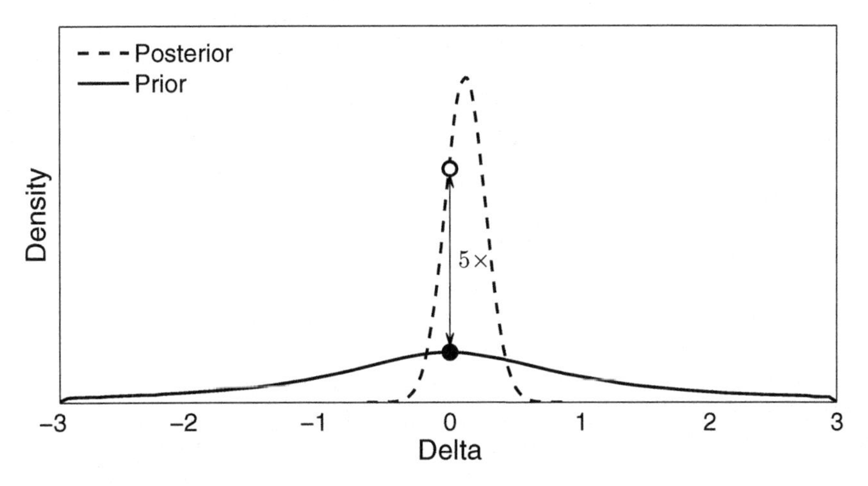

Fig. 8.2 Prior and posterior distributions on effect size δ for the summer and winter data. Markers show the height of the prior and posterior distributions at $\delta = 0$ needed to estimate the Bayes factor between $\mathcal{H}_0 : \delta = 0$ and $\mathcal{H}_1 : \delta \sim \text{Cauchy}(0, 1)$ using the Savage–Dickey method.

Exercises

Exercise 8.1.1 Here we assumed a half-Cauchy prior distribution on the standard deviation `sigma`. Other choices are possible and reasonable. Can you think of a few?

Exercise 8.1.2 Do you think the different priors on `sigma` will lead to substantially different conclusions? Why or why not? Convince yourself by implementing a different prior and studying the result.

Exercise 8.1.3 We also assumed a Cauchy prior distribution on effect size `delta`. Other choices are possible and reasonable. One such choice is the standard Gaussian distribution. Do you think this prior will lead to substantially different conclusions? Why or why not? Convince yourself by implementing the standard Gaussian prior and studying the result.

8.2 Order-restricted one-sample comparison

The Bayes factor computed in the previous section quantified the strength of evidence in favor of $\mathcal{H}_0 : \delta = 0$ versus $\mathcal{H}_1 : \delta \sim \text{Cauchy}(0, 1)$. However, this particular $\mathcal{H}_1$ was not the SMM hypothesis that Dr Smith set out to test. The SMM hypothesis specifically stated that δ should be *negative*. Hence, a more appropriate alternative hypothesis incorporates the constraint $\delta < 0$. This corresponds to a half-Cauchy prior distribution that is defined for negative numbers only,

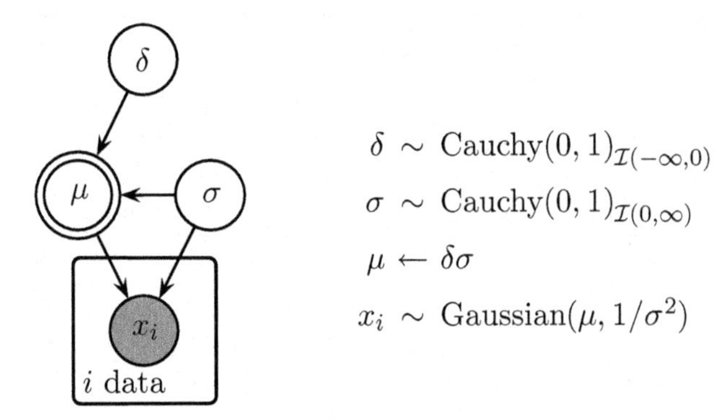

Fig. 8.3 Graphical model for the order-restricted one-sample within-subjects comparison of means.

$\mathcal{H}_2 : \text{Cauchy}(0,1)_{\mathcal{I}(-\infty,0)}$. This alternative hypothesis is called an order-restricted or one-sided hypothesis.

The graphical model for this analysis is shown in Figure 8.3, and it only changes the prior on effect size. The script `OneSampleOrderRestricted.txt` implements the graphical model in WinBUGS:.

```
# One-Sample Order Restricted Comparison of Means
model{
  # Data
  for (i in 1:ndata){
    x[i] ~ dnorm(mu,lambda)
  }
  mu <- delta*sigma
  lambda <- pow(sigma,-2)
  # delta and sigma Come From (Half) Cauchy Distributions
  lambdadelta ~ dchisqr(1)
  delta ~ dnorm(0,lambdadelta)I(,0)
  lambdasigma ~ dchisqr(1)
  sigmatmp ~ dnorm(0,lambdasigma)
  sigma <- abs(sigmatmp)
  # Sampling from Prior Distribution for Delta
  deltaprior ~ dnorm(0,lambdadeltaprior)I(,0)
  lambdadeltaprior ~ dchisqr(1)
}
```

The code `OneSampleOrderRestricted.m` or `OneSampleOrderRestricted.R` again applies the model to the data in Table 8.1. Figure 8.4 plots the prior and the posterior distributions for δ, and shows the key densities for the Savage–Dickey density ratio test at $\delta = 0$ to compute the Bayes factor for $\mathcal{H}_0 : \delta = 0$ versus $\mathcal{H}_2 : \text{Cauchy}(0,1)_{\mathcal{I}(-\infty,0)}$. The data are now about 10 times more likely under $\mathcal{H}_0$ than they are under the order-restricted $\mathcal{H}_2$ associated with SMM. According to the classification scheme proposed by Jeffreys (1961), as presented in Table 7.1, this could be considered "strong evidence" for the null hypothesis.

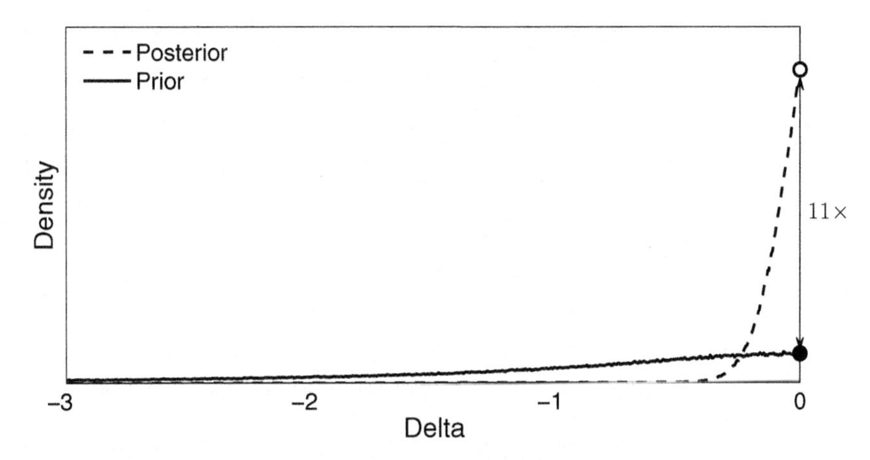

Fig. 8.4 Prior and posterior distributions on effect size δ for the Summer and Winter data. Markers show the height of the prior and posterior distributions at $\delta = 0$ needed to estimate the Bayes factor between $\mathcal{H}_0 : \delta = 0$ and $\mathcal{H}_2 : \text{Cauchy}(0, 1)_{\mathcal{I}(-\infty,0)}$ using the Savage–Dickey method.

Exercises

Exercise 8.2.1 For completeness, estimate the Bayes factor for the summer and winter data between $\mathcal{H}_0 : \delta = 0$ versus $\mathcal{H}_3 : \text{Cauchy}(0, 1)_{\mathcal{I}(0,\infty,)}$, involving the order-restricted alternative hypothesis that assumes the effect is positive.

Exercise 8.2.2 In this example, it matters whether the alternative hypothesis is unrestricted, order-restricted to negative values for δ, or order-restricted to positive values for δ. Why is this perfectly reasonable? Can you think of a situation where the three versions of the alternative hypothesis yield exactly the same Bayes factor?

Exercise 8.2.3 From a practical standpoint, we do not need a new graphical model and WinBUGS script to compute the Bayes factor for $\mathcal{H}_0$ versus the order-restricted $\mathcal{H}_2$. Instead, we can use the original graphical model in Figure 8.1 that implements the unrestricted Cauchy distribution and discard those prior and posterior MCMC samples that are inconsistent with the $\delta < 0$ order-restriction. The Savage–Dicky density ratio test still involves the height of the prior and posterior distributions at $\delta = 0$, but now the samples from these distributions are truncated, respecting the order-restriction, such that they range only from $\delta = -\infty$ to $\delta = 0$. Implement this method in Matlab or R, and check that the same conclusions are drawn from the analysis.

Exercise 8.2.4 Wagenmakers and Morey (2013) describe yet another method to obtain the Bayes factor for order-restricted model comparisons. This method is perhaps the most reliable because it avoids the numerical complications associated with having to estimate the posterior density at a boundary. Go to `http://www.ejwagenmakers.com/papers.html`, download the Wagenmakers

Box 8.1 Estimating densities from samples

Estimating Bayes factors using the Savage–Dickey approach requires estimating the height of the prior and posterior distributions at a specific value of a parameter. Often the height of the prior distribution can be obtained analytically. It is always possible to estimate the required prior and posterior densities from MCMC samples. The simplest approach is by binning. A more advanced approach is to use a non-parametric density estimator (e.g., Stone et al., 1997). In R, one such estimator is included in the `polspline` package. This can be installed by starting R and selecting the `Install Package(s)` option in the `Packages` menu. Once you choose your preferred CRAN mirror, select `polspline` in the `Packages` window and click on `OK`. In Matlab, the Statistics toolbox provides the `ksdensity` function, or you may try the non-parametric density estimators available in the free package developed by Zdravko Botev, `www.mathworks.com/matlabcentral/fileexchange/14034`.

Both binning and density estimation approaches depend on tuning parameters—like the width of the bins—and so some experimentation and tests of robustness are usually required. For example, the Savage–Dickey analysis in Figure 8.2 can give Bayes factors between (at least) about 4.7 and 6.1 using different reasonable density estimation methods. Since the goal is to estimate the Bayes factor, what is important is that the substantive conclusions are trusted, rather than obtaining the exact number. The interpretive framework provided by Table 7.1 is very helpful in this regard.

and Morey paper, and read the introduction with a focus on Equation 1. Implement their suggested method and compare the results to those obtained earlier.

8.3 Two-sample comparison

Often in cognitive science, the comparison of means is based on data from two independent groups, rather than a single group. The two-sample t-test is commonly used as a standard frequentist approach for these sorts of between-subjects designs.

Most textbooks on research methods have their own introductory example, and we consider the one presented by Evans and Rooney (2011, pp. 279–283). Their example involves a between-subjects experiment to test the effect of drinking plain versus oxygenated water. The raw data are presented in Evans and Rooney (2011,

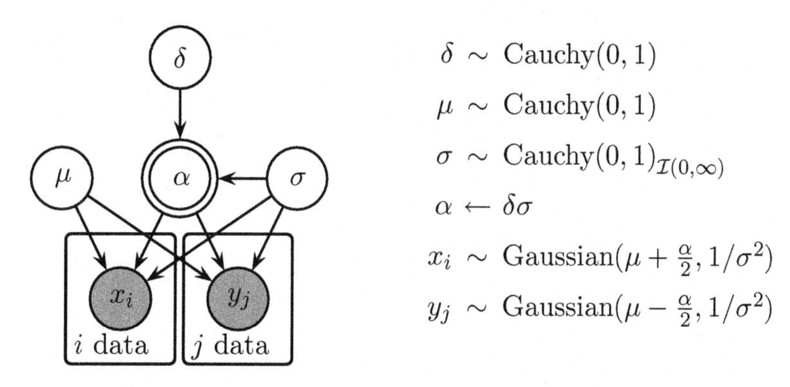

Fig. 8.5 Graphical model for the two-sample between-subjects comparison of means.

Table 13.3), and consists of a set of memory scores for 20 subjects who took plain water, and a set of memory scores for 20 subjects who took oxygenated water. For the plain water (control) group, the mean score is 68.35 with standard deviation 6.38, while for the oxygenated water (treatment or experimental) group, the mean score is 76.65 with standard deviation 4.06. A two-sample t-test gives $t\,(38) = 4.47$, $p < 0.01$.

In our Bayesian approach of making inferences about the means, we rescale the data so that one group has mean 0 and standard deviation 1. This rescaling procedure ensures that the prior distributions for the parameters hold regardless of the scale of measurement. Therefore it does not matter whether, say, response times are measured in seconds or in milliseconds.

The graphical model for the two-sample comparison is shown in Figure 8.5. The variables x and y represent the experimental and control data, respectively. Both x and y follow Gaussian distributions with shared variance σ^2. The mean of x is given by $\mu + \alpha/2$, and the mean of y is given by $\mu - \alpha/2$, so that α is the difference in the means.

Because $\delta = \alpha/\sigma$, α is given by $\alpha = \delta\sigma$. As for the one-sample scenario, the null hypothesis puts all prior mass for δ on a single point, so that $\mathcal{H}_0 : \delta = 0$, whereas the alternative hypothesis assumes that δ follows a Cauchy distribution, so that $\mathcal{H}_1 : \delta \sim \text{Cauchy}(0, 1)$.

The script `TwoSample.txt` implements the graphical model in WinBUGS:

```
# Two-sample Comparison of Means
model{
  # Data
  for (i in 1:n1){
    x[i] ~ dnorm(mux,lambda)
  }
  for (j in 1:n2){
    y[j] ~ dnorm(muy,lambda)
  }
  # Means and precision
  alpha <- delta*sigma
  mux <- mu+alpha/2
```

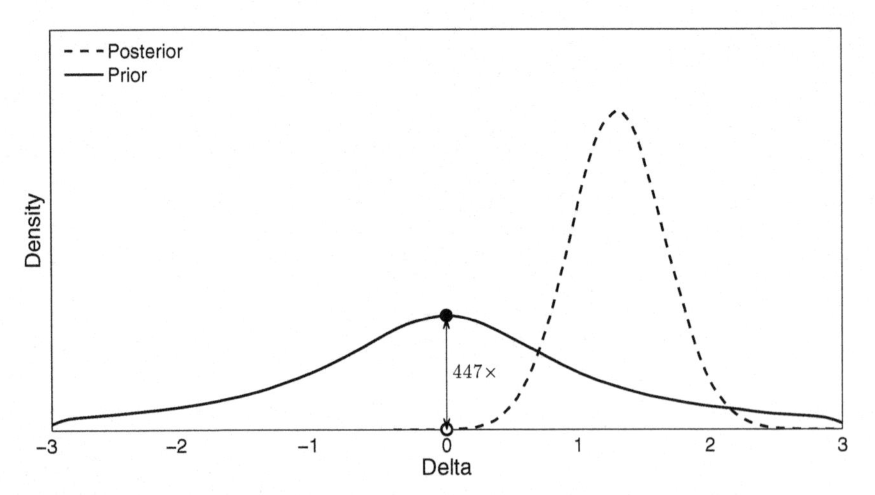

Fig. 8.6 Prior and posterior distributions on effect size δ for the Evans and Rooney (2011) data using the model for a two-sample comparison of means. Markers show the height of the prior and posterior distributions at $\delta = 0$ needed to estimate the Bayes factor between $\mathcal{H}_0 : \delta = 0$ and $\mathcal{H}_1 : \text{Cauchy}(0, 1)$ using the Savage–Dickey method.

```
  muy <- mu-alpha/2
  lambda <- pow(sigma,-2)
  # delta, mu, and sigma Come From (Half) Cauchy Distributions
  lambdadelta ~ dchisqr(1)
  delta ~ dnorm(0,lambdadelta)
  lambdamu ~ dchisqr(1)
  mu ~ dnorm(0,lambdamu)
  lambdasigma ~ dchisqr(1)
  sigmatmp ~ dnorm(0,lambdasigma)
  sigma <- abs(sigmatmp)
  # Sampling from Prior Distribution for Delta
  lambdadeltaprior ~ dchisqr(1)
  deltaprior ~ dnorm(0,lambdadeltaprior)
}
```

The code `TwoSample.m` or `TwoSample.R` applies the model to the data in Evans and Rooney (2011, Table 13.3). Figure 8.6 plots the prior and the posterior distributions for δ, and shows the key densities for the Savage–Dickey density ratio test at $\delta = 0$ to compute the Bayes factor. It is clear that there is a large effect of oxygenated versus plain water on memory performance. The data are now more than 400 times more likely under $\mathcal{H}_1$ than they are under $\mathcal{H}_0$, and this large Bayes factor can be interpreted as providing decisive evidence.

Exercise

Exercise 8.3.1 The two-sample comparison of means outlined above assumes that the two groups have equal variance. How can you extend the model when this assumption is not reasonable?

9 Comparing binomial rates

9.1 Equality of proportions

In their article "After the promise: the STD consequences of adolescent virginity pledges," Brückner and Bearman (2005) analyzed a series of interviews conducted as part of the National Longitudinal Study of Adolescent Health. The focus of the article was on the sexual behavior of adolescents, aged 18–24, who have made a virginity pledge. This is a public or written pledge to remain a virgin until marriage.

Consider the hypothesis that the sexual behavior of pledgers is not very different from that of non-pledgers, except for the fact that pledgers are less likely to use condoms when they first have sex. The Brückner and Bearman (2005) study presents relevant data. Those adolescents who reported using a condom for their first sexual encounter numbered 424 out of 777 ($\approx$ 54.6%) for the pledgers, and 5416 out of 9072 ($\approx$ 59.7%) for non-pledgers. To what extent does a statistical analysis support the assertion that pledgers are less likely than non-pledgers to use a condom?

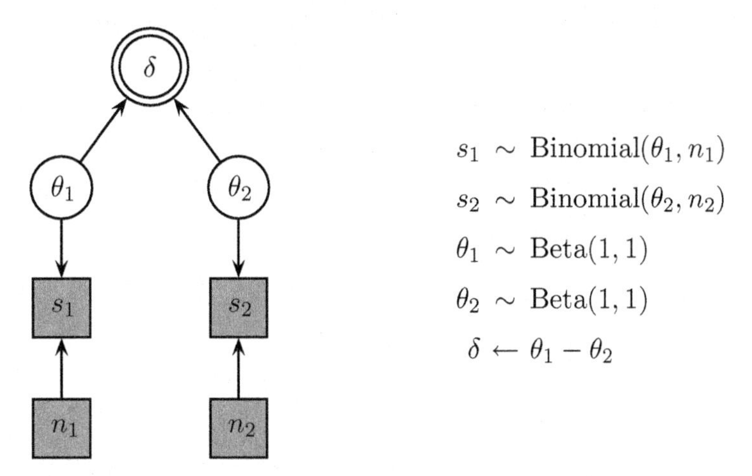

Fig. 9.1 Bayesian graphical model for the comparison of two proportions.

The Bayesian model selection approach for these data is simple and general. We assume that the number of condom users among the pledgers (i.e., $s_1 = 424$ out of $n_1 = 777$) and among the non-pledgers (i.e, $s_2 = 5416$ out of $n_2 = 9072$) are governed by binomial rates θ_1 and θ_2, respectively, which are given uniform priors. The difference between the two rate parameters is $\delta = \theta_1 - \theta_2$. This leads to the

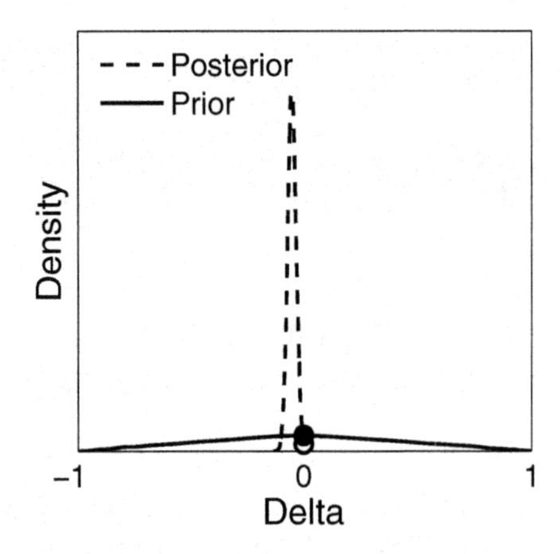

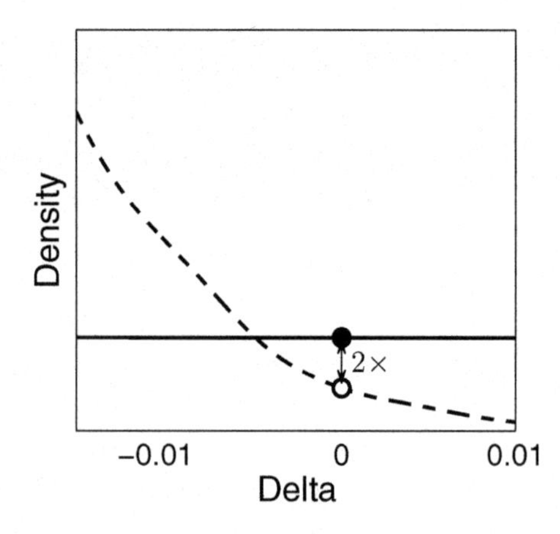

Fig. 9.2 Prior and posterior distributions of the rate difference δ for the pledger data. The left panel shows the distributions across their entire range, and the right panel zooms in on the area that is relevant for the Savage–Dickey test. The markers show the densities at $\delta = 0$ needed to estimate the Bayes factor.

same graphical model we originally considered in Section 3.2, as is shown again in Figure 9.1.

The null hypothesis states that the rates θ_1 and θ_2 are equal, and hence $\mathcal{H}_0 : \delta = 0$. The unrestricted alternative hypothesis states that the rates are free to vary, $\mathcal{H}_1 : \delta \neq 0$. Thus, applying the Savage–Dickey method for finding the Bayes factor requires the prior and posterior distributions for δ.

The script `Pledgers_1.txt` implements the graphical model in WinBUGS:

```
# Pledgers
model{
  # Rates and Difference
  theta1 ~ dbeta(1,1)
  theta2 ~ dbeta(1,1)
  delta <- theta1-theta2
  # Data
  s1 ~ dbin(theta1,n1)
  s2 ~ dbin(theta2,n2)
  # Prior Sampling
  theta1prior ~ dbeta(1,1)
  theta2prior ~ dbeta(1,1)
  deltaprior <- theta1prior-theta2prior
}
```

The code `Pledgers_1.m` or `Pledgers_1.R` calls WinBUGS to draw samples from the posterior and prior of the rate difference δ. The left panel of Figure 9.2 shows the distributions on their entire range, and the right panel zooms in on the relevant region around $\delta = 0$. The critical point $\delta = 0$ is supported about twice as much under the prior as it is under the posterior, and so the Bayes factor is approximately 2 in favor of the alternative hypothesis.

A reasonable interpretation of this Bayes factor is that the data do not provide much evidence in favor of one hypothesis over the other. This seems more conservative than the interpretation that may be drawn from Bayesian parameter estimation, since the Bayesian 95% credible interval for the posterior of δ is approximately $(-0.09, -0.01)$ and does not include 0. The reason for the discrepancy is that the Bayesian hypothesis test punishes $\mathcal{H}_1$ for assigning prior mass to values of δ that yield very low likelihoods (see Berger & Delampady, 1987, for a discussion).

Exercises

Exercise 9.1.1 Because the rate parameters θ_1 and θ_2 both have a uniform prior distribution, the prior distribution for the difference parameter δ can be found analytically as a triangular distribution. What are the advantages of using this result, rather than relying on computational sampling? What are the disadvantages?

Exercise 9.1.2 In the current analysis, we put independent priors on θ_1 and θ_2. Do you think this is plausible? How would you change the model to take into account the possible dependence? How would this affect the outcome of the Bayesian test?

Exercise 9.1.3 This example corresponds to a rare case in which the Bayes factor is available analytically. $BF_{01} = p(D \mid \mathcal{H}_0) / p(D \mid \mathcal{H}_1)$ is given by

$$BF_{01} = \frac{\binom{n_1}{s_1}\binom{n_2}{s_2}}{\binom{n_1+n_2}{s_1+s_2}} \frac{(n_1+1)(n_2+1)}{n_1+n_2+1}.$$

Calculate the Bayes factor analytically, and compare it to the result obtained using the Savage–Dickey method.

Exercise 9.1.4 For the pledger data, a frequentist test for equality of proportions indicates that $p \approx 0.006$. This tells us that when $\mathcal{H}_0$ is true (i.e., the proportions of condom users are equal in the two groups), then the probability is about 0.006 that we would encounter a result at least as extreme as the one that was in fact observed. What conclusions would you draw based on this information? Discuss the usefulness of the Bayes factor and the p-value in answering the scientific question of whether pledgers are less likely than non-pledgers to use a condom.

9.2 Order-restricted equality of proportions

Whether pledgers are less likely than non-pledgers to use condoms has a natural interpretation as involving the order-restriction that the rate is lower for the pledgers than for the non-pledgers. This corresponds to a different alternative hypothesis $\mathcal{H}_2 : \delta < 0$.

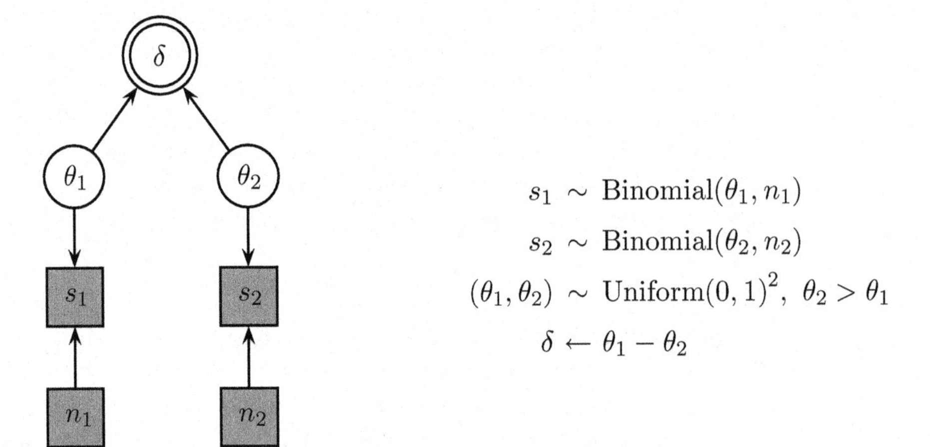

$$s_1 \sim \text{Binomial}(\theta_1, n_1)$$
$$s_2 \sim \text{Binomial}(\theta_2, n_2)$$
$$(\theta_1, \theta_2) \sim \text{Uniform}(0,1)^2, \ \theta_2 > \theta_1$$
$$\delta \leftarrow \theta_1 - \theta_2$$

Fig. 9.3 Bayesian graphical model for the order-restricted comparison of two proportions.

The graphical model for this order-restricted analysis is shown in Figure 9.3, and involves changing the priors on the rates. Since the order-restriction is that $\delta < 0$, the requirement is that $\theta_2 > \theta_1$. This is the same sort of restriction involved in the two-country quiz example in Section 6.4, where it was addressed by using `theta2 ~ dunif(0,1)` and `theta1 ~ dunif(0,theta2)`. While this approximate approach worked satisfactorily for inferring posterior distributions of parameters, an exact approach is needed in the current context of model comparison. This is because the approximate method does not generate a uniform prior distribution of θ_1 and θ_2 over the region of the joint parameter space satisfying the order-restriction, and model selection, unlike parameter estimation, is likely to be sensitive to this mismatch between available information and the implementation of the model.

To make this clear, Figure 9.4 shows samples from the joint prior distribution of (θ_1, θ_2) under the approximate method in the left panel. It is visually clear that the approximate method places too much density in the bottom-left corner of (θ_1, θ_2), because each θ_1 is drawn uniformly, and then θ_1 is constrained to be less than θ_2. The exact distribution is shown in the middle panel, and gives equal density to the valid region in which $\theta_2 > \theta_1$. The impact of the approximation on the prior for δ is shown in the right panel of Figure 9.4, and involves too much prior density being given to values near zero.

The script `Pledgers_2.txt` implements the graphical model in WinBUGS:

```
# Pledgers, Order Constrained Rates
model{
  # Order Constrained Rates
  thetap[1:2] ~ dmnorm(mu[],TI[,])
  theta1 <- phi(cos(angle)*thetap[1]-sin(angle)*abs(thetap[2]))
  theta2 <- phi(sin(angle)*thetap[1]+cos(angle)*abs(thetap[2]))
  # Data
  s1 ~ dbin(theta1,n1)
  s2 ~ dbin(theta2,n2)
```

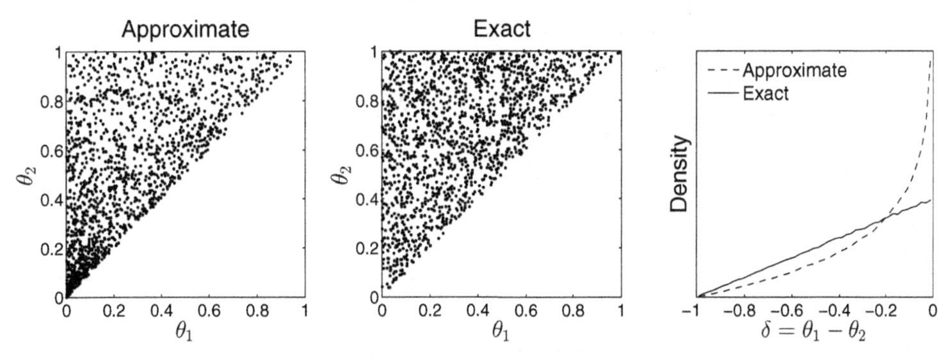

Fig. 9.4 Samples from approximate (left panel) and exact (middle panel) approaches to implementing the order-constrained prior on (θ_1, θ_2) in which $\theta_2 > \theta_1$. The right panel shows the implied prior distributions on $\delta = \theta_1 - \theta_2$ arising from both methods.

```
  # Difference
  delta <- theta1-theta2
  # Prior Sampling
  thetapprior[1:2] ~ dmnorm(mu[],TI[,])
  theta1prior <- phi(cos(angle)*thetapprior[1]-sin(angle)*abs(thetapprior[2]))
  theta2prior <- phi(sin(angle)*thetapprior[1]+cos(angle)*abs(thetapprior[2]))
  deltaprior  <- theta1prior-theta2prior
  # Constants
  angle <- 45*3.1416/180
  TI[1,1] <- 1
  TI[1,2] <- 0
  TI[2,1] <- 0
  TI[2,2] <- 1
  mu[1] <- 0
  mu[2] <- 0
}
```

An exact method is used for the order-constrained prior, involving jointly drawing samples from a bivariate standard Gaussian, rotating these samples by 45 degrees, and then transforming the rotated samples into rates that lie in the unit square.

The code `Pledgers_2.m` or `Pledgers_2.R` draws samples from the posterior and prior of the rate difference δ under the order-constrained prior. The left panel of Figure 9.5 shows the distributions on their entire range, and the right panel zooms in on the relevant region around $\delta = 0$. It is clear that order-restriction has the effect of doubling the density of the prior, but has little effect on the posterior. This means that the critical point $\delta = 0$ is now supported about four times as much under the prior as it is under the posterior, and so the Bayes factor is approximately 4 in favor of the alternative hypothesis.

Exercise

Exercise 9.2.1 Consider an order-restricted test of $\mathcal{H}_0 : \delta = 0$ versus $\mathcal{H}_3 : \delta > 0$. What do you think the result will be? Check your intuition by implementing the appropriate graphical model, and estimating the Bayes factor.

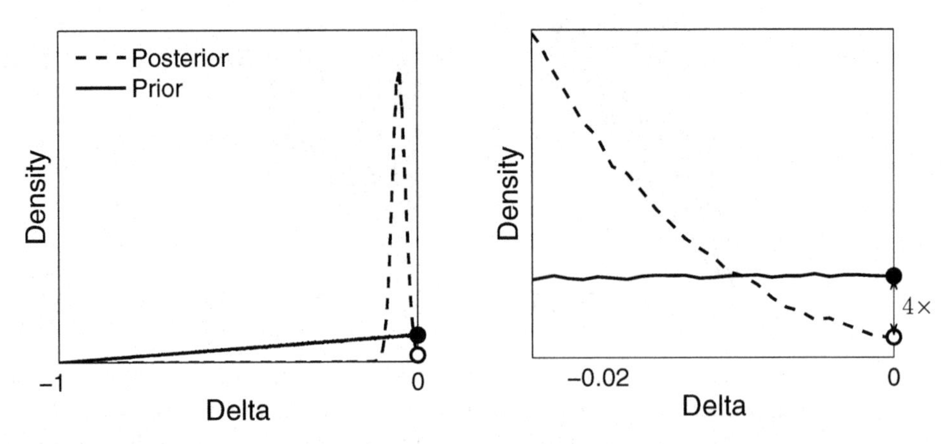

Fig. 9.5 Prior and posterior distributions of the rate difference δ for the pledger data, under the order-restricted analysis. The left panel shows the distributions across their entire range, and the right panel zooms in on the area that is relevant for the Savage–Dickey test. The markers show the densities at $\delta = 0$ needed to estimate the Bayes factor.

Fig. 9.6 Example pair of similar pictures used in Experiment 3 from Zeelenberg et al. (2002).

9.3 Comparing within-subject proportions

In their article "Priming in implicit memory tasks: Prior study causes enhanced discriminability, not only bias," Zeelenberg, Wagenmakers, and Raaijmakers (2002) reported three experiments in two-alternative forced-choice perceptual identification. In the test phase of each experiment, a stimulus (e.g., a picture of a clothespeg) is briefly presented and masked. Immediately after the mask the subject is confronted with two choice options, examples of which are shown in Figure 9.6. One is the target (e.g., a picture of the clothespeg) and a similar foil alternative (e.g., a picture of a stapler). The subject's goal is to identify the target.

Prior to the test phase, the Zeelenberg et al. (2002) experiments featured a study phase, in which subjects studied a subset of the choice alternatives that would also be presented in the later test phase. Two conditions were critical: the "study-neither" condition, in which neither choice alternative was studied, and the "study-both" condition, in which both choice alternatives were studied. In the first two experiments reported by Zeelenberg et al. (2002), subjects choose the target stimulus more often in the study-both condition than in the study-neither condition. This *both-primed benefit* suggests that prior study leads to enhanced discriminability, not just a bias to prefer the studied alternative (e.g., Ratcliff & McKoon, 1997).

Here we focus on statistical inference for Experiment 3 from Zeelenberg et al. (2002). In the study phase of this experiment, all 74 subjects were presented with 21 pairs of similar pictures, as in the example in Figure 9.6. In the test phase, all participants had to identify briefly presented target pictures among a set of two alternatives. The test phase was composed of 42 pairs of similar pictures, 21 of which had been presented in the study phase.

In order to assess the evidence in favor of the both-primed benefit, the authors carried out a standard analysis and computed a one-sample t-test:

> Mean percentage of correctly identified pictures was calculated for each participant. When neither the target nor the foil had been studied, 71.5% of the pictures were correctly identified. When both the target and the foil had been studied, 74.7% of the pictures were correctly identified. The difference between the study-both and study-neither conditions was significant, $t(73) = 2.19$, $p < .05$.

Our Bayes factor test of the both-primed benefit is shown in Figure 9.7. The model assumes that the number of correct choices made by the ith subject is binomially distributed with accuracy rate parameters θ_i^b and θ_i^n in the study-both and study-neither conditions, respectively. We allow for individual differences between the accuracy of subjects, and individual differences in the impact of the both-primed benefit.

Individual differences in the study-neither condition are modeled as having a Gaussian distribution, with the rates themselves given by a probit transformation. This is conceptually the same idea as the logit transformation approach in Section 6.6, with probit and logit transformations being quantitatively slightly different. The probit transform is shown in Figure 9.8, and is the inverse cumulative distribution function of the standard Gaussian distribution. This means, for example, that a probit rate of $\phi = 0$ maps on to a rate of $\theta = \Phi(0) = 0.5$, and a probit rate of $\phi = 1.96$ maps on to a rate of $\theta = \Phi(1.96) = 0.975$. The standard Gaussian distribution on the probit scale, $\phi \sim \text{Gaussian}(0, 1)$, corresponds to a standard uniform distribution on the rate scale, $\theta \sim \text{Uniform}(0, 1)$. Formally, the ith subject has $\phi_i^n \sim \text{Gaussian}(\mu, 1/\sigma^2)$ and their accuracy rate is then given by $\theta_i^n = \Phi(\phi_i^n)$, transforming a real number into a probability.

The model in Figure 9.7 uses the probit scale to implement the potential both-primed benefit. The benefit for the ith subject is drawn from a Gaussian distribution with mean μ_α and and standard deviation σ_α. This effect additively changes the study-neither into the study-both accuracy on the probit scale, so that $\phi_i^b = \phi_i^n + \alpha_i$. The model incorporates a parameter δ that quantifies the critical effect size, $\delta = \mu_\alpha / \sigma_\alpha$. Effect size is a dimensionless quantity, and this makes it relatively easy

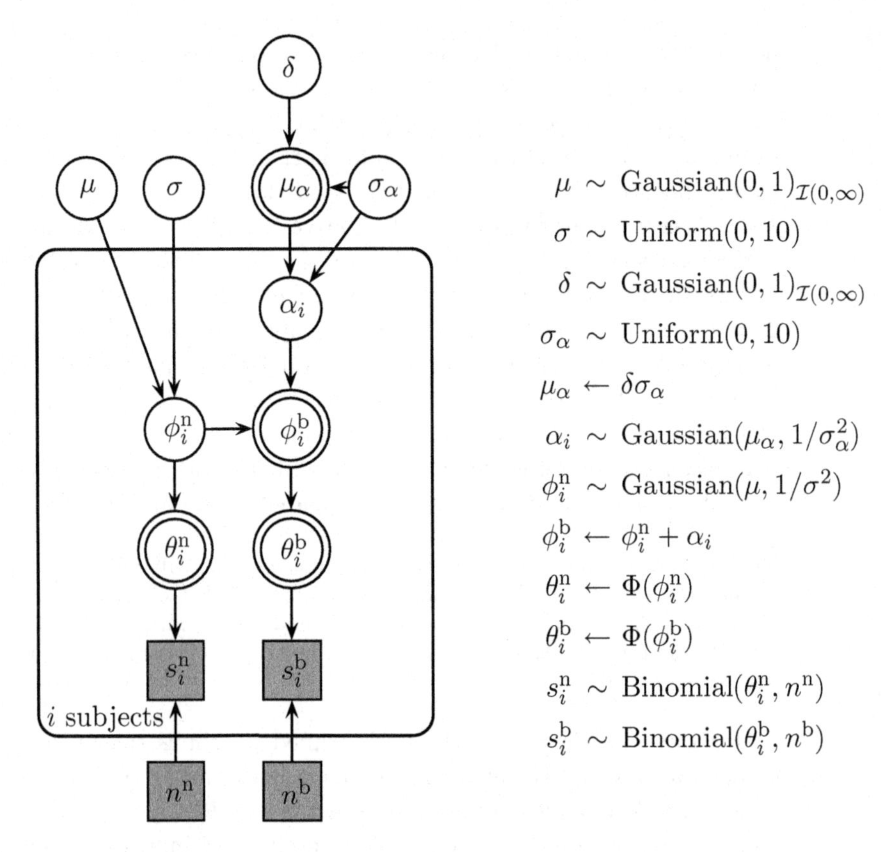

Fig. 9.7 Graphical model for the analysis of the Zeelenberg et al. (2002) data.

to define a principled prior, such as the Cauchy distribution (i.e., a t distribution with one degree of freedom) and the standard Gaussian distribution (e.g., Gönen, Johnson, Lu, & Westfall, 2005; Rouder et al., 2009). The latter prior is known as the "unit information prior," as it carries as much information as a single observation (Kass & Wasserman, 1995). Our model uses the standard Gaussian distribution prior.

For the parameters that are not the focus of the statistical test (i.e., μ_ϕ, σ_ϕ, and σ_α) relatively uninformative priors are used. The prior for the group mean of the study-neither condition, μ_ϕ, is a standard Normal truncated to be greater than zero, which on the rate scale corresponds to a uniform distribution from 0.5 to 1. For σ_ϕ and σ_α, the model uses uniform priors over a large enough range from 0 to 10.

With this account of the data-generating model in place, we can now turn to hypothesis testing. The null hypothesis states that there is no both-primed benefit, and hence the effect size is zero: $\mathcal{H}_0 : \delta = 0$. The alternative, order-restricted hypothesis states that there is a both-primed benefit, and hence $\mathcal{H}_1 : \delta > 0$.

The script **`Zeelenberg.txt`** implements the graphical model in WinBUGS:

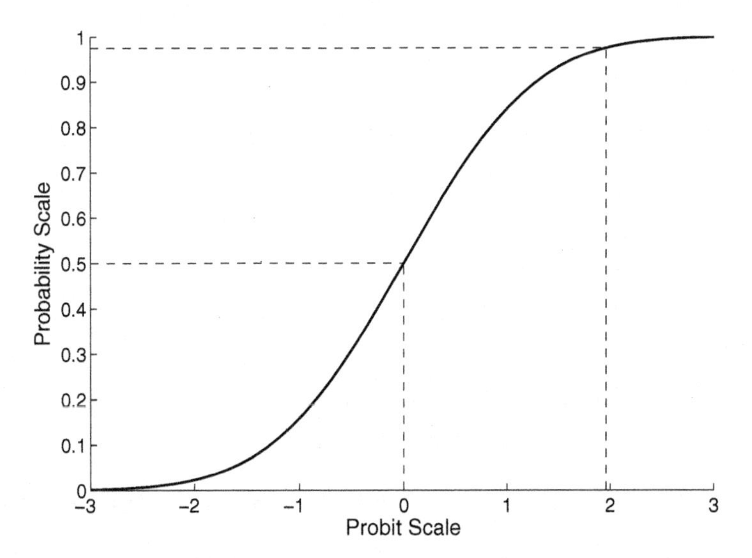

Fig. 9.8 The probit transformation.

```
# Zeelenberg
model{
  for (i in 1:ns){
    # Data
    sb[i] ~ dbin(thetab[i],nb)
    sn[i] ~ dbin(thetan[i],nn)
    # Probit Transformation
    thetab[i] <- phi(phib[i])
    thetan[i] <- phi(phin[i])
    # Individual Parameters
    phin[i] ~ dnorm(mu,lambda)
    alpha[i] ~ dnorm(mualpha,lambdaalpha)
    phib[i] <- phin[i]+alpha[i]
  }
  # Priors
  mu ~ dnorm(0,1)I(0,)
  sigma ~ dunif(0,10)
  lambda <- pow(sigma,-2)
  # Priming Effect
  sigmaalpha ~ dunif(0,10)
  lambdaalpha <- pow(sigmaalpha,-2)
  delta ~ dnorm(0,1)I(0,)
  mualpha <- delta*sigmaalpha
  # Sampling from Prior Distribution for Delta
  deltaprior ~ dnorm(0,1)I(0,)
}
```

The code `Zeelenberg.m` or `Zeelenberg.R` draws samples from the posterior and the prior for the effect size δ, and Figure 9.9 shows the results of applying the Savage–Dickey method. The critical effect size of $\delta = 0$ is supported about 4 times as much under the prior as it is under the posterior. That is, the data have decreased

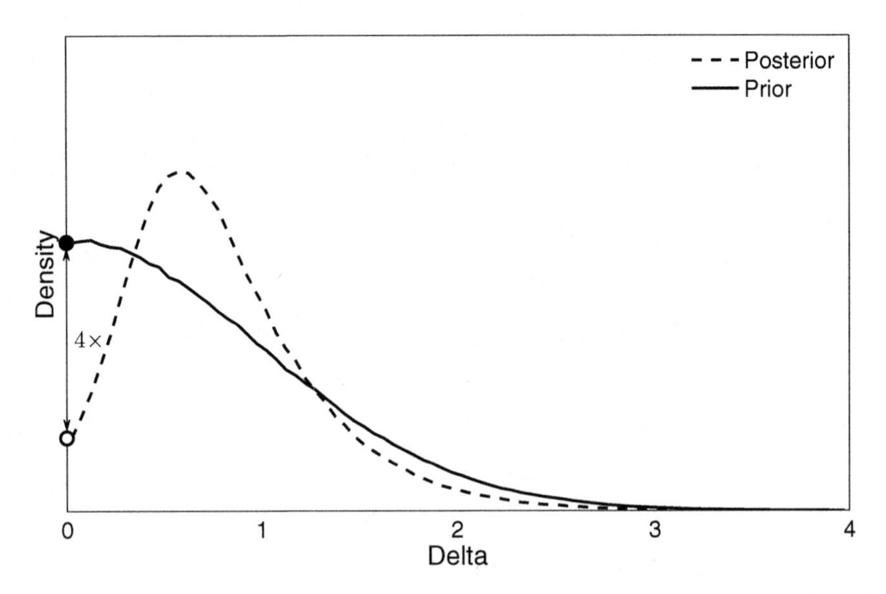

Fig. 9.9 Prior and posterior distributions of the effect size δ for the Zeelenberg et al. (2002) data. The markers show the densities at $\delta = 0$ needed to estimate the Bayes factor.

the support for $\delta = 0$ by a factor of 4. Thus, the Bayes factor is about 4 in favor of the alternative hypothesis, and a reasonable conclusion might be that the data weakly support the assertion that there is a both-primed benefit.

Exercise

Exercise 9.3.1 The Zeelenberg data can also be analyzed using the Bayesian t-test discussed in Chapter 8. Think of a few reasons why this might not be such a good idea. Then, despite your reservations, apply the Bayesian t-test anyway. How do the results differ? Why?

9.4 Comparing between-subject proportions

In their article "How specific are executive functioning deficits in Attention Deficit Hyperactivity Disorder and autism?", Geurts, Verté, Oosterlaan, Roeyers, and Sergeant (2004) studied the performance of children with ADHD and autism on a range of cognitive tasks. Here we focus on a small subset of the data and consider the question whether children who develop typically (i.e., "normal controls") outperform children with ADHD on the Wisconsin Card Sorting Test (WCST). The WCST requires people to learn, by trial and error, to sort cards according to an implicit rule. The complication is that, over the course of the experiment, the sorting rule sometimes changes. This means that in order to avoid too many mistakes,

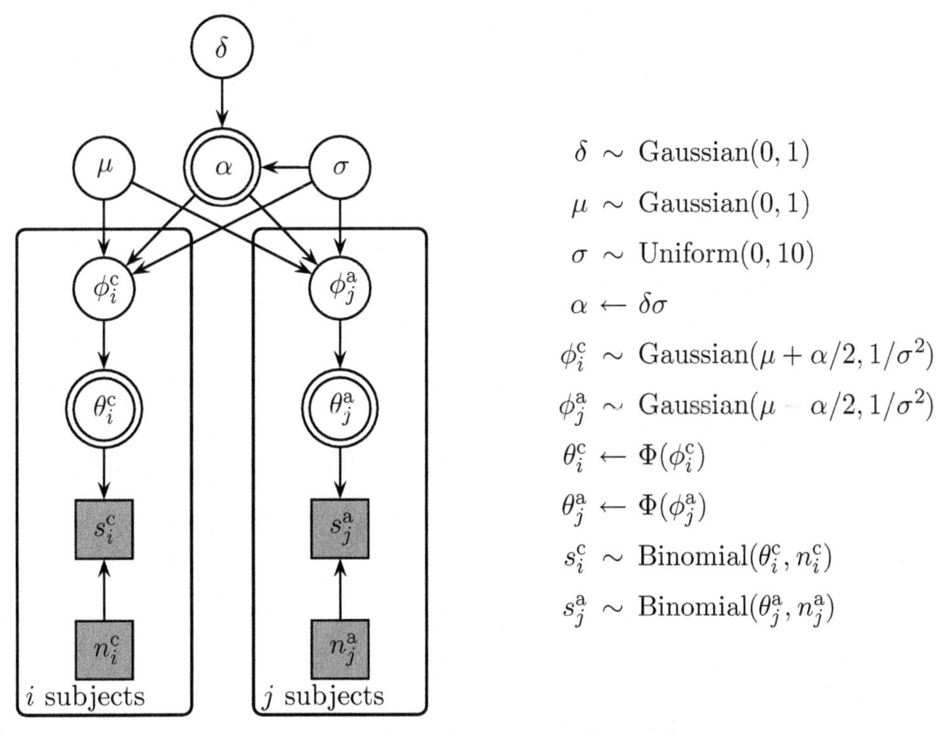

Fig. 9.10 Graphical model for the analysis of the Geurts et al. (2004) data.

people have to suppress the tendency to perseverate, and discover and adopt the new rule. Because of these task demands, performance on the WCST is thought to quantify cognitive flexibility or set shifting ability.

The experiment of interest contains data from 26 normal controls and 52 children with ADHD. Each child performed the WCST, and the measure of interest is the number of correctly sorted cards relative to the total number of sorting opportunities. The WCST provides a maximum of 128 cards to sort, but, depending on a child's performance, this number could also be lower. Overall, the group of normal controls sorted the cards correctly on 65.4% of the cases, and the group of ADHD children sorted the cards correctly on 66.9% of the cases. The null hypothesis states that the probability of sorting cards correctly does not differ between the normal controls and the ADHD children. A between-subjects frequentist t-test on the proportion of correctly sorted cards does not allow one to reject the null hypothesis, $t(40.2) = 0.37$, $p = 0.72$. But this statistic does not quantify the evidence in favor of the null hypothesis.

The key difference between this example and the Zeelenberg example in Section 9.3 is that the experimental design is now between-subjects. Figure 9.10 shows the graphical model for this design. There are now separate plates for the control and ADHD groups. Within these groups the ith or jth child, respectively, has a success rate of θ_i^c and θ_j^a. These account for the observed data, which are the number of correctly sorted cards s_i^c and s_j^a out of the number of attempts n_i^c and n_j^a.

Individual variation is again modeled by a Gaussian distribution on the probit scale. The means of the control and ADHD groups are $\mu+\alpha/2$ and $\mu-\alpha/2$, so that they differ by α, and both Gaussians have the same standard deviation σ. Thus, on the probit scale, $\phi_i^c \sim \text{Gaussian}(\mu+\alpha/2, 1/\sigma^2)$ and $\phi_i^a \sim \text{Gaussian}(\mu-\alpha/2, 1/\sigma^2)$, with the transformations $\theta_i^c = \Phi(\phi_i^c)$ and $\theta_j^a = \Phi(\phi_j^a)$.

As before, the parameters that are not the focus of the statistical test are given relatively uninformative priors. Also as before, a parameter $\delta = \alpha/\sigma$ is used to quantify effect size, and is given a "unit information" standard normal prior. The null hypothesis states that normal controls and ADHD children perform the same on the WCST, and hence the effect size is zero: $\mathcal{H}_0 : \delta = 0$. The unrestricted alternative hypothesis states that there is a difference in performance, and hence $\mathcal{H}_1 : \delta \neq 0$.

The script `Geurts.txt` implements the graphical model in WinBUGS:

```
# Geurts
model{
  for (i in 1:nsc){
    kc[i] ~ dbin(thetac[i],nc[i])
    thetac[i] <- phi(phic[i])
    phic[i] ~ dnorm(muc,lambda)
  }
  for (j in 1:nsa){
    ka[j] ~ dbin(thetaa[j],na[j])
    thetaa[j] <- phi(phia[j])
    phia[j] ~ dnorm(mua,lambda)
  }
  muc <- mu+alpha/2
  mua <- mu-alpha/2
  # Priors
  mu ~ dnorm(0,1)
  sigma ~ dunif(0,10)
  alpha <- delta*sigma
  lambda <- pow(sigma,-2)
  delta ~ dnorm(0,1)
  # Sampling from Prior Distribution for Delta
  deltaprior ~ dnorm(0,1)
}
```

The code `Geurts.m` or `Geurts.R` draws samples from the posterior and the prior for the effect size δ. The results are shown in Figure 9.11. The ADHD children performed slightly better than the normal controls, and this is reflected in a posterior distribution for δ. This posterior is slightly asymmetric around zero, and assigns more mass to negative than to positive values of δ. The Bayesian 95% credible interval for δ is approximately $(-0.54, 0.42)$.

Figure 9.11 shows that the data have made the value $\delta = 0$ about four times more likely than it was before. Thus, the data support the claim, with a Bayes factor of about 4, that normal controls and ADHD children perform equally well on the WCST, when compared to the claim that these groups perform differently.

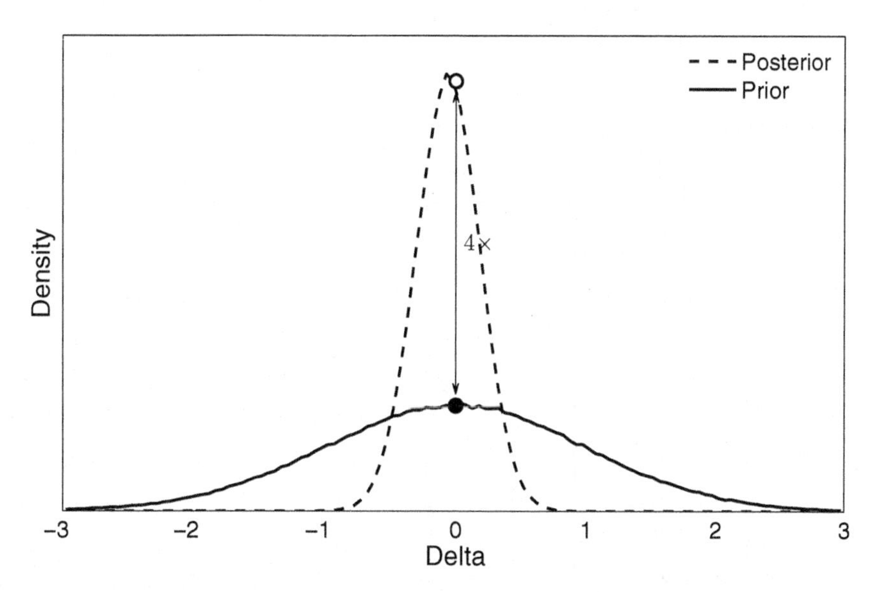

Fig. 9.11 Prior and posterior distributions of the effect size δ for the Geurts et al. (2004) data. The markers show the densities at $\delta = 0$ needed to estimate the Bayes factor.

Exercises

Exercise 9.4.1 A between-subjects frequentist t-test on the proportion of correctly sorted cards does not allow one to reject the null hypothesis, $t\,(40.2) = 0.37$, $p = 0.72$. In what way does the Bayesian approach improve upon the frequentist inference?

Exercise 9.4.2 In what way is the model of the data in Figure 9.10 superior to the statistical model assumed by the t-test?

9.5 Order-restricted between-subjects comparison

A natural modification of the alternative hypothesis for the Geurts et al. (2004) data is to impose the order-restriction corresponding to the assumption that normal controls perform better than ADHD children. This alternative hypothesis is $\mathcal{H}_2 : \delta > 0$. This hypothesis seems reasonable to entertain because of its a priori plausibility, but the data suggest that, if anything, the reverse is true. What can we expect when we test $\mathcal{H}_0 : \delta = 0$ versus $\mathcal{H}_2 : \delta > 0$?

First, note that the posterior for δ is not far from being symmetric around zero. If it were completely symmetric, the height of both the prior and the posterior is multiplied by 2, so that their ratio stays the same. Second, the posterior for δ is not quite symmetric around zero, and assigns slightly more mass to values that are inconsistent with $\mathcal{H}_2$. This will slightly increase the support for $\mathcal{H}_0$ over $\mathcal{H}_2$. These

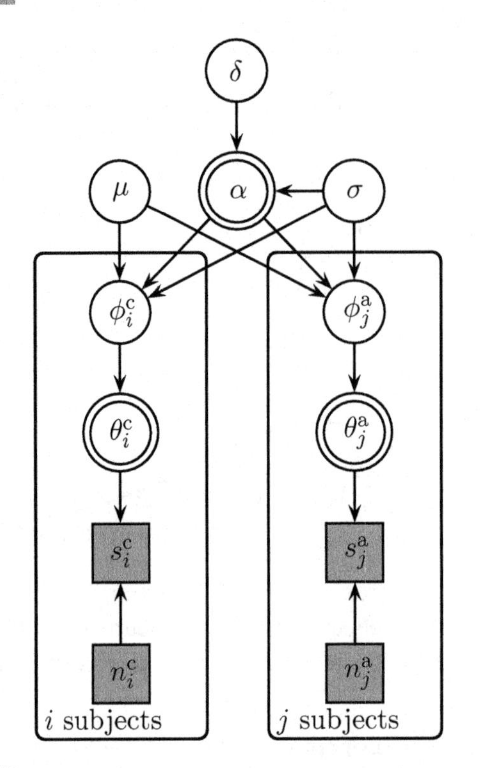

$$\delta \sim \text{Gaussian}(0,1)_{\mathcal{I}(0,\infty)}$$
$$\mu \sim \text{Gaussian}(0,1)$$
$$\sigma \sim \text{Uniform}(0,10)$$
$$\alpha \leftarrow \delta\sigma$$
$$\phi_i^c \sim \text{Gaussian}(\mu + \alpha/2, 1/\sigma^2)$$
$$\phi_j^a \sim \text{Gaussian}(\mu - \alpha/2, 1/\sigma^2)$$
$$\theta_i^c \leftarrow \Phi(\phi_i^c)$$
$$\theta_j^a \leftarrow \Phi(\phi_j^a)$$
$$s_i^c \sim \text{Binomial}(\theta_i^c, n_i^c)$$
$$s_j^a \sim \text{Binomial}(\theta_j^a, n_j^a)$$

Fig. 9.12 Graphical model for the analysis of the Geurts et al. (2004) data.

two considerations lead us to expect that the evidence in favor of $\mathcal{H}_0$ over $\mathcal{H}_2$ will be slightly larger than that of $\mathcal{H}_0$ over $\mathcal{H}_1$. First, note that the posterior for δ is not far from being symmetric around zero. If it were completely symmetric, the height of both the prior and the posterior is multiplied by 2, so that their ratio stays the same. Second, the posterior for δ is not quite symmetric around zero, and assigns slightly more mass to values that are inconsistent with $\mathcal{H}_2$. This will slightly increase the support for $\mathcal{H}_0$ over $\mathcal{H}_2$. These two considerations lead us to expect that the evidence in favor of $\mathcal{H}_0$ over $\mathcal{H}_2$ will be slightly larger than that of $\mathcal{H}_0$ over $\mathcal{H}_1$.

Figure 9.12 shows the modified graphical model for the order-restricted analysis, which simply restricts the prior on δ to positive values.

The script `GeurtsOrderRestricted.txt` implements the graphical model in WinBUGS:

```
# Geurts, Order Restricted
model{
  for (i in 1:nsc){
    kc[i] ~ dbin(thetac[i],nc[i])
    thetac[i] <- phi(phic[i])
    phic[i] ~ dnorm(muc,lambda)
  }
  for (j in 1:nsa){
    ka[j] ~ dbin(thetaa[j],na[j])
```

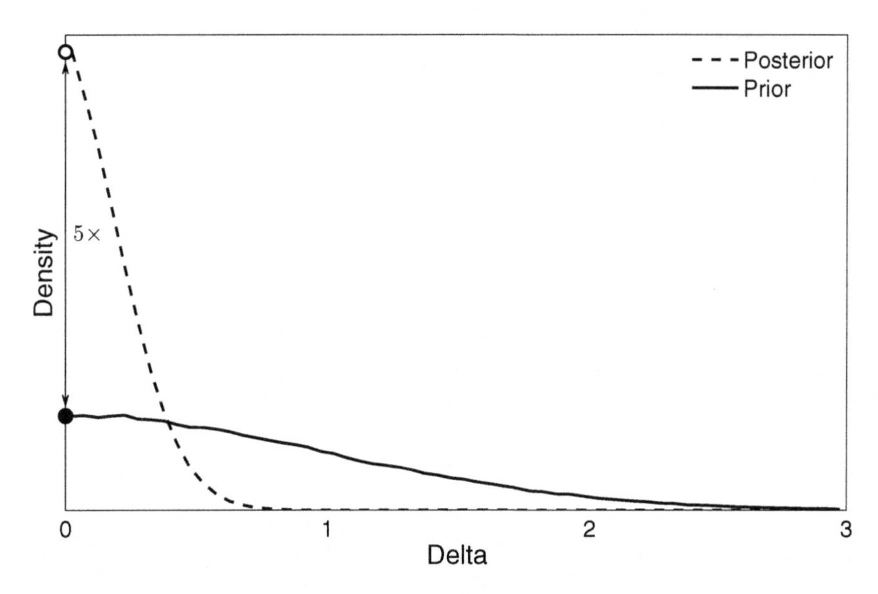

Fig. 9.13 Prior and posterior distributions of the effect size δ for the Geurts et al. (2004) data. The markers show the densities at $\delta = 0$ needed to estimate the Bayes factor.

```
    thetaa[j] <- phi(phia[j])
    phia[j] ~ dnorm(mua,lambda)
  }
  muc <- mu+alpha/2
  mua <- mu-alpha/2
  # Priors
  mu ~ dnorm(0,1)
  sigma ~ dunif(0,10)
  alpha <- delta*sigma
  lambda <- pow(sigma,-2)
  delta ~ dnorm(0,1)I(0,)
  # Sampling from Prior Distribution for Delta
  deltaprior ~ dnorm(0,1)
}
```

The code `GeurtsOrderRestricted.m` or `GeurtsOrderRestricted.R` draws samples from the posterior and the prior for δ. Figure 9.13 shows the results, confirming the expectation that the order-restriction slightly increases the evidence in favor of $\mathcal{H}_0$. The Bayes factor in favor of the null hypothesis is now about 5. Thus, the data support the assertion that normal controls and children with ADHD perform similarly on the WCST, although the evidence is not overwhelming.

Exercises

Exercise 9.5.1 For the order-restricted comparison of $\mathcal{H}_0 : \delta = 0$ versus $\mathcal{H}_2 : \delta > 0$, what is the maximum support in favor of $\mathcal{H}_0$ that could possibly be obtained, given the present number of subjects, and given that the average rate of correct card sorts is 65%?

Exercise 9.5.2 What is the maximum support for the earlier unrestricted test of $\mathcal{H}_0 : \delta = 0$ versus $\mathcal{H}_1 : \delta \neq 0$?

PART IV

CASE STUDIES

Fisher and Jeffreys were occupied with very different problems. Fisher studied biological problems, where one had no prior information and no guiding theory ..., and the data taking was very much like drawing from Bernoulli's urn. Jeffreys studied problems of geophysics, where one had a great deal of cogent prior information and a highly developed guiding theory ..., and the data taking procedure had no resemblance to drawing from an urn ... As science progressed to more and more complicated problems of inference, the shortcomings of the orthodox methods have become more and more troublesome ... Scientists, engineers, biologists, and economists with good Bayesian training are now finding for themselves the correct solutions appropriate for their problems, which can adapt effortlessly to many kinds of prior information, thus achieving a flexibility unknown in orthodox statistics.

Jaynes, 2003, pp. 496–497

10 Memory retention

Finding a lawful relationship between memory retention and time is one of the oldest cognitive modeling question, going back to Ebbinghaus in the 1880s. The usual experiment involves giving people many items of information on a list, and then testing their ability to remember items from the list after different periods of time have elapsed. Various mathematical functions, usually with psychological interpretations, have been proposed as describing the relationship between time and the level of retention (Rubin & Wenzel, 1996; Rubin, Hinton, & Wenzel, 1999).

We consider a simplified version of the exponential decay model. The model assumes that the probability that an item will be remembered after a period of time t has elapsed is $\theta_t = \exp(-\alpha t) + \beta$, with the restriction $0 < \theta_t < 1$. The α parameter corresponds to the rate of decay of information. The β parameter corresponds to a baseline level of remembering that is assumed to remain even after very long time periods. Our analyses using this model are based on fictitious data from a potential memory retention study, to help illustrate key modeling points.

Table 10.1 Memory retention data for 4 subjects and 10 time intervals.

	Time Interval									
Subject	1	2	4	7	12	21	35	59	99	200
1	18	18	16	13	9	6	4	4	4	?
2	17	13	9	6	4	4	4	4	4	?
3	14	10	6	4	4	4	4	4	4	?
4	?	?	?	?	?	?	?	?	?	?

These data are given in Table 10.1, and relate to 4 subjects tested on 18 items at 10 time intervals: 1, 2, 4, 7, 12, 21, 35, 59, 99, and 200. The number of items tested and the first 9 time intervals are those used by Rubin et al. (1999). Each datum in Table 10.1 counts the number of correct memory recalls for each subject at each time interval. Included in Table 10.1 are missing data, shown by "?" symbols, so that the prediction and generalization properties of models can be tested. All of the subjects have missing data for the final time period of 200, which tests the ability of models to generalize to new measurements. For Subject 4, there are no data at all, which tests the ability of models to generalize to new subjects.

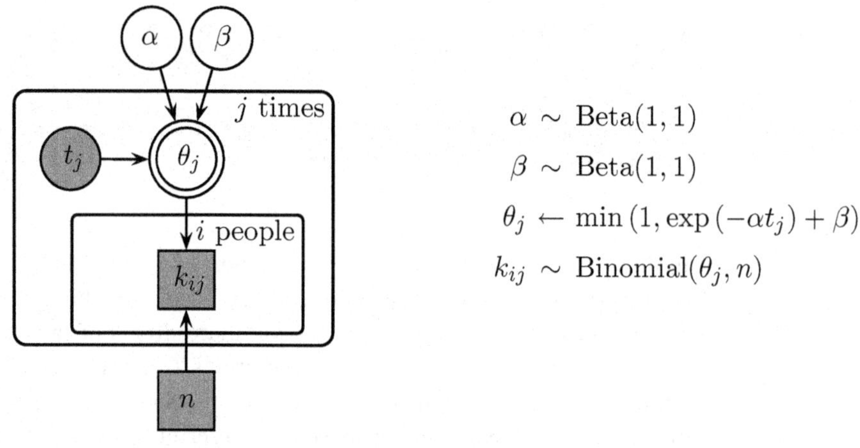

Fig. 10.1 Graphical model for the exponential decay model of memory retention, assuming no individual differences.

10.1 No individual differences

The graphical model for our first attempt to account for the data is shown in Figure 10.1. The model assumes that every subject has the same retention curve, and so there is one true value for the α and β parameters. The outer plate corresponds to the different time periods with values given by the observed t_j variable. Together with the α and β parameters, the time period defines the probability θ_j that the jth item will be remembered.

The inner plate corresponds to the subjects. Each has the same probability of recall at any given time period, but their experimental data, given by the success counts k_{ij}, vary, and are binomially distributed according to the success rate and number of trials.

The script `Retention_1.txt` implements the graphical model in WinBUGS. Note that the code calculates the success rate for each subject at each interval separately, and so is more elaborate than it needs to be:

```
# Retention With No Individual Differences
model{
  # Observed and Predicted Data
  for (i in 1:ns){
    for (j in 1:nt){
      k[i,j] ~ dbin(theta[i,j],n)
      predk[i,j] ~ dbin(theta[i,j],n)
    }
  }
  # Retention Rate At Each Lag For Each Subject Decays Exponentially
  for (i in 1:ns){
    for (j in 1:nt){
      theta[i,j] <- min(1,exp(-alpha*t[j])+beta)
    }
```

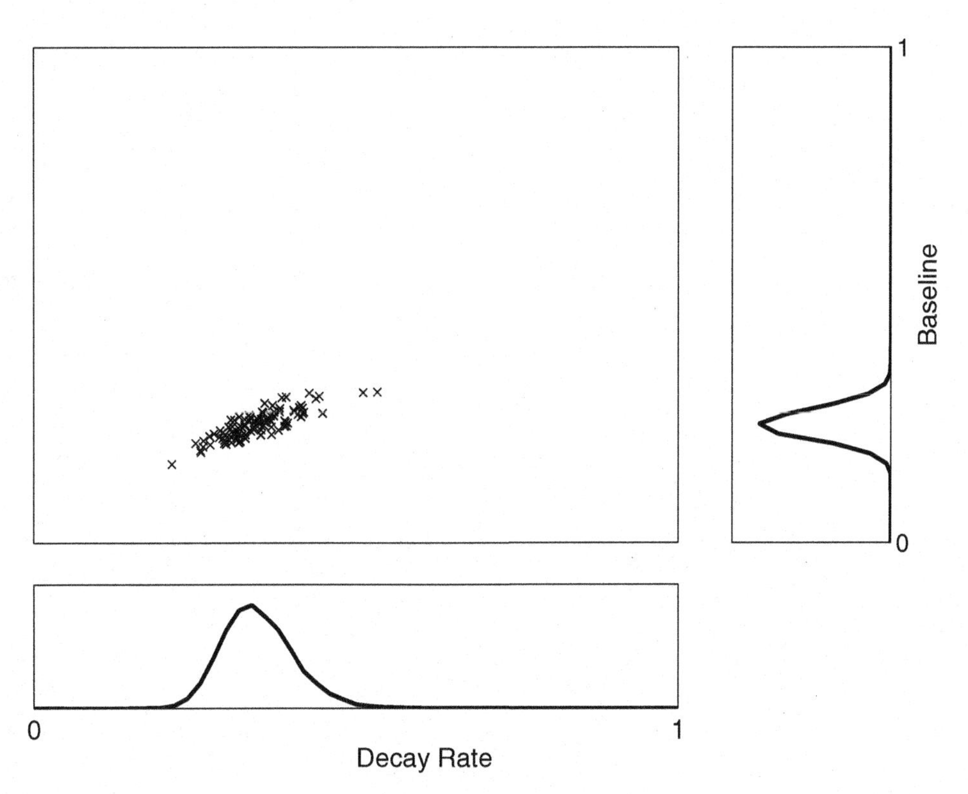

Fig. 10.2 The joint and marginal posterior distributions over the decay and baseline parameters, for the memory retention model that assumes no individual differences.

```
  }
  # Priors
  alpha ~ dbeta(1,1)
  beta ~ dbeta(1,1)
}
```

The code `Retention_1.m` or `Retention_1.R` applies the model to the data, and produces analysis of the posterior and posterior predictive distributions.

The joint posterior distribution of α and β is shown in the main panel of Figure 10.2, as a two-dimensional scatter-plot. Each of the points in the scatter-plot corresponds to a posterior sample selected at random from those available. The marginal distributions of both α and β are shown below and to the right, and are based on all samples. The marginals show the distribution of each parameter, conditioned on the data, considered independently of the other parameter (i.e., averaged across the other parameter).

It is clear from Figure 10.2 that the joint posterior carries more information than the two marginal distributions. If the joint posterior were independent, it would be just the product of the two marginals. But the joint posterior shows a mild relationship, with larger values of α generally corresponding to larger values of β.

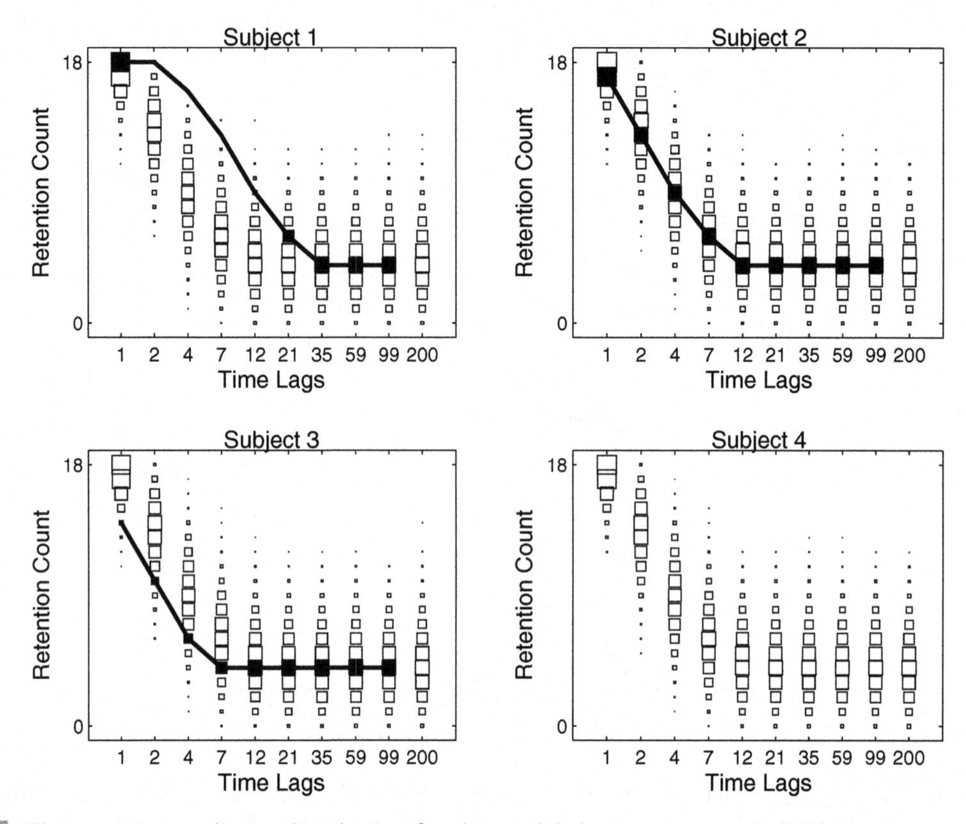

Fig. 10.3 The posterior predictive distribution for the model that assumes no individual differences.

This can be interpreted psychologically as meaning that it is uncertain whether the parameters are a relatively higher baseline coupled with a relatively higher decay rate, or a relatively lower baseline coupled with a relatively lower decay rate.

Figure 10.3 shows the posterior predictive distribution over the number of successful retentions at each time interval. For each subject, at each interval, the squares show the posterior mass given to each possible number of items recalled. These correspond to the model's predictions about observed behavior in the retention experiment, based on what the model has learned from the data. Also shown, by the black squares and connecting lines, are the actual observed data for each subject, where available.

The obvious feature of Figure 10.3 is that the current model does not meet a basic requirement of descriptive adequacy. For both Subjects 1 and 3 the model gives little posterior mass to the observed data at many time periods. It describes a steeper rate of decay than shown by the data of Subject 1, and a shallower rate of decay than shown by the data of Subject 3. After evaluating the model using the posterior predictive analysis, we can conclude that the modeling assumption of no individual differences is inappropriate. It is important to understand that

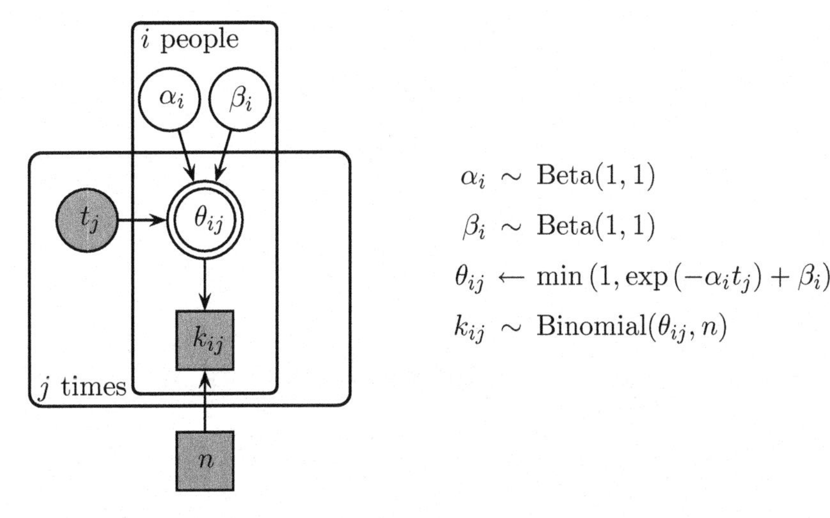

Fig. 10.4 Graphical model for the exponential decay model of memory retention, assuming full individual differences.

this conclusion neuters the usefulness of the posterior distribution over parameters shown in Figure 10.2. The posterior distribution is conditioned on the assumption that the model is appropriate, and is not relevant when the model is fundamentally deficient.

Exercise

Exercise 10.1.1 Why is the posterior predictive distribution for all four subjects the same? Are there any (real or fabricated) data that could make the model predict different patterns of retention for different subjects? What if there were massive qualitative differences, such as one subject remembering everything, and the other two remembering nothing?

10.2 Full individual differences

A revised graphical model that does accommodate individual differences is shown in Figure 10.4. The change from the previous model is that the ith subject now has their own α_i and β_i parameters, and that the probability of retention for an item θ_{ij} now changes for both subjects and retention intervals.

The script `Retention_2.txt` implements the graphical model in WinBUGS:

```
# Retention With Full Individual Differences
model{
  # Observed and Predicted Data
  for (i in 1:ns){
```

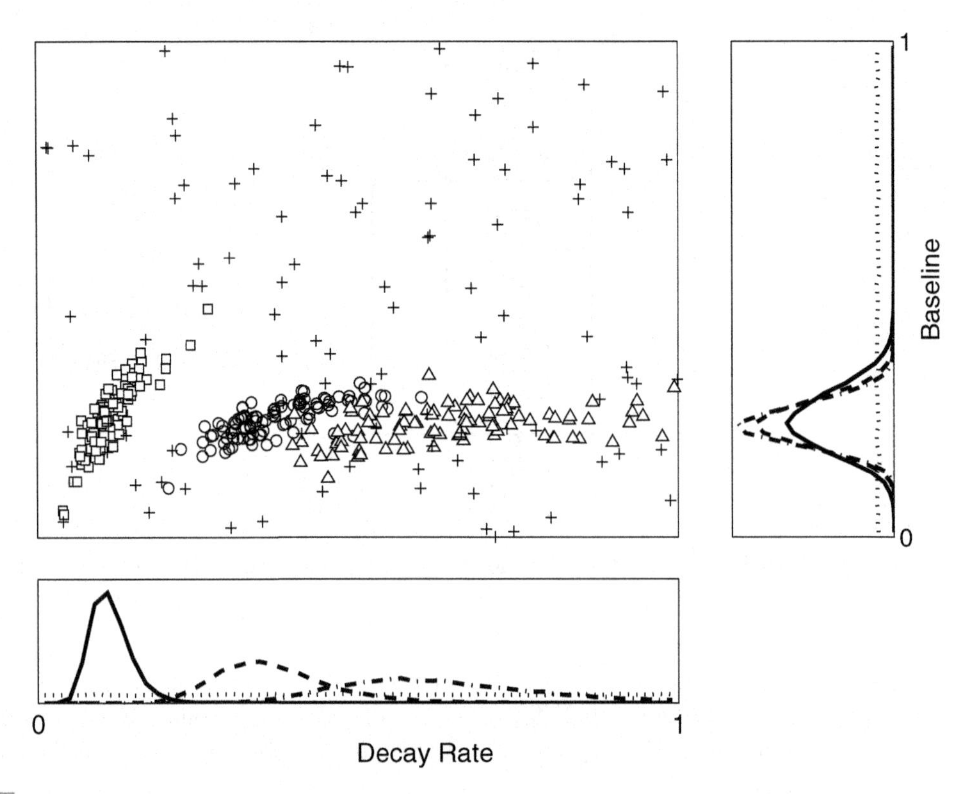

Fig. 10.5 The joint and marginal posterior distributions over the decay and baseline parameters, for the memory retention model that assumes full individual differences.

```
    for (j in 1:nt){
      k[i,j] ~ dbin(theta[i,j],n)
      predk[i,j] ~ dbin(theta[i,j],n)
    }
  }
  # Retention Rate At Each Lag For Each Subject Decays Exponentially
  for (i in 1:ns){
    for (j in 1:nt){
      theta[i,j] <- min(1,exp(-alpha[i]*t[j])+beta[i])
    }
  }
  # Priors For Each Subject
  for (i in 1:ns){
    alpha[i] ~ dbeta(1,1)
    beta[i] ~ dbeta(1,1)
  }
}
```

The code `Retention_2.m` or `Retention_2.R` applies the model to the same data, and again produces analysis of the posterior and posterior predictive distributions.

The joint posterior distributions for each subject are shown in the main panel of Figure 10.5. Each point in the scatter-plot corresponds to a posterior sample,

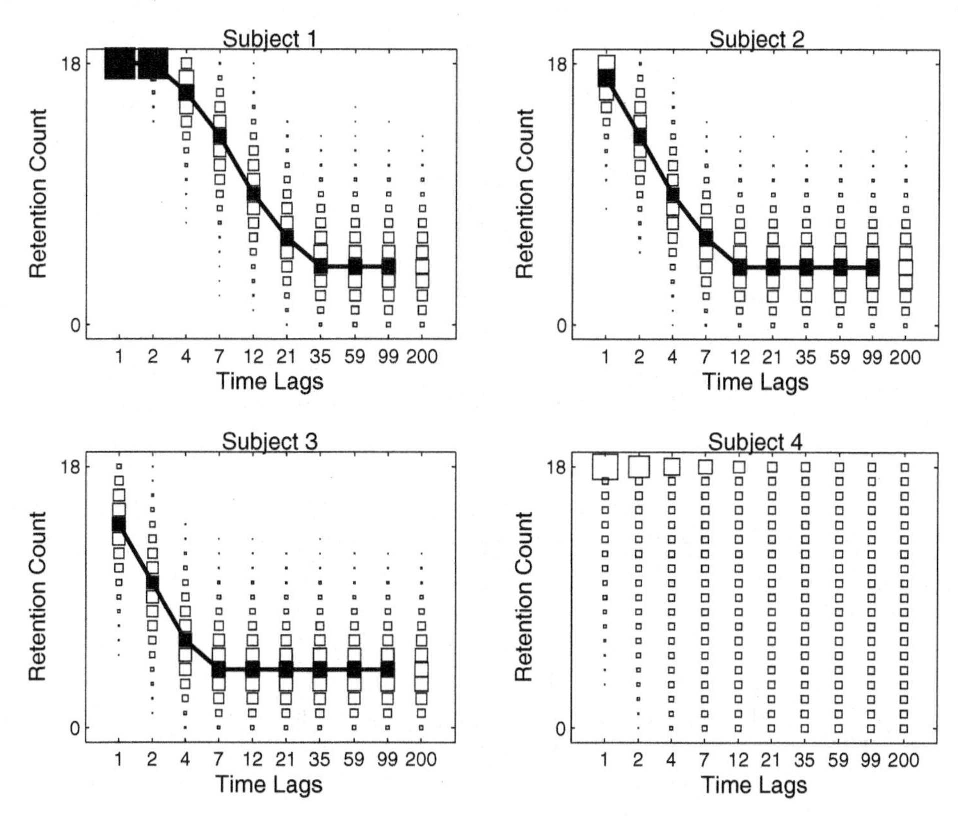

Fig. 10.6 The posterior predictive distribution for the model that assumes full individual differences.

with different markers representing different subjects. The first, second, third, and fourth subjects use square, circular, triangular, and cross markers, respectively. The marginal distributions are shown below and to the right, and use different line styles to represent the subjects.

Figure 10.6 shows the same analysis of the posterior predictive distributions over the number of successful retentions at each time interval, for each subject. It is clear that allowing for individual differences lets the model achieve a basic level of descriptive adequacy for Subjects 1, 2, and 3. The posteriors in Figure 10.5 show that different values for the α decay parameter are used for each of these subjects, corresponding to our intuitions from the earlier analysis.

The weakness in the current model is evident in its predictions for Subject 4. Because each subject is assumed to have decay and baseline parameters that are different, the only information the model has about the new subject is the priors for the α and β parameters. The relationships between parameters for subjects that are visually evident in Figure 10.5 are not formally captured by the model. This means, as shown in Figure 10.5, the posteriors for Subject 4 are just the priors, and so the posterior predictive distribution for Subject 4, as shown in Figure 10.6, does not have any useful structure. In this way, the model fails a basic test of generalizability, since it does not make sensible predictions for the behavior of subjects other than those for whom data are available. Intuitively, one might want to predict that Subject 4 will be likely to have model parameters consistent with regularities in the inferred parameters for Subjects 1, 2, and 3.

Exercise

Exercise 10.2.1 What are the relative strengths and weaknesses of the full individual differences model and the no individual differences model? Think about this, because the hierarchical approach we consider next could be argued to combine the best features of both of these approaches.

10.3 Structured individual differences

The relationship between the parameters of different subjects, visually evident in Figure 10.5, can be captured using a hierarchical model. A graphical model implementing this approach is shown in Figure 10.7. The key change is that now the α_i and β_i parameters for each subject are modeled as coming from Gaussian distributions. The over-arching Gaussian distribution models this group-level structure for each parameter. This group structure itself has parameters, in the form of means μ_α for the decay and μ_β for the baseline, and precisions λ_α for the decay and λ_β for the baseline. In this way, the individual differences between subjects are given structure.

Each α_i and β_i parameter is independently sampled, so they can be different, but they are sampled from the same distribution, so they have a relationship to one another. This means that inferences made for one subject influence predictions made for another. Since the means and precisions of the group-level distributions are common to all subjects, what is learned about them from one subject affects what is known about another. In addition, because they are sampled from over-arching distributions, the α_i and β_i parameters at the individual subject level no longer have priors explicitly specified, but inherit them from the priors on the means and precisions of the group-level Gaussian distributions.

The script `Retention_3.txt` implements the graphical model in WinBUGS:

```
# Retention With Structured Individual Differences
model{
  # Observed and Predicted Data
  for (i in 1:ns){
    for (j in 1:nt){
      k[i,j] ~ dbin(theta[i,j],n)
      predk[i,j] ~ dbin(theta[i,j],n)
    }
  }
  # Retention Rate At Each Lag For Each Subject Decays Exponentially
```

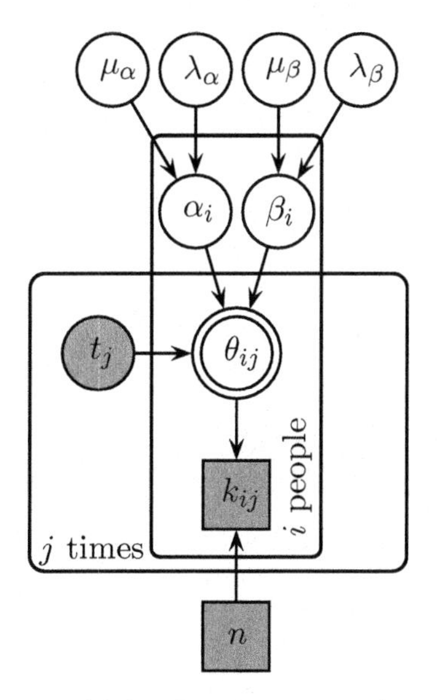

$$\mu_\alpha \sim \text{Beta}(1,1)$$
$$\lambda_\alpha \sim \text{Gamma}(.001,.001)$$
$$\mu_\beta \sim \text{Beta}(1,1)$$
$$\lambda_\beta \sim \text{Gamma}(.001,.001)$$
$$\alpha_i \sim \text{Gaussian}(\mu_\alpha,\lambda_\alpha)_{\mathcal{I}(0,1)}$$
$$\beta_i \sim \text{Gaussian}(\mu_\beta,\lambda_\beta)_{\mathcal{I}(0,1)}$$
$$\theta_{ij} \leftarrow \min\left(1,\exp\left(-\alpha_i t_j\right)+\beta_i\right)$$
$$k_{ij} \sim \text{Binomial}(\theta_{ij},n)$$

Fig. 10.7 Graphical model for the exponential decay model of memory retention, assuming structured individual differences.

```
  for (i in 1:ns){
    for (j in 1:nt){
      theta[i,j] <- min(1,exp(-alpha[i]*t[j])+beta[i])
    }
  }
  # Parameters For Each Subject Drawn From Gaussian Group Distributions
  for (i in 1:ns){
    alpha[i] ~ dnorm(alphamu,alphalambda)I(0,1)
    beta[i] ~ dnorm(betamu,betalambda)I(0,1)
  }
  # Priors For Group Distributions
  alphamu ~ dbeta(1,1)
  alphalambda ~ dgamma(.001,.001)I(.001,)
  alphasigma <- 1/sqrt(alphalambda)
  betamu ~ dbeta(1,1)
  betalambda ~ dgamma(.001,.001)I(.001,)
  betasigma <- 1/sqrt(betalambda)
}
```

The code `Retention_3.m` or `Retention_3.R` applies the model to the data, and again produces an analysis of the posterior and posterior predictive distributions.

The joint and marginal posterior distributions for this model are shown in Figure 10.8 using the same markers and lines as before. For Subjects 1, 2, and 3, these distributions are extremely similar to those found using the full individual differences model. The important difference is for Subject 4, who now has sensible posterior distributions for both parameters. For the decay parameter α there is still considerable uncertainty, consistent with the range of values seen for the first

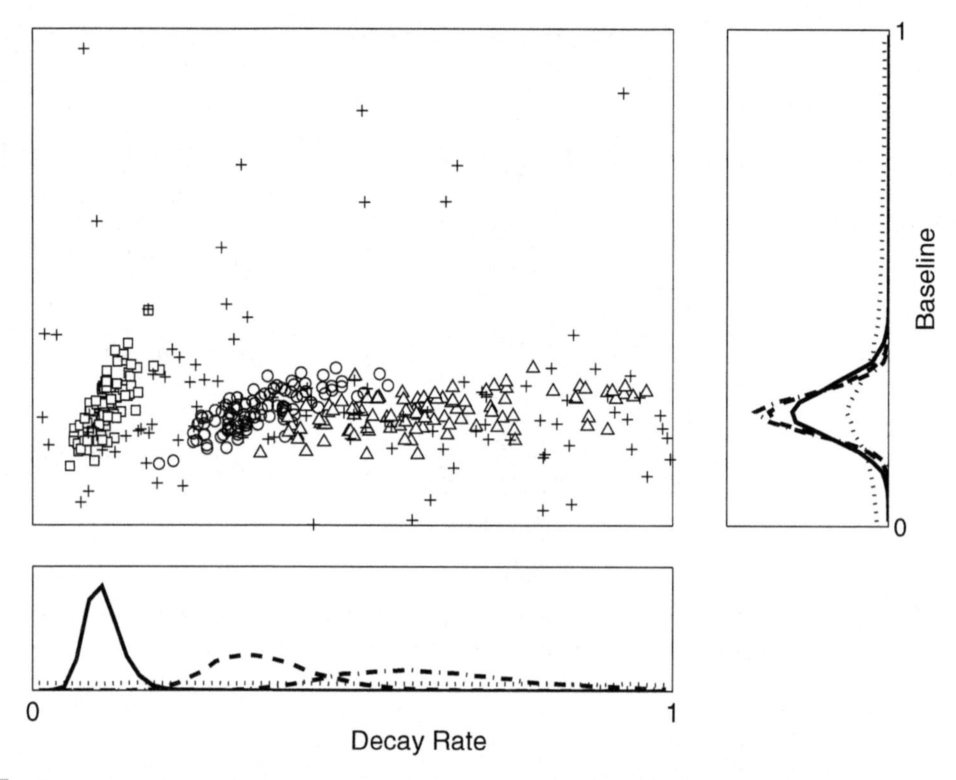

Fig. 10.8 The joint and marginal posterior distributions over the decay and baseline parameters, for the memory retention model that assumes structured individual differences.

three subjects, but for the baseline parameter β, Subject 4 now has a much more constrained posterior distribution.

The posterior predictive distributions for each subject under the hierarchical model are shown in Figure 10.9. The predictions remain useful for the first three subjects, and are now also appropriate for Subject 4. The structured prediction for Subject 4, from whom no data have yet been collected, comes directly from the nature of the hierarchical model. Based on the data from Subjects 1, 2, and 3, inferences are made about the means and precisions of the group distributions for the two parameters of the retention model. The new Subject 4 has values sampled from the Gaussians with these parameters, producing the sensible parameter distributions in Figure 10.8 that lead to the sensible predictive distributions in Figure 10.9.

Psychologically, hierarchical models are powerful because they are able to represent knowledge at different levels of abstraction in a cognitive process (Lee, 2011a). Just as the data have been assumed to be generated by the decay and baseline parameters combining in a memory process for individual subjects, the hierarchical model assumes that those parameters themselves are generated by more abstract latent parameters that describe group distributions across subjects. In other words,

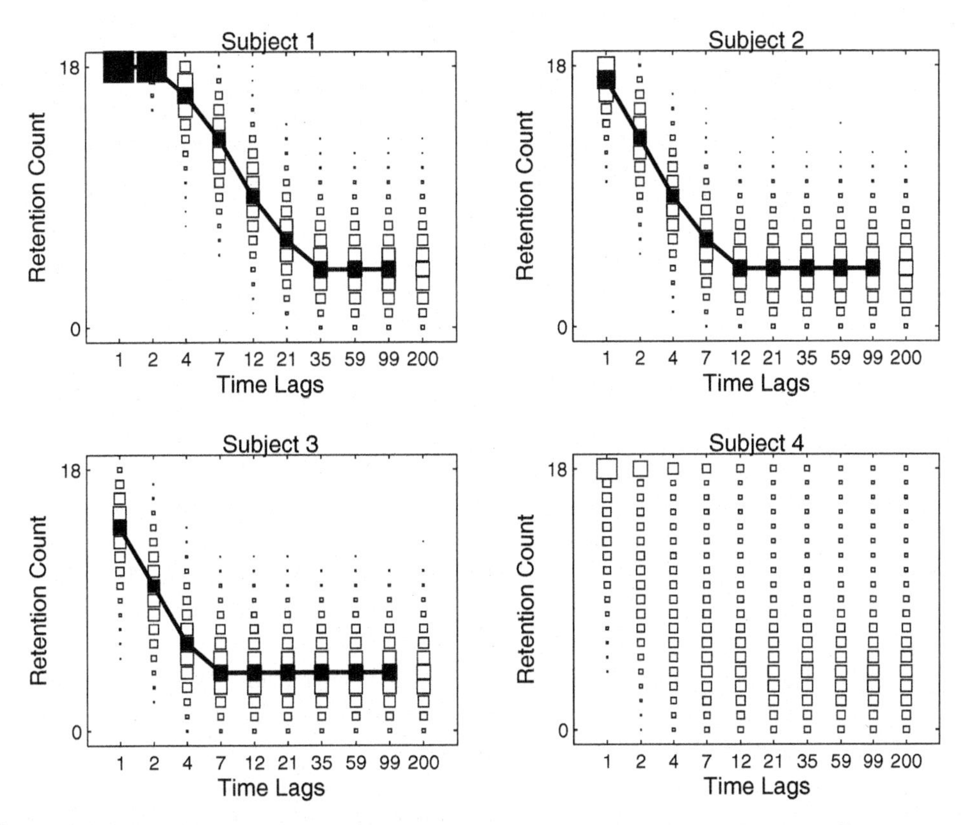

Fig. 10.9 The posterior predictive for the model that assumes structured individual differences.

a hierarchical model lets a theory of memory retention be combined with a theory of individual differences, to provide a more complete account of behavioral data from multiple subjects.

Exercises

Exercise 10.3.1 Think of a psychological model and data, in a different context from the current memory retention example, where the hierarchical approach might be useful.

Exercise 10.3.2 Develop a modified model that does not require you to truncate the rate scale when sampling the α_i decay rates and β_i baselines for each subject. The truncation is not only theoretically inelegant, but technically problematic as it is implemented in WinBUGS, which does not distinguish between censoring and truncation. Implement your modified model and see whether it leads to different conclusions than the ones presented here.

11 Signal detection theory

11.1 Signal detection theory

Signal detection theory (SDT: see D. M. Green & Swets, 1966; MacMillan & Creelman, 2004, for detailed treatments) is a very general, useful, and widely employed method for drawing inferences from data in psychology. It is particularly applicable to two-alternative forced choice experiments, although it can be applied to any situation that can be conceived as a 2×2 table of counts.

Table 11.1 gives the basic data and terminology for SDT. There are "signal" trials and "noise" trials, and "yes" responses and "no" responses. When a "yes" response is given for a signal trial, it is called a "hit." When a "yes" response is given for a noise trial, it is called a "false alarm." When a "no" response is given for a signal trial, it is called a "miss." When a "no" response is given for a noise trial, it is called a "correct rejection."

The basic data for an SDT analysis are just the counts of hits, false alarms, misses, and correct rejections. It is common to consider just the hit and false alarm counts which, together with the total number of signal and noise trials, completely describe the data.

Table 11.1 Basic signal detection theory data and terminology.

	Signal trial	Noise trial
Yes response	Hit	False alarm
No response	Miss	Correct rejection

The key assumptions of SDT are shown in Figure 11.1, and involve representation and decision-making assumptions. Representationally, the idea is that signal and noise trials can be represented as values along a uni-dimensional "strength" construct. Both types of trials are assumed to produce strengths that vary according to a Gaussian distribution along this dimension. The signal strengths are assumed to be greater, on average, than the noise strengths, and so the signal strength distribution has a greater mean. In the most common equal-variance form of SDT, both the distributions are assumed to have the same variance. The decision-making assumption of SDT is that yes and no responses are produced by comparing the

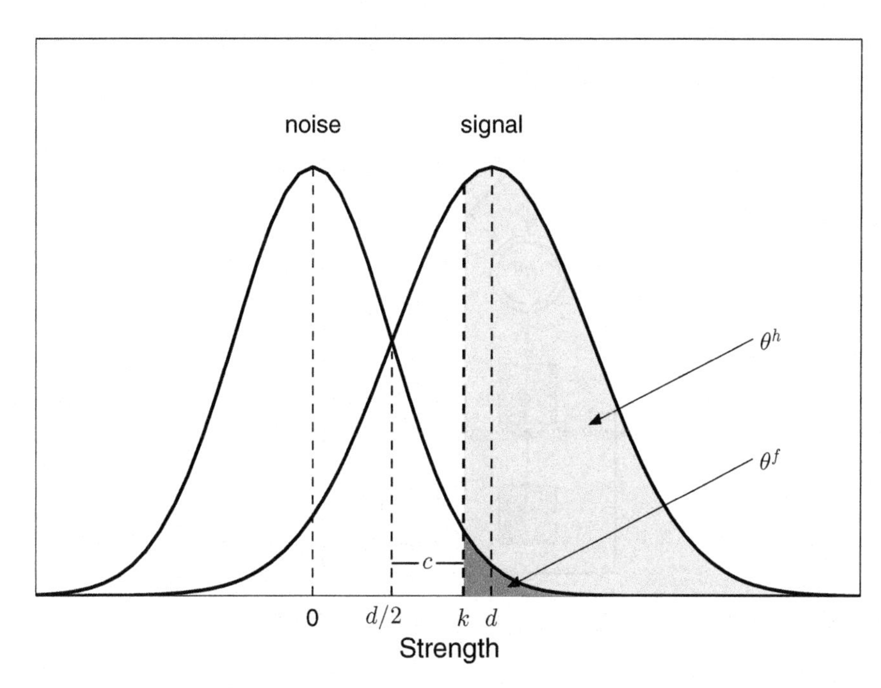

Fig. 11.1 Equal-variance Gaussian signal detection theory framework.

strength of the current trial to a fixed criterion. If the strength exceeds the criterion a "yes" response is made, otherwise a "no" response is made.

Figure 11.1 provides a formal version of the equal-variance SDT model. Since the underlying strength scale has arbitrary units, the variances are fixed to one, and the mean of the noise distribution is set to zero. The mean of the signal distribution is d. This makes d a measure of the *discriminability* of the signal trials from the noise trials, because it corresponds to the distance between the two distributions.

The strength value $d/2$ is special, because it is the criterion value that maximizes the probability of a correct classification when signal and noise trials are equally likely to occur. In this sense, using a criterion of $d/2$ corresponds to unbiased responding. The actual criterion used for responding is denoted k, and the distance between this criterion and the unbiased criterion is denoted c. This makes c a measure of *bias*, because it corresponds to how different the actual criterion is from the unbiased one. Positive values of c correspond to a bias towards saying no, and so to an increase in correct rejections at the expense of an increase in misses. Negative values of c correspond to a bias towards saying yes, and so to an increase in hits at the expense of an increase in false alarms.

The SDT model, with its representation and decision-making assumptions, makes predictions about hit rates and false alarm rates, and so maps naturally onto the counts in Table 11.1. In Figure 11.1, the hit rate, θ^h, is shown as the proportion of the signal distribution above the criterion k. Similarly, the false alarm rate, θ^f, is the proportion of the noise distribution above the criterion k.

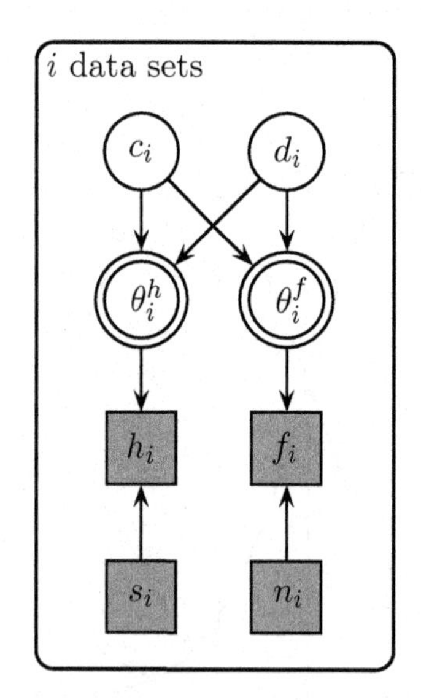

$$d_i \sim \text{Gaussian}(0, \tfrac{1}{2})$$
$$c_i \sim \text{Gaussian}(0, 2)$$
$$\theta_i^h \leftarrow \Phi(\tfrac{1}{2}d_i - c_i)$$
$$\theta_i^f \leftarrow \Phi(-\tfrac{1}{2}d_i - c_i)$$
$$h_i \sim \text{Binomial}(\theta_i^h, s_i)$$
$$f_i \sim \text{Binomial}(\theta_i^f, n_i)$$

Fig. 11.2 Graphical model for signal detection theory.

The usefulness of SDT is that, through this relationship, it is possible to take the sort of data in Table 11.1 and convert the counts of hits and false alarms into psychologically meaningful measures of discriminability and bias. Discriminability is a measure of how easily signal and noise trials can be distinguished. Bias is a measure of how the decision-making criterion being used relates to the optimal criterion.

A graphical model for inferring discriminability and bias from hit and false alarm counts for a number of data sets is shown in Figure 11.2. The hit rates θ_i^h and false alarm rates θ_i^f for the ith data set follow from the geometry of Figure 11.1. They are functions of their associated discriminabilities d_i and biases c_i, using the cumulative standard Gaussian distribution function $\Phi(\cdot)$. The observed counts of hits h_i and false alarms f_i are binomially distributed according to the hits and false alarm rates, and the number of signal trials s_i and noise trials n_i. The priors for discriminability and bias are both Gaussian distributions, constructed to correspond to uniform prior distributions over the hit and false alarm rates.[1]

The script `SDT_1.txt` implements the graphical model in WinBUGS:

```
# Signal Detection Theory
model{
  for (i in 1:k){
  # Observed counts
    h[i] ~ dbin(thetah[i],s[i])
    f[i] ~ dbin(thetaf[i],n[i])
```

[1] The proof relies on the probability integral transform theorem (e.g., Angus, 1994), which says that if X is a continuous real-valued random variable with strictly increasing cumulative distribution function F_X, then $F_X(X)$ is uniformly distributed on $(0, 1)$. This means, for example, that if $X \sim \text{Gaussian}(0, 1)$ then $\Phi(X) \sim \text{Uniform}(0, 1)$.

```
    # Reparameterization Using Equal-Variance Gaussian SDT
    thetah[i] <- phi(d[i]/2-c[i])
    thetaf[i] <- phi(-d[i]/2-c[i])
    # These Priors over Discriminability and Bias Correspond
    # to Uniform Priors over the Hit and False Alarm Rates
    d[i] ~ dnorm(0,0.5)
    c[i] ~ dnorm(0,2)
  }
}
```

The code SDT_1.m or SDT_1.R applies the model to make inferences for three illustrative data sets. Figure 11.3 shows the results produced, plotting the posterior distributions for discriminability, bias, hit rate, and false alarm for each data set.

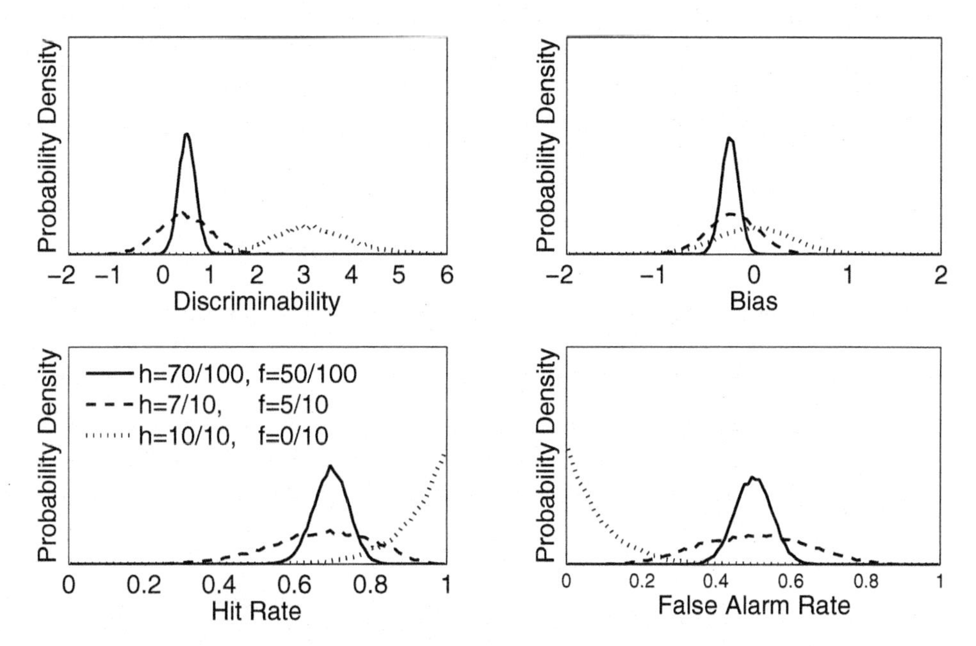

Fig. 11.3 Posterior distributions for discriminability, bias, hit rate, and false alarm rate using three illustrative data sets.

In the first data set, 70 hits and 50 false alarms are observed in 100 target and 100 noise trials. Because of the large number of trials, there is relatively little uncertainty surrounding the hit and false alarm rates, with narrow posteriors centered on 0.7 and 0.5, respectively. Discriminability and bias are also known with some certainty, centered on about 0.5 and -0.25 respectively.

In the second data set, 7 hits and 5 false alarms are observed in 10 target and 10 noise trials. These are the same proportions of hits and false alarms as the first situation, but based on many fewer samples. Accordingly, the posterior distributions have (essentially) the same means, but show much greater uncertainty.

In the third data set, perfect performance is observed, with 10 hits and no false alarms in 10 target and 10 noise trials. The modal hit and false alarm rates are 1.0 and 0.0, but other possibilities have some density. Discriminability is certain to be

Table 11.2 Recognition memory for odors reported by Lehrner et al. (1995).

	Control group		Group I		Group II	
	Old odor	New odor	Old odor	New odor	Old odor	New odor
Old resp.	148	29	150	40	150	51
New resp.	32	151	30	140	40	139

large, although the exact value is not clear. These data provide no information to help estimate bias, and so it retains its prior distribution. This outcome contrasts favorably with traditional frequentist analyses, which have to employ ad hoc edge corrections to avoid both discriminability and bias being undefined when either no hits or no false alarms are observed.

Exercises

Exercise 11.1.1 Do you feel that the priors on discriminability and bias are plausible, a priori? Why or why not? Try out some alternative priors and study the effect that this has on your inference for the data sets discussed above.

Exercise 11.1.2 Lehrner, Kryspin-Exner, and Vetter (1995) report data on the recognition memory for odors of three groups of subjects. Group I had 18 subjects, all with positive HIV antibody tests, and CD-4 counts of 240–700/mm^3. Group II had 19 subjects, all also with positive HIV antibody tests, but with CD-4 counts of 0–170/mm^3. The CD-4 count is a measure of the strength of the immune system, with a normal range being 500–700/mm^3, so Group II subjects had weaker immune systems. Group III had 18 healthy subjects and functioned as a control group. The odor recognition task involved each subject being presented with 10 common household odors to memorize, with a 30-sec. interval between each presentation. After an interval of 15 min., a total of 20 odors were presented to subjects. This test set comprised the 10 previously presented odors, and 10 new odors, presented in a random order. Subjects had to decide whether each odor was "old" or "new." The signal detection data that resulted are shown in Table 11.2.[2] Analyze these three data sets using signal detection theory to infer the discriminability and bias of the recognition performance for each group. What conclusions do you draw from this analysis? What, if anything, can you infer about individual differences between the subjects in the same groups?

[2] One or two of the counts might be out by one, because these data have been recovered from hit and false alarm rates truncated at two decimal places.

11.2 Hierarchical signal detection theory

We now consider a hierarchical extension of SDT, applied to a different problem where individual subject data are available. This allows us to model possible individual differences using a hierarchical extension of the basic SDT model in the previous case study. The idea is that different subjects have different discriminabilities and biases that are drawn from group-level Gaussian distributions.

The data come from the empirical evaluation, presented by Heit and Rotello (2005), of a conjecture made by Rips (2001) that inductive and deductive reasoning can be unified within a signal detection theory framework. The conjecture involves considering the strength of an argument as a uni-dimensional construct, but allowing different criteria for induction and deduction. The criterion separates between "weak" and "strong" arguments in the inductive case, and 'invalid" and "valid" arguments in the deductive case, with the deductive criterion being more extreme. Under this conception, deduction is simply a more stringent form of induction. Accordingly, empirical evidence for or against the SDT model has strong implications for the many-threaded contemporary debate over the existence of different kinds of reasoning systems or processes (e.g., Chater & Oaksford, 2000; Heit, 2000; Parsons & Osherson, 2001; Sloman, 1998).

In their study, Heit and Rotello (2005) tested the inductive and deductive judgments of 80 participants on eight arguments. They used a between-subjects design, so that 40 subjects were asked induction questions about the arguments (i.e., whether the conclusion was "plausible"), while the other 40 participants were asked deduction questions (i.e., whether the conclusion was "necessarily true"). These decisions made by participants have a natural characterization in term of the hit and false alarm counts.

A graphical model for analyzing the Heit and Rotello (2005) data is shown in Figure 11.4. It uses SDT to infer the discriminability d_i and bias c_i from hit θ_i^h and false alarm counts θ_i^f and for the ith subject. Individual differences are modeled hierarchically by assuming the individual discriminabilities and biases come from Gaussian group-level distributions, with means and precisions given by μ_d, μ_c, λ_d, and λ_c, respectively.

The script `SDT_2.txt` implements the graphical model in WinBUGS:

```
# Hierarchical Signal Detection Theory
model{
  for (i in 1:k){
    # Observed counts
    h[i] ~ dbin(thetah[i],s)
    f[i] ~ dbin(thetaf[i],n)
    # Reparameterization Using Equal-Variance Gaussian SDT
    thetah[i] <- phi(d[i]/2-c[i])
    thetaf[i] <- phi(-d[i]/2-c[i])
    # Discriminability and Bias
    c[i] ~ dnorm(muc,lambdac)
    d[i] ~ dnorm(mud,lambdad)
```

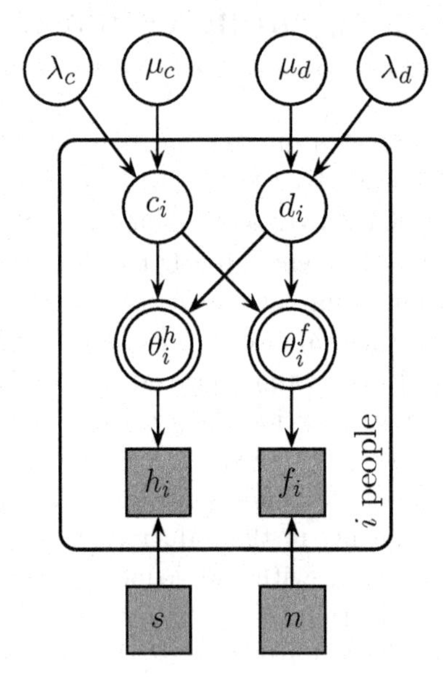

$$
\begin{aligned}
\mu_d, \mu_c &\sim \text{Gaussian}(0, .001) \\
\lambda_d, \lambda_c &\sim \text{Gamma}(.001, .001) \\
d_i &\sim \text{Gaussian}(\mu_d, \lambda_d) \\
c_i &\sim \text{Gaussian}(\mu_c, \lambda_c) \\
\theta_i^h &\leftarrow \Phi(\tfrac{1}{2}d_i - c_i) \\
\theta_i^f &\leftarrow \Phi(-\tfrac{1}{2}d_i - c_i) \\
h_i &\sim \text{Binomial}(\theta_i^h, s) \\
f_i &\sim \text{Binomial}(\theta_i^f, n)
\end{aligned}
$$

Fig. 11.4 Graphical model for hierarchical signal detection theory.

```
  }
  # Priors
  muc ~ dnorm(0,.001)
  mud ~ dnorm(0,.001)
  lambdac ~ dgamma(.001,.001)
  lambdad ~ dgamma(.001,.001)
  sigmac <- 1/sqrt(lambdac)
  sigmad <- 1/sqrt(lambdad)
}
```

The code `SDT_2.m` or `SDT_2.R` applies the model to the Heit and Rotello (2005) data. It applies the graphical model separately to the individual data for each experimental condition. Having done this, it produces a display of the joint posterior over the group means for both discriminability and bias, for both experimental conditions, as shown in Figure 11.5.

Exercises

Exercise 11.2.1 Of key interest for testing the Rips (2001) conjecture is how the group-level means for bias and (especially) discriminability differ between the induction and deduction conditions. What conclusion do you draw about the Rips (2001) conjecture base on the current analysis of the Heit and Rotello (2005) data?

Exercise 11.2.2 Heit and Rotello (2005) used standard significance testing methods on their data to reject the null hypothesis that there was no difference between discriminability for induction and deduction conditions. Their analy-

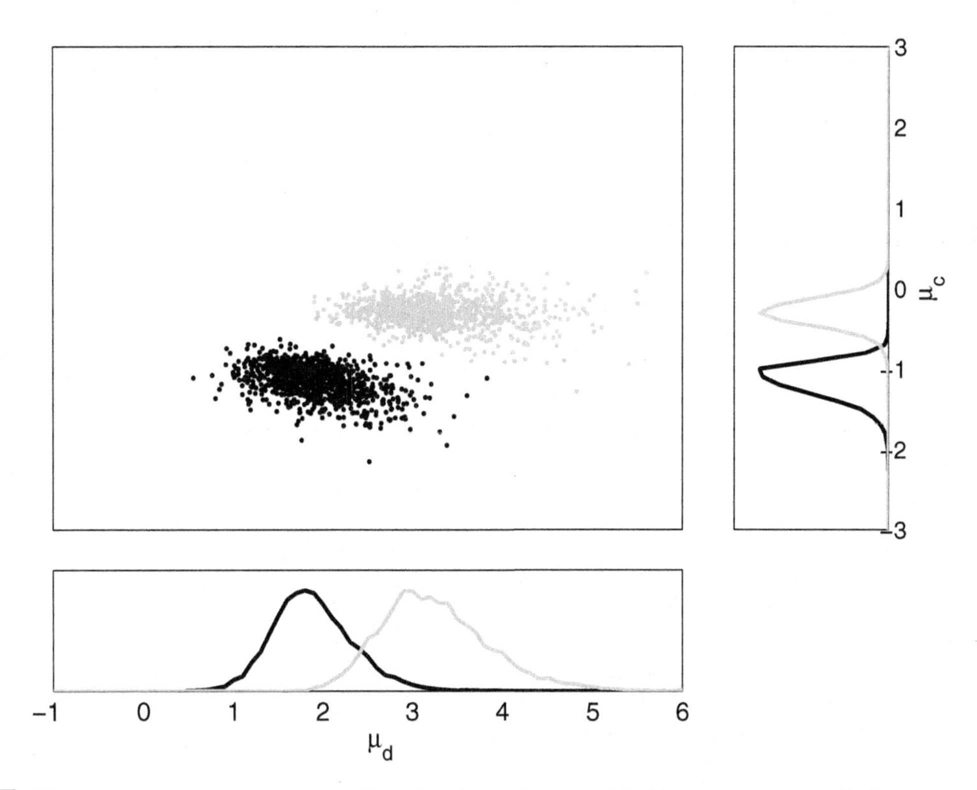

Fig. 11.5 The joint posterior over μ_d and μ_c for the induction (dark) and deduction (light) conditions.

sis involved calculating the mean discriminabilities for each participant, using edge-corrections where perfect performance was observed. These sets of discriminabilities gave means of 0.93 for the deduction condition and 1.68 for the induction condition. By calculating via the t statistic, and so assuming associated Gaussian sampling distributions, and observing that the p-value was less than 0.01, Heit and Rotello (2005) rejected the null hypothesis of equal means. According to Heit and Rotello (2005), this finding of different discriminabilities provided evidence against the criterion-shifting uni-dimensional account offered by SDT. Is this consistent with your conclusions from the Bayesian analysis?

Exercise 11.2.3 Re-run the analysis without discarding burn-in samples. This can be done by setting `nburnin` to 0 in the code `SDT_2.m` or `SDT_2.R`. The result should look something like Figure 11.6. Notice the strange set of samples leading from zero to the main part of the sampled distribution. Explain why these samples exist, and why they suggest burn-in is important in this analysis.

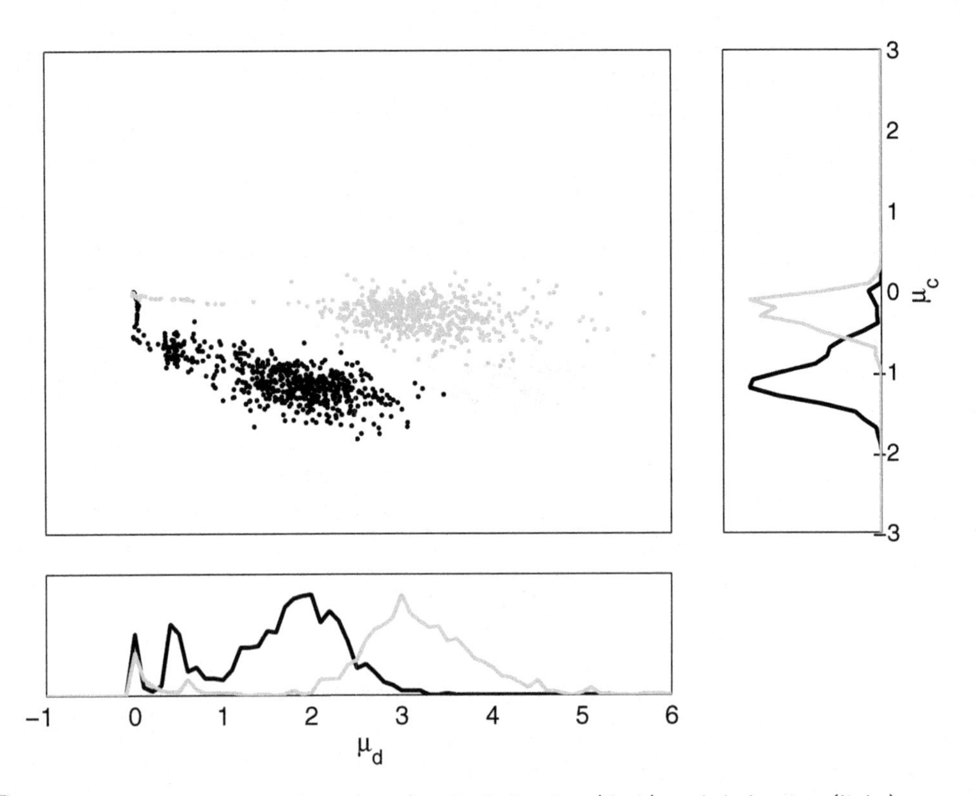

Fig. 11.6 The joint posterior over μ_d and μ_c for the induction (dark) and deduction (light) conditions, without discarding burn-in samples.

11.3 Parameter expansion

WITH DORA MATZKE

Even after the introduction of a burn-in period, there is a sampling issue with the way the chains of the hierarchical variance parameter σ_c behave. As shown in Figure 11.7, the σ_c chains can get stuck near zero, with the samples around 2500 in the induction condition providing an especially good example. This undesirable sampling behavior is fairly common in complicated hierarchical Bayesian models.

The problem is as follows. Suppose that σ_c happens to be estimated near zero. As a result, the individual bias parameters c_i will be pooled toward their population mean μ_c. On the next MCMC iteration, σ_c will be estimated again near zero because it depends on the current values of the c_i parameters. Eventually, the chain of σ_c will break out of the "zero variance trap." However, this may require several iterations and the chain may become trapped again later in sampling.

A good way to enable the sampling process to escape the trap is to use a technique known as parameter expansion (e.g., Gelman, 2004; Gelman & Hill, 2007; Liu & Wu, 1999). This technique involves augmenting the original model with redun-

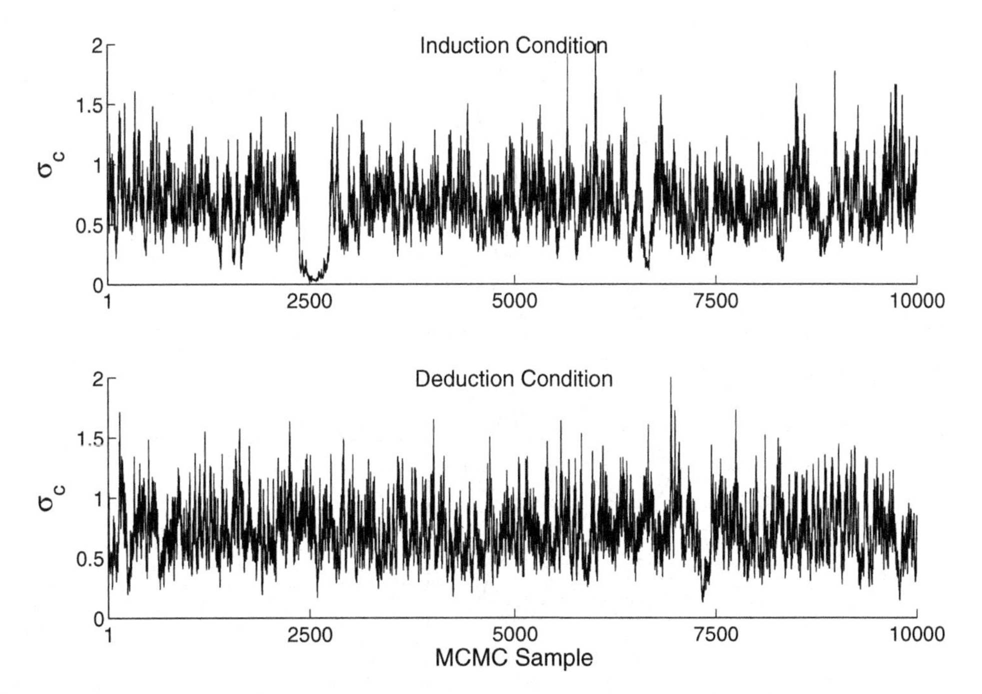

Fig. 11.7 MCMC chains of the σ_c parameter of the hierarchical signal detection model.

dant multiplicative parameters. Specifically, we can extend the hierarchical signal detection model with two multiplicative parameters, say ξ_c and ξ_d. The role of these additional parameters is to rescale the original c_i and d_i parameters and their corresponding standard deviations σ_c and σ_d.

The graphical model that implements this parameter-expanded model is shown in Figure 11.8. Note that the new model is equivalent to the original hierarchical signal detection model, because it simply reparameterizes the original model. In the parameter-expanded model, c_i and σ_c are rescaled by multiplying by ξ_c. The c_i parameter is now expressed in terms of μ_c, ξ_c, and δ_{c_i}, and σ_c is expressed in terms of $|\xi_c|\,\sigma_c^{\text{new}}$. Similarly the d_i and σ_d parameters are rescaled by multiplying them with ξ_d. The d_i parameter is now expressed in terms of μ_d, ξ_d, and δ_{d_i}, and σ_d is expressed in terms of $|\xi_d|\,\sigma_d^{\text{new}}$.

The rationale behind parameter expansion is that updating the ξ_c and ξ_d parameters includes an additional random component in the sampling process. This component causes the samples of $\sigma_c = |\xi_c|\,\sigma_c^{\text{new}}$ and $\sigma_d = |\xi_d|\,\sigma_d^{\text{new}}$ to be less dependent on the previous iteration and prevents the chains getting trapped near zero regardless of how small their previous values were. In order to draw inferences under the original model, however, the parameters from the expanded model must be transformed back to their original scale. For example, inferences about the original σ_c parameter should be based on samples for $|\xi_c|\,\sigma_c^{\text{new}}$ instead of σ_c^{new}.

The script `SDT_3.txt` implements the graphical model in WinBUGS:

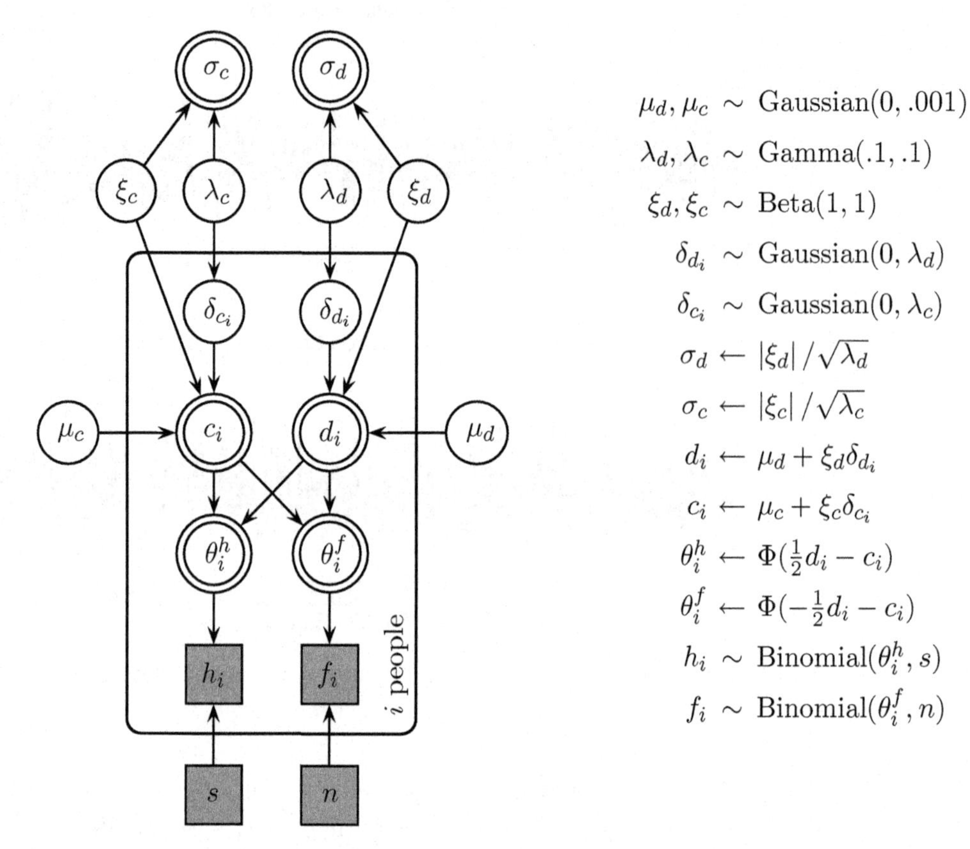

Fig. 11.8 Graphical model for the parameter-expanded hierarchical signal detection theory.

```
# Hierarchical SDT With Parameter Expansion
model{
  for (i in 1:k){
    # Observed counts
    h[i] ~ dbin(thetah[i],s)
    f[i] ~ dbin(thetaf[i],n)
    # Reparameterization Using Equal-Variance Gaussian SDT
    thetah[i] <- phi(d[i]/2-c[i])
    thetaf[i] <- phi(-d[i]/2-c[i])
    # Discriminability and Bias
    c[i] <- muc + xic*deltac[i]
    d[i] <- mud + xid*deltad[i]
    deltac[i] ~ dnorm(0,lambdac)
    deltad[i] ~ dnorm(0,lambdad)
  }
  # Priors
  muc ~ dnorm(0,0.001)
  mud ~ dnorm(0,0.001)
  xic ~ dbeta(1,1)
  xid ~ dbeta(1,1)
```

```
  lambdac ~ dgamma(.1,.1)
  lambdad ~ dgamma(.1,.1)
  sigmacnew <- 1/sqrt(lambdac)
  sigmadnew <- 1/sqrt(lambdad)
  sigmac <- abs(xic)*sigmacnew
  sigmad <- abs(xid)*sigmadnew
}
```

The code SDT_3.m or SDT_3.R applies the parameter-expanded model to the Heit and Rotello (2005) data. After you run the code, WinBUGS should display MCMC chains similar to those shown in Figure 11.9. The chains for the variance parameter σ_c now seem to have escaped the zero variance trap.

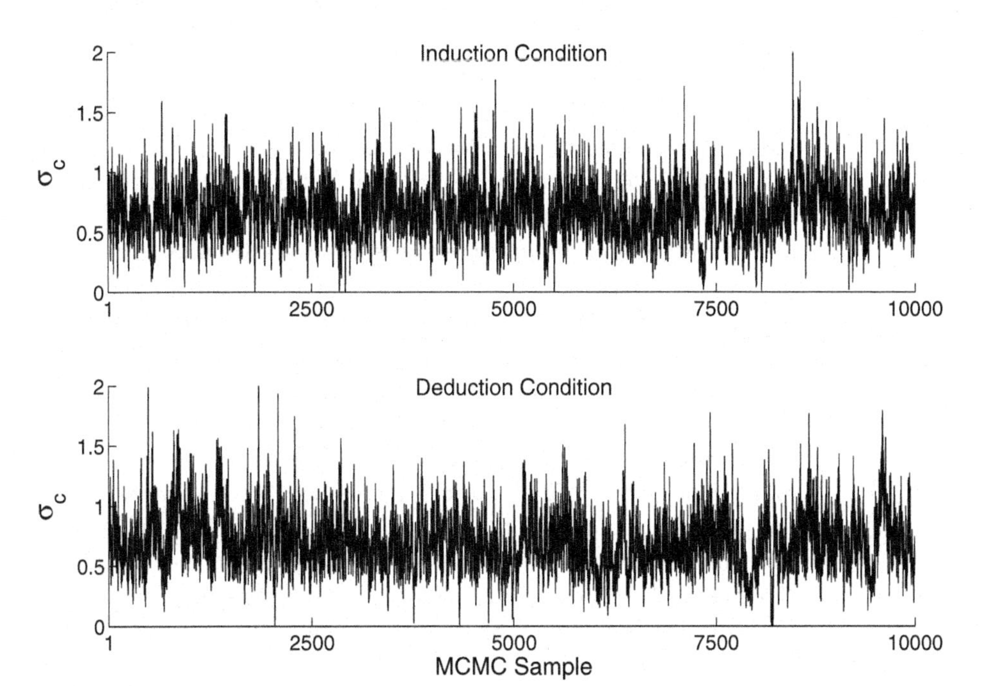

Fig. 11.9 MCMC chains of the σ_c parameter of the parameter-expanded hierarchical signal detection model.

Exercise

Exercise 11.3.1 Experiment with different priors for the unscaled precision, such as dgamma(0.1,0.1), dgamma(0.01,0.01), or dgamma(0.001,0.001), and for the scaling parameters, such as dunif(0,1), dunif(0,2), or dnorm(0,1). How does the prior for the scaled standard deviations change when you change the scaling factor?

12 Psychophysical functions

WITH JORAM VAN DRIEL

12.1 Psychophysical functions

Psychophysics is concerned with measuring how external physical stimuli cause internal psychological sensations. In a typical psychophysical experiment, subjects are repeatedly confronted with two similar stimuli, such as two sounds, two weights, two smells, two lines, or two time intervals. One stimulus—the standard stimulus—always has the same intensity, whereas the other stimulus—the test stimulus—varies in intensity from trial to trial. On each trial of the experiment, the subject's task is to detect which of the two stimuli is more intense: louder, heavier, stronger-smelling, more tilted, or longer lasting. The more similar the stimuli, the more difficult it is for the subject to discriminate between them.

The relation between task difficulty and performance usually follows a sigmoid or S-shaped curve, as shown in Figure 12.1. The x-axis represents differences in stimulus intensity between the test stimulus and the standard. The y-axis represents the probability of a response indicating that the test stimulus has higher intensity. The curve linking these physical and psychological measures is known as a psychophysical function, and is used to define several values of interest. The "point of subjective equality" (PSE) represents that difference in intensity for which the participant chooses the correct response 50% of the time, which is not necessarily the point where the two stimuli are equally intense physically. The "just noticeable difference" (JND) is the intensity threshold at which the subject "just" notices a difference in intensity between two stimuli. The y-value of the psychometric function that is often used to quantify the JND is the 84% point (Ernst, 2005). Thus, here we define the JND to be the difference in stimulus intensity that makes classification performance rise from 50% to 84%.

In this chapter we illustrate the basics of psychophysics and the psychometric function with an experiment on time perception (e.g., Ivry, 1996) that allows inferences about JNDs.[1] In the experiment, 8 subjects each completed 3 blocks of 80 trials. Every trial featured two beeps of different duration. The standard beep always lasted 300 ms, and the test beep was of variable duration. The subject's task was to indicate whether the duration of the test beep was shorter or longer than

[1] This experiment was carried out in the lab of Rich Ivry at UC Berkeley (June–July, 2009). We thank Rich Ivry for letting us present these data here.

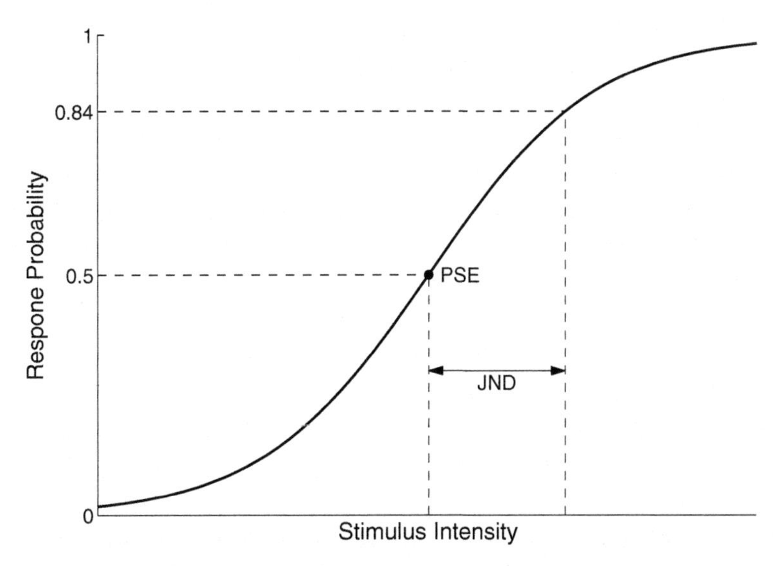

Fig. 12.1 A psychophysical function, showing the S-shaped relationship between stimulus intensity and choice behavior. Important theoretical measures known as the point of subjective equality (PSE) and just noticeable difference (JND) are highlighted.

the standard interval. The duration of the test beep depended on the accuracy of the subject, using an adaptive staircase routine. Figure 12.2 shows, for each test interval, the proportion of trials on which the subject responded longer, and shows the typical S-shape of a psychometric function.

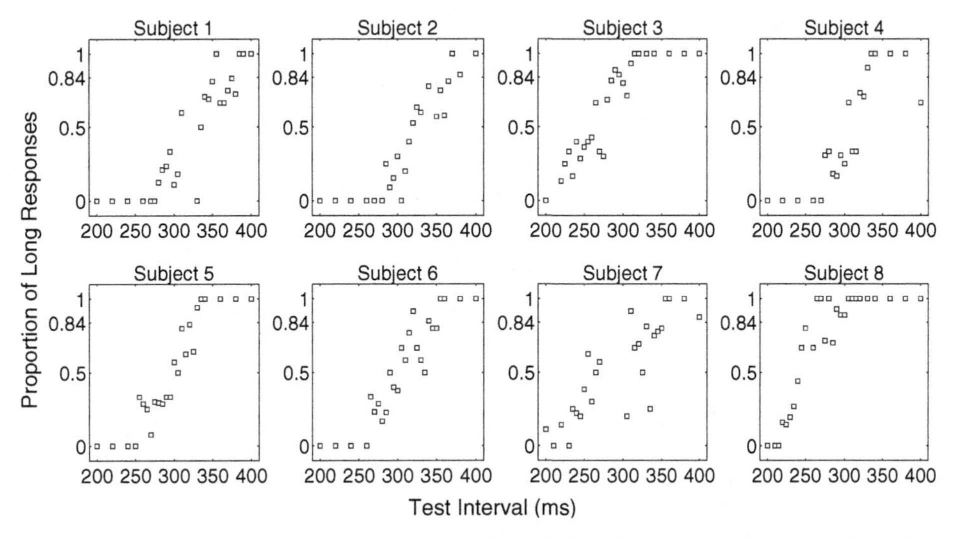

Fig. 12.2 Data for all 8 subjects, showing the proportion of "long" responses as a function of test interval duration.

To make inferences about JNDs, however, requires a psychological model of the data-generating process. Consider the ith subject who makes n_{ij} duration discrimination judgments for the jth interval pair, where the test interval has duration x_{ij}. The number of times this subject classifies this test interval as longer is denoted by r_{ij}. Thus, for test duration x_{ij} we observe r_{ij} long responses out of n_{ij} trials, and we assume this process is determined by a binomial rate parameter θ_{ij}. The rate parameter θ_{ij} is given, in turn, by the psychometric function evaluated at the intensity x_{ij} of the test stimulus.

The psychometric function can be modeled by distributions such as the Gumbel or Weibull (Kuss, Jäkel, & Wichmann, 2005), but here we use the logistic function with parameters α_i and β_i, so that

$$\theta_{ij} = \frac{1}{1 + \exp\left\{-\left[\alpha_i + \beta_i(x_{ij} - \bar{x}_i)\right]\right\}}. \tag{12.1}$$

Note that the stimulus intensity values have been mean-centered. For each subject, the mean interval duration $\bar{x}_i$ is subtracted from each of the test interval durations x_{ij}. Implementing this function in WinBUGS is conveniently done using the `logit` transformation, so that Equation 12.1 is represented as

$$\text{logit}\left(\theta_{ij}\right) = \alpha_i + \beta_i\left(x_{ij} - \bar{x}_i\right). \tag{12.2}$$

A graphical model for the psychophysical model is shown in Figure 12.3. The model is hierarchical in the sense that values for the parameters α_i and β_i that define the psychometric function for the ith subject are drawn from group-level Gaussian distributions.

The script `Psychophysical_1.txt` implements the graphical model in WinBUGS:

```
# Logistic Psychophysical Function
model{
  for (i in 1:nsubjs){
    for (j in 1:nstim[i]){
      r[i,j] ~ dbin(thetalim[i,j],n[i,j])
      logit(thetalim[i,j]) <- lthetalim[i,j]
      lthetalim[i,j] <- min(999,max(-999,ltheta[i,j]))
      ltheta[i,j] <- alpha[i]+beta[i]*(x[i,j]-xmean[i])
    }
    beta[i] ~ dnorm(mub,lambdab)
    alpha[i] ~ dnorm(mua,lambdaa)
  }
  # Priors
  mub ~ dnorm(0,.001)
  mua ~ dnorm(0,.001)
  sigmab ~ dunif(0,1000)
  sigmaa ~ dunif(0,1000)
  lambdab <- pow(sigmab,-2)
  lambdaa <- pow(sigmaa,-2)
}
```

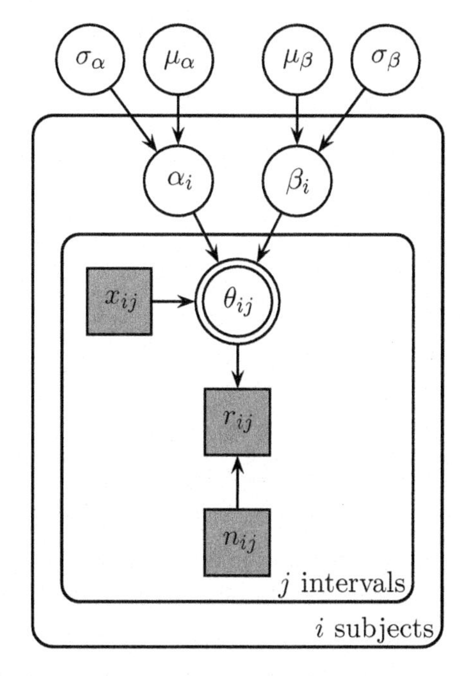

$$
\begin{aligned}
r_{ij} &\sim \text{Binomial}(\theta_{ij}, n_{ij}) \\
\text{logit}(\theta_{ij}) &\leftarrow \alpha_i + \beta_i(x_{ij} - \bar{x}_i) \\
\alpha_i &\sim \text{Gaussian}(\mu_\alpha, \sigma_\alpha) \\
\beta_i &\sim \text{Gaussian}(\mu_\beta, \sigma_\beta) \\
\mu_\alpha &\sim \text{Gaussian}(0, 0.001) \\
\mu_\beta &\sim \text{Gaussian}(0, 0.001) \\
\sigma_\alpha &\sim \text{Uniform}(0, 1000) \\
\sigma_\beta &\sim \text{Uniform}(0, 1000)
\end{aligned}
$$

Fig. 12.3 Graphical model for the psychometric model of duration discrimination.

The code `Psychophysical_1.m` or `Psychophysical_1.R` applies the model to the data. Figure 12.4 shows the data and the psychometric functions corresponding to the expected posterior values of the α_i and β_i parameters.

Exercises

Exercise 12.1.1 What do you think is the function of the `thetalim` construction in the WinBUGS script?

Exercise 12.1.2 The sigmoid curves in Figure 12.4 are single lines derived from point estimates. How can you visualize the uncertainty in the psychometric function?

Exercise 12.1.3 Figure 12.4 shows the PSE for each subject. Compare subject 2 with subject 8. How do they differ in their perception of the intervals?

Exercise 12.1.4 One of the aims of the analysis is to use the psychometric function to infer the JND. In Figure 12.4 the JND is indicated by the difference on the x-axis between the dashed lines corresponding to the 50% and 84% points on the y-axis. The JNDs from Figure 12.4 are point estimates. Plot posterior distributions for the JND, and interpret the results. Which subjects are better at perceiving differences in time, and how certain are your conclusions?

Exercise 12.1.5 Look closely at the data points that are used to fit the psychometric functions. Are all of them close to the sigmoid curve? How do you think possible outliers would influence the function, and the inferred JND?

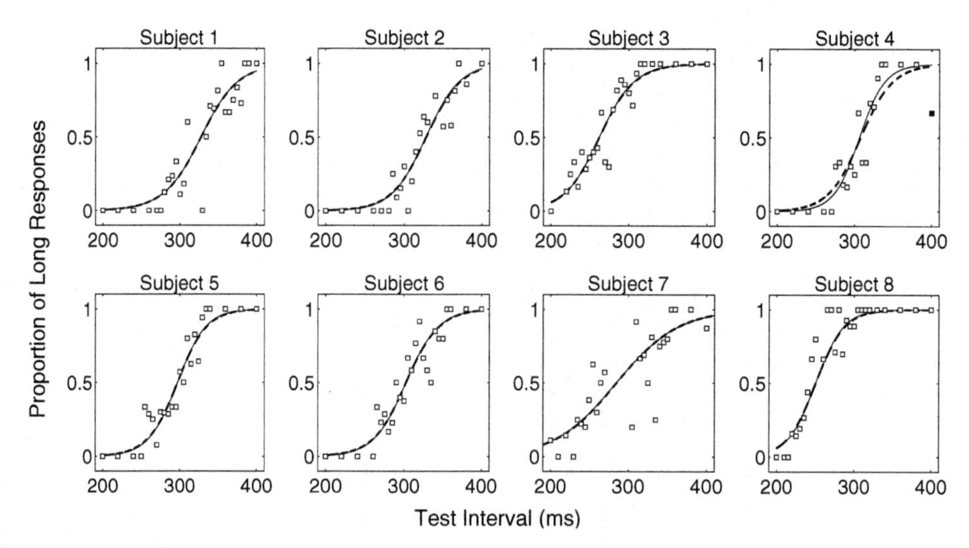

Fig. 12.4 Psychophysical functions corresponding to expected posterior parameter values for each of the 8 subjects. Dashed lines represent the JND between the 50% (PSE) and the 84% points.

12.2 Psychophysical functions under contamination

Experimental data are rarely a clean reflection of the psychological process of interest. Attentional lapses can produce contaminant data that affect inference, and this means that it is often a good idea to account for contaminants explicitly, using a mixture-model approach. To deal with potential contaminants in the current data, we extend the model by equipping it with a separate contaminant process, using the graphical model shown in Figure 12.5.

For the ith subject and jth stimulus pair, the binary variable z_{ij} determines the nature of responses. When $z_{ij} = 0$, responses r_{ij} are generated from the psychophysical function, with parameters α_i and β_i, as before. When $z_{ij} = 1$, responses are generated by a separate process with success rate π_{ij} that is uniformly distributed. The contaminant model also assumes that the ith subject has a probability ϕ_i of producing contaminant behavior for any stimulus interval. These probabilities are assumed to be drawn from a group-level Gaussian distribution, using a probit transformation.

The script `Psychophysical_2.txt` implements the graphical model in WinBUGS:

```
# Logistic Psychophysical Function with Contaminants
model{
  for (i in 1:nsubjs){
    for (j in 1:nstim[i]){
      z[i,j] ~ dbern(phi[i])
      z1[i,j] <- z[i,j] + 1
```

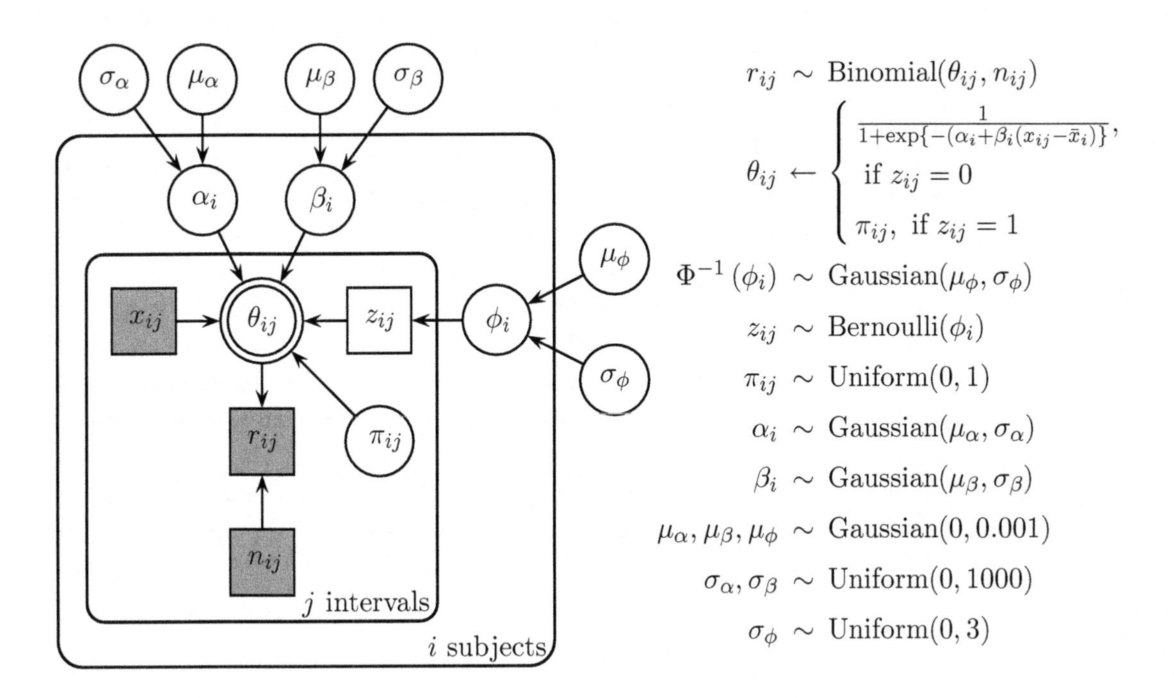

Fig. 12.5 Graphical model for the psychometric model of duration discrimination, including a contaminant process.

```
      thetalim[i,j,2] <- pi[i,j]
      pi[i,j] ~ dbeta(1,1)
      r[i,j] ~ dbin(thetalim[i,j,z1[i,j]],n[i,j])
      logit(thetalim[i,j,1]) <- lthetalim[i,j]
      lthetalim[i,j] <- min(999,max(-999,ltheta[i,j]))
      ltheta[i,j] <- alpha[i]+beta[i]*(x[i,j]-xmean[i])
    }
    phi[i] <- phi(probitphilim[i])
    probitphilim[i] <- min(5, max(-5,probitphi[i]))
    probitphi[i] ~ dnorm(mup,lambdap)
    beta[i] ~ dnorm(mub,lambdab)
    alpha[i] ~ dnorm(mua,lambdaa)
  }
  # Priors
  mub ~ dnorm(0,.001)
  mua ~ dnorm(0,.001)
  sigmab  ~ dunif(0,1000)
  sigmaa  ~ dunif(0,1000)
  lambdab <- pow(sigmab,-2)
  lambdaa <- pow(sigmaa,-2)
  mup ~ dnorm(0,1)
  sigmap ~ dunif(0,3)
  lambdap <- pow(sigmap,-2)
}
```

Box 12.1 Modeling contamination

There are always many useful models of any psychological process, or people's behavior in any task. Developing models is a creative exercise, and there is never a single best or correct answer. Sometimes simple statistical accounts of behavior can suffice, and sometimes much richer accounts based on psychological theory are more useful. The same holds for developing contaminant models. These are also models of psychological processes; just not the psychological processes that were intended to be applied to the task. Often simple statistical models of contaminant behavior—as in the psychophysical model proposed here—will suffice. But sometimes richer cognitive contaminant models might be more useful, capturing more complicated strategies which people used to complete the task. Zeigenfuse and Lee (2010) present case studies involving recognition memory and sequential decision-making, using the same latent-mixture approach, but incorporating more elaborate contaminant models.

The code `Psychophysical_2.m` or `Psychophysical_2.R` applies the model to the same data as before. Figure 12.6 shows the psychophysical functions inferred by both models. The square markers representing each interval are shaded according to the mean value of the posterior for z_{ij}, ranging from white, for responses certainly from the psychophysical function, to black, for responses certainly from the contaminant process.

Figure 12.7 shows the posterior distributions of the JNDs estimated by the model both without and with accounting for contaminants.

Exercise

Exercise 12.2.1 How did the inclusion of the contaminant process change the inference for the psychophysical functions, and the key JND and PSE properties?

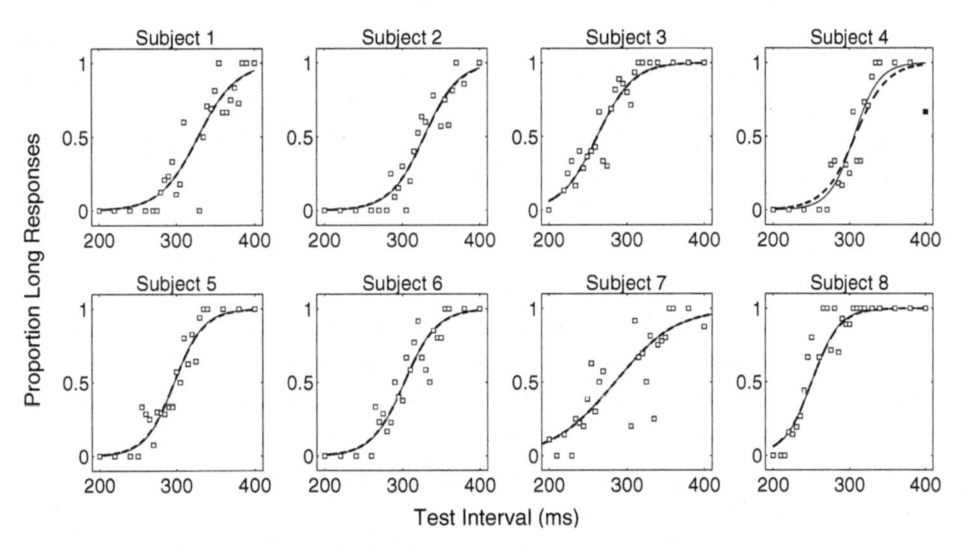

Fig. 12.6 Psychophysical functions corresponding to expected posterior parameter values, using the model including a contaminant process, for each of the 8 subjects. Square markers representing data are colored to represent how certain they are to be generated by the psychophysical process (lighter) or the contaminant process (darker).

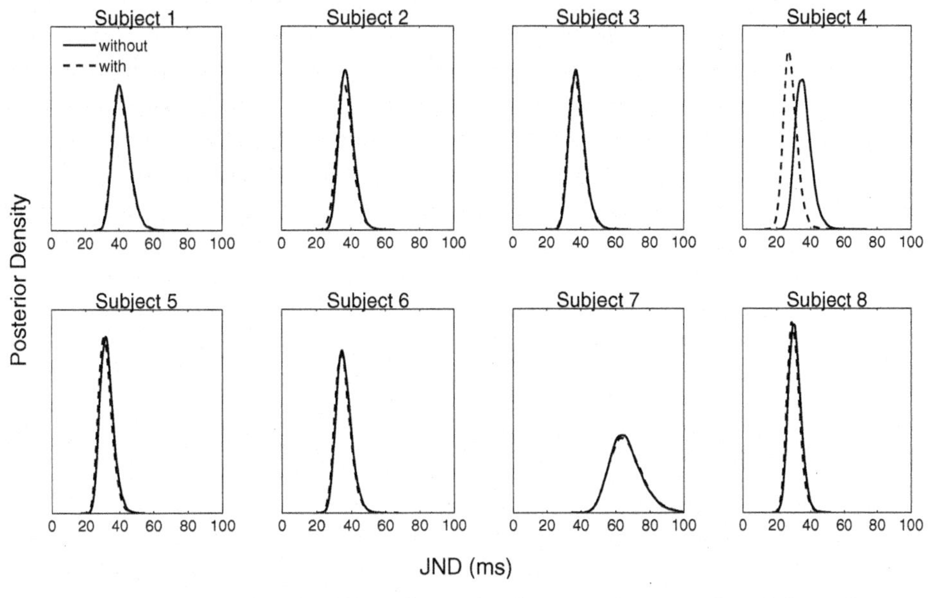

Fig. 12.7 Posterior distributions for the JNDs, for each of the 8 subjects, inferred from the psychophysical model without (solid line) and with (dashed line) the inclusion of a contaminant process.

13 Extrasensory perception

More than a hundred years ago, some of the world's most renowned scientists were curious whether or not people can look into the future (precognition), move objects with their mind (telekinesis), or transmit messages by thought (telepathy). These so-called extrasensory abilities were studied by members of the *Society for Psychical Research*, a society that included intellectual heavyweights such as William James, Carl Jung, and Alfred Wallace. Even as late as 1950, the great Alan Turing argued that his famous test for artificial intelligence should be carried out in a telepathy-proof room.

How times have changed. Scientific work on extrasensory perception, or ESP, is now conducted only by a few self-pronounced academic mavericks, on a Quixotic mission to demonstrate to the world that the phenomenon is real. In 2011, the debate on the existence of ESP was re-ignited when reputable social psychologist Dr Daryl Bem published nine ESP experiments with over 1000 subjects (Bem, 2011). On the basis of these data, Bem argued that people are able to look into the future. In Bem's first experiment, for example, subjects had to guess whether a picture was going to appear on the left or the right side of the computer screen. The location of the picture was random, and this means that, on average, subjects cannot do better than a chance rate of 50% correct—unless, of course, people can look into the future. Bem (2011) found that people guessed the upcoming location of the pictures with above-chance accuracy of 53.1%. Interestingly, this effect occurred only for erotic pictures, and was absent for neutral pictures, romantic but not erotic pictures, negative pictures, and positive pictures. It was also found that the effect was largest for extravert women.

Do the Bem studies show that people can look into the future? Hardly. The Bem studies have been criticized on several grounds (Wagenmakers, Wetzels, Borsboom, & van der Maas, 2011). Here we use Bayesian parameter estimation and model

Box 13.1 Turing on telepathy

"I assume that the reader is familiar with the idea of extra-sensory perception, and the meaning of the four items of it, *viz.* telepathy, clairvoyance, precognition and psycho-kinesis. These disturbing phenomena seem to deny all our usual scientific ideas. How we should like to discredit them! Unfortunately the statistical evidence, at least for telepathy, is overwhelming." (Turing, 1950, p. 453).

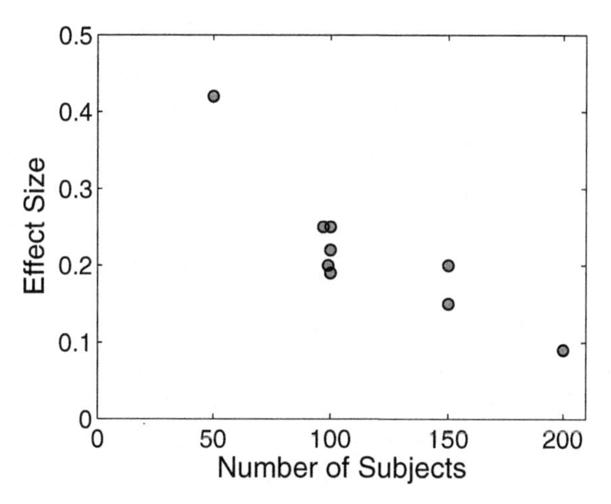

Fig. 13.1 The relationship between the number of subjects and the effect size for the experiments reported by Bem (2011).

selection methods to explore one of the criticisms, namely that the original Bem results had been obtained by optional stopping. We also analyze data from a replication experiment and assess the evidence for stable individual differences in ability, and for the effect of extraversion.

13.1 Evidence for optional stopping

When researchers report p-values they often do not realize that they have to specify a sampling plan in advance of data collection. When you state that you are going to test 100 subjects, you are not allowed to take sneak peeks at the data and stop whenever the result is significant (e.g., $p < 0.05$); nor are you allowed to test more than 100 subjects in case the outcome of your test is ambiguous (e.g., $p = 0.09$). The reason for this requirement is that researchers who take sneak peeks at their data can achieve any desired p-value, no matter how low, even if the null hypothesis is exactly true. For this reason, the optional stopping procedure is also known as "sampling to a foregone conclusion."

For a single study it can be very difficult to determine whether or not the results were due to optional stopping. Whenever a researcher reports multiple studies, however, a diagnostic tool is to plot the number of subjects or observations against the effect size (Hyman, 1985). Figure 13.1 shows this relation for the experiments reported by Bem.[1]

The negative relation between the number of subjects and effect size suggests that the results are contaminated by optional stopping. When the effect size is

[1] We thank Ray Hyman for attending us to this regularity.

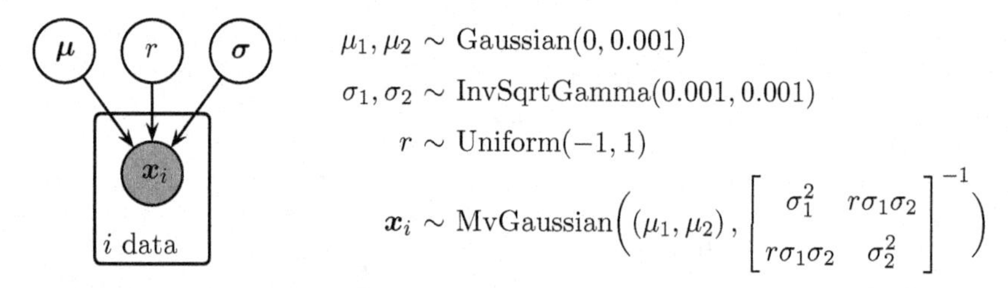

Fig. 13.2 Graphical model for inferring a correlation coefficient.

large, the researcher can afford to stop early. When the effect size is small, more subjects need to be tested before the result reaches significance.

How strong is the evidence that there is a negative association between sample size and effect size? Within the Bayesian framework, two natural ways to address this question come from the parameter estimation and model selection perspectives on inference. Parameter estimation involves inferring the posterior distribution of the correlation coefficient, as covered in Section 5.1. Model selection involves comparing, for example, the hypothesis that the correlation is zero to the hypothesis that the correlation is some other value. This can be done by applying the Savage–Dickey density ratio test, as covered in Section 7.6.

A graphical model for inferring the correlation coefficient is shown again in Figure 13.2. The script `Correlation_1.txt` implements the graphical model in WinBUGS:

```
# Pearson Correlation
model{
  # Data
  for (i in 1:n){
    x[i,1:2] ~ dmnorm(mu[],TI[,])
  }
  # Priors
  mu[1] ~ dnorm(0,.001)
  mu[2] ~ dnorm(0,.001)
  lambda[1] ~ dgamma(.001,.001)
  lambda[2] ~ dgamma(.001,.001)
  r ~ dunif(-1,1)
  # Reparameterization
  sigma[1] <- 1/sqrt(lambda[1])
  sigma[2] <- 1/sqrt(lambda[2])
  T[1,1] <- 1/lambda[1]
  T[1,2] <- r*sigma[1]*sigma[2]
  T[2,1] <- r*sigma[1]*sigma[2]
  T[2,2] <- 1/lambda[2]
  TI[1:2,1:2] <- inverse(T[1:2,1:2])
}
```

The graphical model is used to infer the posterior distribution of the correlation coefficient, assuming the prior distribution is $r \sim \text{Uniform}(-1, 1)$. In other words, all values of the correlation coefficient are deemed equally likely a priori. In hypothesis testing or model selection terms, this corresponds to the alternative hypothesis

Box 13.2 Flexibility of Bayesian data collection

"The rules governing when data collection stops are irrelevant to data interpretation. It is entirely appropriate to collect data until a point has been proven or disproven, or until the data collector runs out of time, money, or patience." (Edwards et al., 1963, p. 193) ... "if you set out to collect data until your posterior probability for a hypothesis which is unknown to you is true has been reduced to .01, then 99 times out of 100 you will never make it, no matter how many data you, or your children after you, may collect." (Edwards et al., 1963, p. 239)

$\mathcal{H}_1$. Under the null hypothesis $\mathcal{H}_0$, the assumption is that there is no correlation. Thus, using the Savage-Dickey density ratio method, the Bayes factor is simply the height of the prior divided by the height of the posterior, evaluated at the point of test $r = 0$.

The code `OptionalStopping.m` or `OptionalStopping.R` applies the graphical model to the data from Figure 13.1, plots the posterior distribution, and applies the Savage–Dickey method.

The results are shown in Figure 13.3, with the left panel showing the data again for convenience. The right panel shows the prior (horizontal dotted line) and posterior (solid line) distribution for the correlation coefficient. The expected value of the posterior is about -0.77, and the mode is near the frequentist value of -0.87 shown by the broken vertical line.

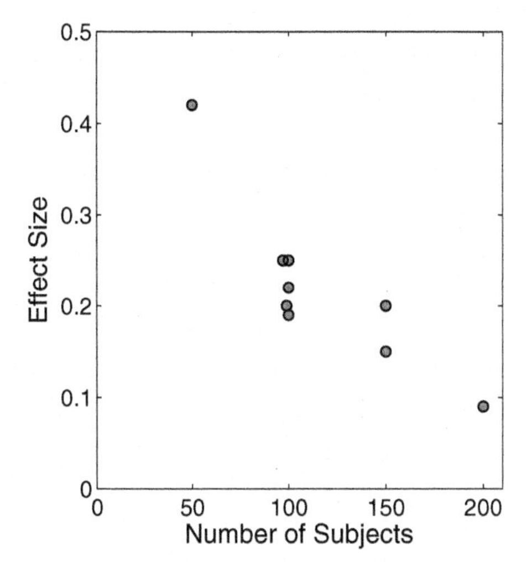

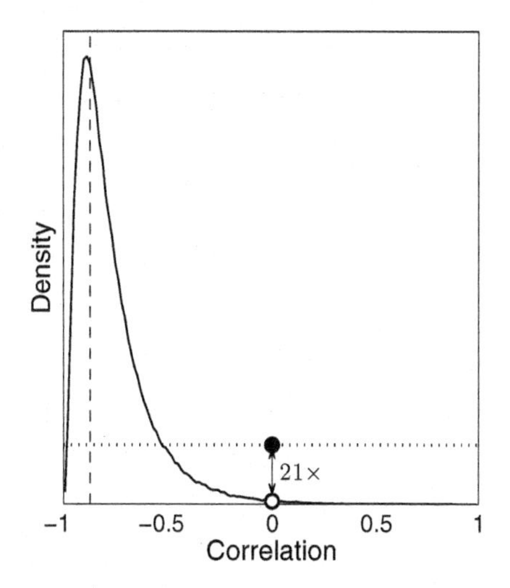

Fig. 13.3 Raw data (left panel), and Bayesian analysis (right panel) for the correlation between sample size and effect size in the Bem (2011) experiments.

Box 13.3 **Boring Bayes**

"What is the principal distinction between Bayesian and classical statistics? It is that Bayesian statistics is fundamentally boring. There is so little to do: just specify the model and the prior, and turn the Bayesian handle. There is no room for clever tricks or an alphabetic cornucopia of definitions and optimality criteria. I have heard people who should know better use this 'dullness' as an argument against Bayesianism. One might as well complain that Newton's dynamics, being based on three simple laws of motion and one of gravitation, is a poor substitute for the richness of Ptolemy's epicyclic system." (Dawid, 2000, p. 326)

The right panel of Figure 13.3 also shows the density of the prior and posterior at the point of test $r = 0$ by black and white circles, respectively. The density of the prior is about 21 times greater than that of the posterior at this point, so the Bayes factor is about 21 in favor of the alternative hypothesis that the correlation is not zero.

Exercises

Exercise 13.1.1 What does the Bayesian analysis tell you about the association between sample size and effect size in the Bem (2011) studies?

Exercise 13.1.2 Section 5.2 considered extending the correlation model in Figure 13.2 to incorporate uncertainty about the measures being related. Could that extension be usefully applied here?

Exercise 13.1.3 A classical p-value test on the Pearson product-moment correlation coefficient yields $r = -0.87$, 95% CI $= [-0.97, -0.49]$, $p = 0.002$. What conclusions would you draw from this analysis, and how do they compare to the conclusions you drew from the Bayesian analysis?

Exercise 13.1.4 We do not need to compute the Savage–Dickey density ratio on the original scale. For example, there are good arguments first to transform the posterior samples using the Fisher z-transform, so that $z = \text{arctanh}\,(r)$. Try using this transformation. What difference do you observe?

13.2 Evidence for differences in ability

Wagenmakers, Wetzels, Borsboom, van der Maas, and Kievit (2012) conducted a replication of Bem's (2011) original experiment, which differed in a few details. Because Wagenmakers et al. (2012) wanted to maximize the probability of finding an effect, they tested only women, and included only neutral and erotic pictures.

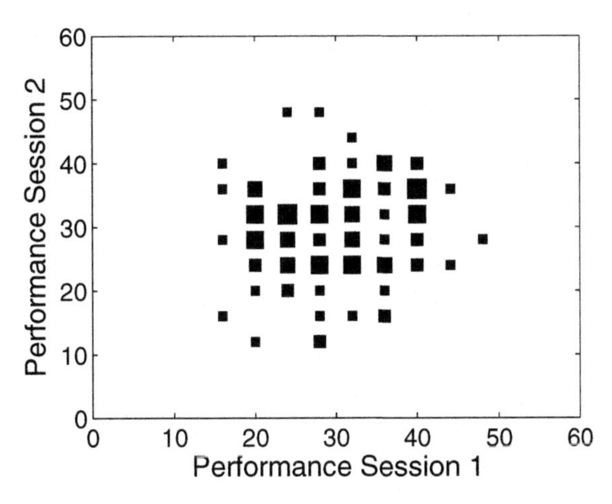

Fig. 13.4 Performance of 100 subjects in two sessions, each with 60 trials, in correctly identifying the hidden location of erotic pictures. The size of each point indicates the number of subjects with that combination of correct predictions.

Another difference was that they included two consecutive sessions, reasoning that "If participants have ESP, this trait should be related from session 1 to session 2. In other words, individual differences in ESP express themselves statistically as a positive correlation between performance on erotic pictures for session 1 and session 2."

Figure 13.4 shows the ability of all 100 subjects, on two sessions of 60 trials, to identify the hidden locations of erotic pictures. The visual impression is that there is no systematic association between performance on session 1 and session 2.

As before, it makes sense to make inferences about the correlation coefficient, and test alternative hypotheses. Since the hypothesis being tested is specifically about the possibility of a positive correlation, the alternative hypothesis $\mathcal{H}_1$ now states that the correlation is positive, and so uses the prior distribution $r \sim \text{Uniform}(0, 1)$.

Another difference in this example is that it is clear how to model the uncertainty in psychological variables that generate the behavioral measures. Performance on the two sessions is simply a count of the number of correct responses for each person, assumed to be generated by an underlying ability. Thus, if the ith person has k_{i1} correct answers in the first session out of $n = 60$ trials, this performance is related to an underlying rate of correctly responding θ_{i1} on the first session by $k_{i1} \sim \text{Binomial}(\theta_{i1}, n)$. In this way, by providing a complete account of the probabilistic process by which the observed data are generated, the inherent uncertainty in the underlying abilities for people and sessions is naturally taken into account.

Figure 13.5 shows a graphical model for inferring the correlation coefficient between performance on the first and second session, and for modeling the subjects' behavior in making correct decisions from underlying abilities. The script `Ability.txt` implements the graphical model in WinBUGS:

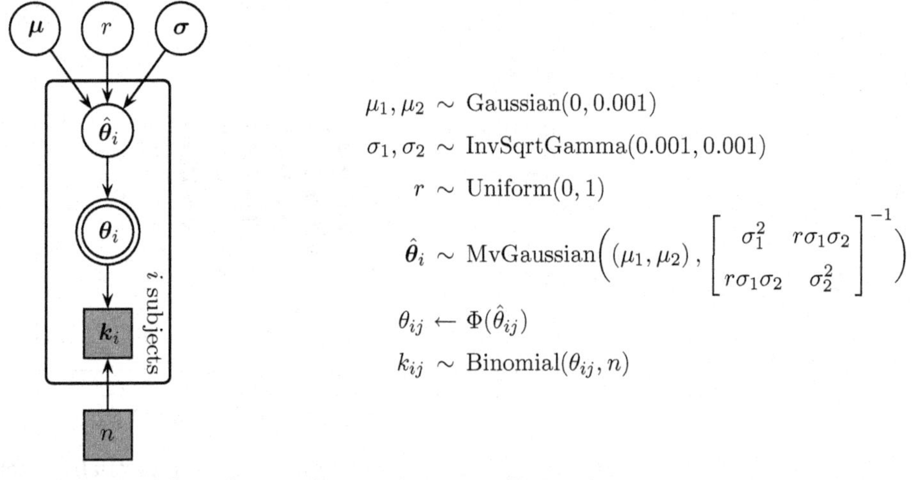

Fig. 13.5 Graphical model for inferring the correlation coefficient between performance across subjects on the first and second session of the ESP replication experiment.

```
# Ability Correlation for ESP Replication
model{
  # Data
  for (i in 1:nsubjs){
    thetap[i,1:2] ~ dmnorm(mu[],TI[,])
    for (j in 1:2){
      theta[i,j] <- phi(thetap[i,j])
      k[i,j] ~ dbin(theta[i,j],ntrials)
    }
  }
  # Priors
  mu[1] ~ dnorm(0,.001)
  mu[2] ~ dnorm(0,.001)
  lambda[1] ~ dgamma(.001,.001)
  lambda[2] ~ dgamma(.001,.001)
  r ~ dunif(0,1)
  # Reparameterization
  sigma[1] <- 1/sqrt(lambda[1])
  sigma[2] <- 1/sqrt(lambda[2])
  T[1,1] <- 1/lambda[1]
  T[1,2] <- r*sigma[1]*sigma[2]
  T[2,1] <- r*sigma[1]*sigma[2]
  T[2,2] <- 1/lambda[2]
  TI[1:2,1:2] <- inverse(T[1:2,1:2])
}
```

The code `Ability.m` or `Ability.R` applies the graphical model to the data from Figure 13.4, plots the posterior distribution, and applies the Savage–Dickey method.

The results are shown in Figure 13.6. The left panel shows the inferred abilities, for each subject on each session. The circles show the expected value of the abilities for each subject, and the lines connect this expectation to a sample of points from the joint posterior. The right panel shows the prior (horizontal dotted line) and

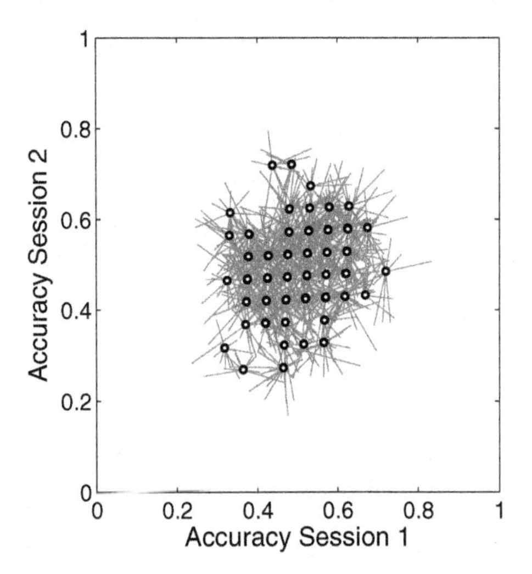

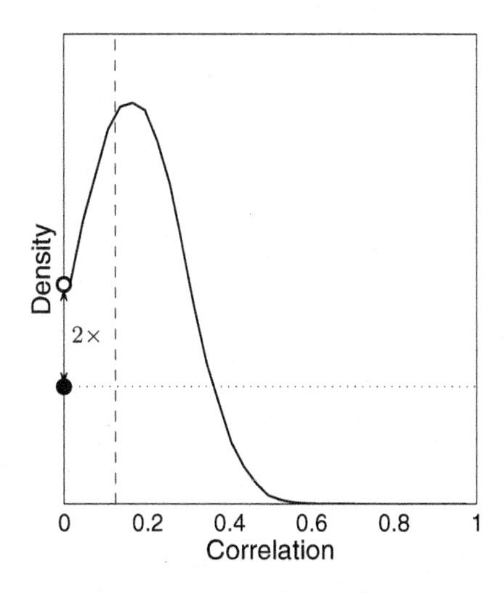

Fig. 13.6 Inferred abilities (left-hand panel) and correlation analysis (right-hand panel) for the relationship between sample size and effect size in ESP replication experiments.

posterior (solid line) distribution for the correlation coefficient. The expected value of the posterior is about 0.18. The mode is near the frequentist value of 0.12 shown by the broken vertical line, but does not correspond as closely as Figure 13.3.

The right panel of Figure 13.6 also shows the density of the prior and posterior at the point of test $r = 0$ by black and white circles, respectively. The density of the posterior is about 2 times greater than that of the prior at this point, so the Bayes factor is about 2 in favor of the null hypothesis that the correlation is zero. This is an evidence level that Jeffreys (1961) called "not worth more than a bare mention."

Exercises

Exercise 13.2.1 Suppose that the alternative hypothesis does not assume a positive correlation between the abilities of subjects over the two sessions, but instead allows for any correlation, so that the prior is $r \sim \text{Uniform}\big(-1, 1\big)$. Intuitively, what is the value of the Bayes factor in this case?

Exercise 13.2.2 A classical analysis yields $r = 0.12$, 95% CI $= [-0.08, 0.31]$, $p = 0.23$. This non-significant p-value, however, fails to indicate whether the data are ambiguous or whether there is evidence in favor of $\mathcal{H}_0$. How does the Bayes factor resolve this ambiguity?

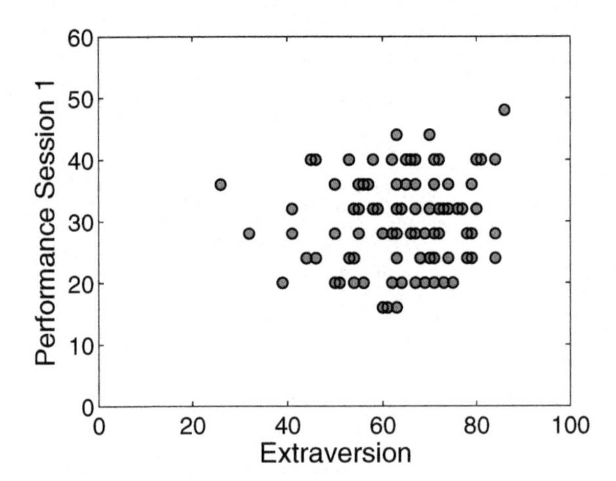

Fig. 13.7 The extraversion score, and performance on the first session, for 100 subjects in the ESP replication experiment

13.3 Evidence for the impact of extraversion

Wagenmakers et al. (2012), following a suggestion of Bem (2011), also considered the possibility of there being a positive correlation between performance and extraversion across subjects. Figure 13.7 shows the data for the extraversion score of each subject, and their performance on the first session. The visual impression is that there is no strong correlation.

The performance on the first session can be modeled as before, but modeling the inherent uncertainty in extraversion requires additional assumptions. One approach, also considered in Section 5.2, is to treat each extraversion score as the mean of a Gaussian distribution with some standard deviation. The assumed value of the standard deviation then corresponds to the assumed precision of the psychometric instrument used to generate the observed score.

A graphical model that represents this approach is shown in Figure 13.8. For the ith subject, the counts of correct predictions k_i is modeled as before, with $k_i \sim \text{Binomial}(\theta_{i1}, n)$. Their extraversion score x_i is modeled as $x_i \sim \text{Gaussian}(\theta_{i2}, \lambda^x)$, where $\theta_{i2} = 100\Phi(\hat{\theta}_{i2})$ is the underlying true extraversion on a 0–100 scale, and λ^x is the precision of the Gaussian. Note that the prior on the correlation coefficient has reverted to $r \sim \text{Uniform}(-1, 1)$.

The script `Extraversion.txt` implements the graphical model in WinBUGS:

```
# Extraversion Correlation for ESP Replication
model{
  # Data
  for (i in 1:nsubjs){
    thetap[i,1:2] ~ dmnorm(mu[],TI[,])
    theta[i,1] <- phi(thetap[i,1])
    k[i] ~ dbin(theta[i,1],ntrials)
```

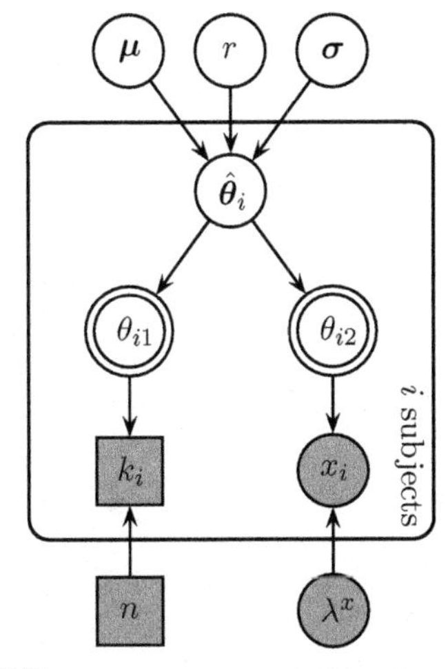

$$\mu_1, \mu_2 \sim \text{Gaussian}(0, 0.001)$$
$$\sigma_1, \sigma_2 \sim \text{InvSqrtGamma}(0.001, 0.001)$$
$$r \sim \text{Uniform}(-1, 1)$$
$$\hat{\boldsymbol{\theta}}_i \sim \text{MvGaussian}\left((\mu_1, \mu_2), \begin{bmatrix} \sigma_1^2 & r\sigma_1\sigma_2 \\ r\sigma_1\sigma_2 & \sigma_2^2 \end{bmatrix}^{-1}\right)$$
$$\theta_{i1} \leftarrow \Phi(\hat{\theta}_{i1})$$
$$\theta_{i2} \leftarrow 100\Phi(\hat{\theta}_{i2})$$
$$k_i \sim \text{Binomial}(\theta_{i1}, n)$$
$$x_i \sim \text{Gaussian}(\theta_{i2}, \lambda^x)$$

Fig. 13.8 Graphical model for inferring the correlation coefficient between performance across subjects on the first and second blocks of the ESP replication experiment.

```
    theta[i,2] <- 100*phi(thetap[i,2])
    x[i] ~ dnorm(theta[i,2],lambdax)
  }
  # Priors
  mu[1] ~ dnorm(0,.001)
  mu[2] ~ dnorm(0,.001)
  lambda[1] ~ dgamma(.001,.001)
  lambda[2] ~ dgamma(.001,.001)
  r ~ dunif(-1,1)
  # Reparameterization
  sigma[1] <- 1/sqrt(lambda[1])
  sigma[2] <- 1/sqrt(lambda[2])
  T[1,1] <- 1/lambda[1]
  T[1,2] <- r*sigma[1]*sigma[2]
  T[2,1] <- r*sigma[1]*sigma[2]
  T[2,2] <- 1/lambda[2]
  TI[1:2,1:2] <- inverse(T[1:2,1:2])
}
```

The code `Extraversion.m` or `Extraversion.R` applies the graphical model to the data from Figure 13.7, plots the posterior distribution, and applies the Savage-Dickey method. Note that the code makes the assumption about the extraversion test precision on the standard deviation scale, which seems an easier one to express the uncertainty of measurement, and converts it to a precision to supply to the graphical model.

The results when $\lambda^x = 1/9$—that is, when the standard deviation is 3 for the extraversion measure—are shown in Figure 13.9. The left panel shows the inferred ability on the first session, and underlying level of extraversion, for each subject. The circles show the expected values, and the lines connect this expectation to a sample of points from the joint posterior. The right panel shows the prior (horizontal

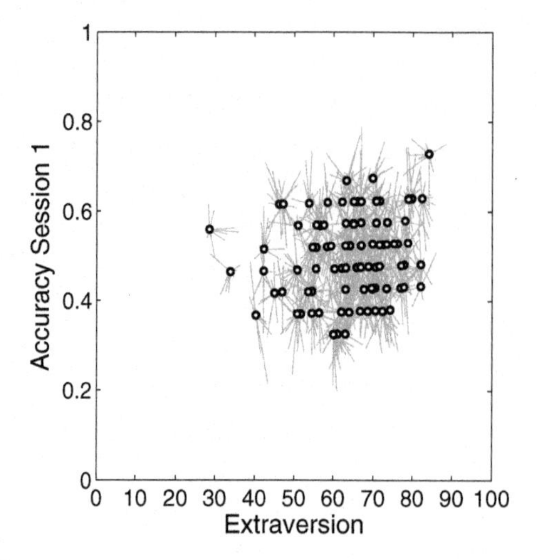

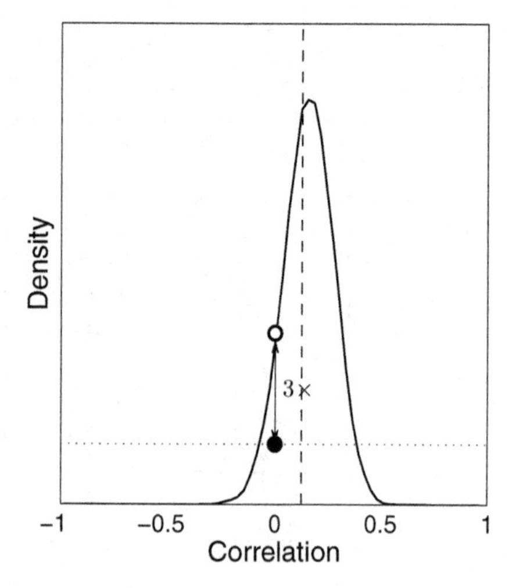

Fig. 13.9 Inferred extraversion measures and abilities (left-hand panel), and correlation analysis (right-hand panel) for the relationship between extraversion and performance in the ESP replication experiment.

dotted line) and posterior (solid line) distribution for the correlation coefficient. The expected value of the posterior is about 0.16. The mode is near the frequentist value of 0.12 shown by the broken vertical line.

The right panel of Figure 13.9 also shows the density of the prior and posterior at the point of test $r = 0$ by black and white circles, respectively. The density of the posterior is about 3 times greater than that of the prior at this point, so the Bayes factor is about 3 in favor of the null hypothesis that the correlation is zero.

Exercises

Exercise 13.3.1 What do you conclude about whether or not the correlation is zero, based on the Bayes factor?

Exercise 13.3.2 Try more extreme assumptions about the accuracy with which extraversion is measured, by setting $\lambda^x = 1$ and $\lambda^x = 1/100$. How does the Bayes factor change in response to this change in available information?

14 Multinomial processing trees

WITH DORA MATZKE

14.1 Multinomial processing model of pair-clustering

Consider a free recall task in which people study a list of words—table, dog, brick, pencil, cat, news, doctor, keys, nurse, soccer—and, after a short delay, are asked to recall the words in any order. An interesting property of this list is that it contains some pairs of semantically related words, like dog and cat, or doctor and nurse, as well as some words, called "singletons," like table or soccer, that are not semantically related to another on the list.

A standard finding is that semantically related words are often recalled consecutively, even when they are not adjacent in the study list. For example, a person may recall soccer, cat, dog, table, doctor, and nurse. The finding that semantically related items are often recalled consecutively can be taken as evidence for the idea that they were stored and retrieved as a cluster.

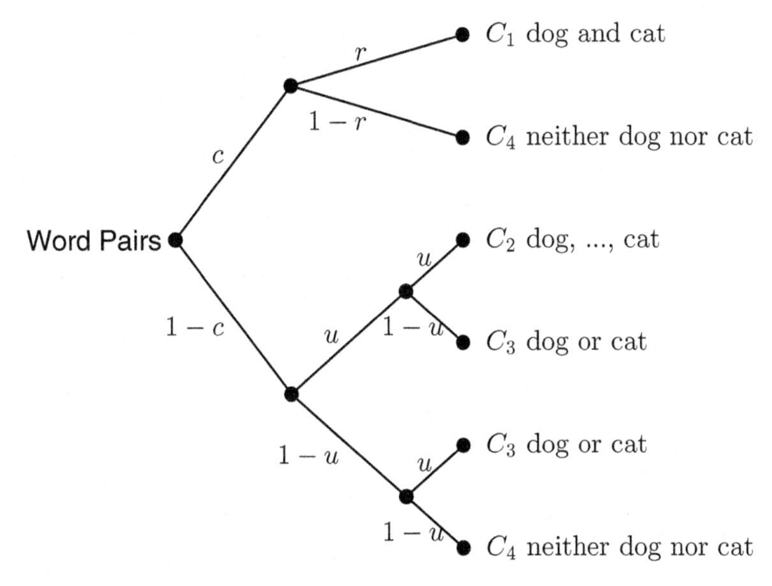

Fig. 14.1 Multinomial processing tree model for pair-clustering effects in recall from memory, with cluster-storage c, cluster-retrieval r, and unique storage-retrieval u parameters.

Multinomial processing trees (MPTs: Batchelder & Riefer, 1980, 1986; Chechile, 1973; Chechile & Meyer, 1976) provide one approach to modeling these memory effects. An MPT model assumes that observed behavior arises from a sequence of cognitive events, able to be represented by a rooted tree architecture such as the one shown in Figure 14.1. In this MPT model, the focus is on the recall of word pairs, but more general models can be developed that also account for singletons.

The model in Figure 14.1 considers four categories of response behavior for word pairs. In the first category, C_1, both words in a word pair are recalled consecutively. In the second category, C_2, both words in a word pair are recalled, but not consecutively. In the third category, C_3, only one word of a word pair is recalled. In the fourth category, C_4, neither word in a word pair is recalled.

The pair-clustering MPT model describes a simple sequence of cognitive processes that can produce these four behavioral outcomes, controlled by three parameters. The cluster-storage parameter c is the probability that a word pair is clustered and stored in memory. The cluster-retrieval parameter r is the conditional probability that a word pair is retrieved from memory, given that is was clustered. The unique storage-retrieval parameter u is the conditional probability that a member of a word pair is stored and retrieved from memory, given that the word pair was not stored as a cluster.

Under the MPT model for the pair-clustering paradigm, the probabilities of the four response categories are as follows:

$$\begin{aligned}
\Pr\left(C_{11} \mid c, r, u\right) &= cr \\
\Pr\left(C_{12} \mid c, r, u\right) &= (1-c)\,u^2 \\
\Pr\left(C_{13} \mid c, r, u\right) &= 2u\,(1-c)\,(1-u) \\
\Pr\left(C_{14} \mid c, r, u\right) &= c\,(1-r) + (1-c)\,(1-u)^2 .
\end{aligned}$$

For example, the MPT model states that the probability of category C_1—that is, consecutively recalling semantically related study words—requires that the words are first stored in memory as a pair, with probability c, and then retrieved as a pair, with probability r. The probabilities for other categories can also be decomposed as products and sums of the MPT parameters. In this way, the MPT model provides an account of the number of times words are recalled according to the different categories of behavior, controlled by the cognitive processes represented by the tree and its parameters.

Table 14.1 Category counts for Trials 1, 2, and 6 from Riefer et al. (2002).

	C_1	C_2	C_3	C_4
Trial 1	45	24	97	254
Trial 2	106	41	107	166
Trial 6	243	64	65	48

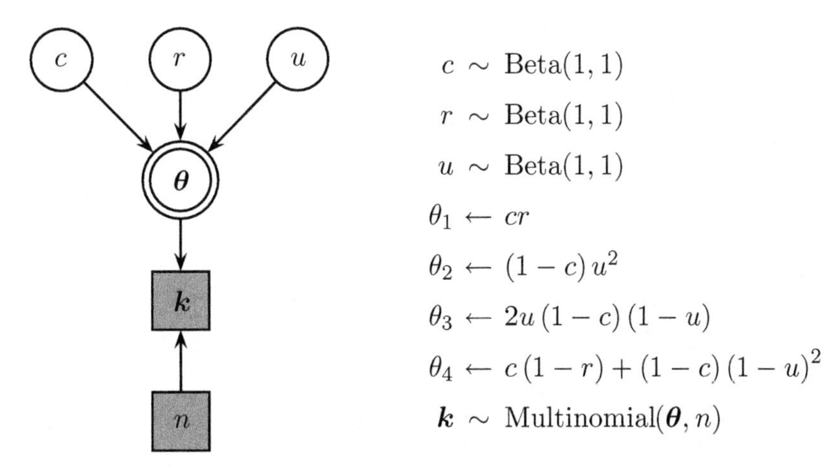

Fig. 14.2 Graphical model for the pair-clustering MPT model of aggregated data.

Most analyses of MPT models rely on category response data that are aggregated over subjects and items (e.g., Hu & Batchelder, 1994). Our applications use a subset of the free recall data reported in Riefer et al. (2002). We analyze the free recall performance of 21 subjects responding to 20 categorically related word pairs in a series of six study-test trials. Each trial featured exactly the same word materials, and it therefore seems reasonable to expect that all three model parameters will increase over trials. Hence, we focus on performance in the first, second, and sixth session. Table 14.1 shows the aggregated data for each of the four categories of behavior, in all three of these trials.

A graphical model for the MPT account of these data is shown in Figure 14.2. The c, r, and u parameters are given uniform priors, and generate probabilities $\boldsymbol{\theta} = (\theta_1, \ldots, \theta_4)$ for each of the four categories. The aggregated count data k for any trial thus follow $k \sim \text{Multinomial}(\boldsymbol{\theta}, n)$, where n is the total number of behaviors over all subjects and word pairs.

The script MPT_1.txt implements the graphical model in WinBUGS:

```
# Multinomial Processing Tree
model{
  # MPT Category Probabilities for Word Pairs
  theta[1] <- c * r
  theta[2] <- (1-c) * pow(u,2)
  theta[3] <- (1-c) * 2 * u * (1-u)
  theta[4] <- c * (1-r) + (1-c) * pow(1-u,2)
  # Data
  k[1:4] ~ dmulti(theta[1:4],n)
  # Priors
  c ~ dbeta(1,1)
  r ~ dbeta(1,1)
  u ~ dbeta(1,1)
}
```

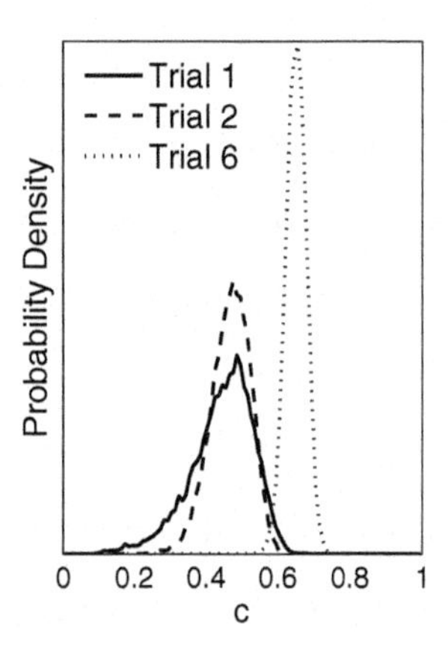

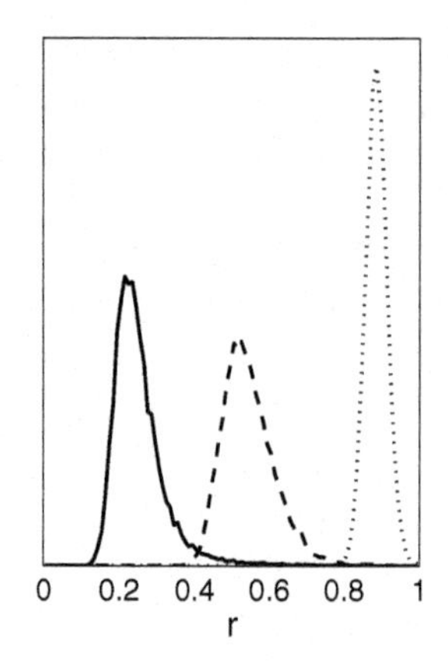

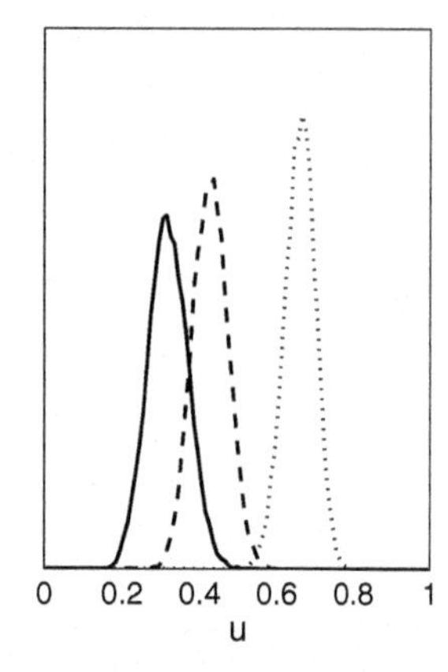

Fig. 14.3 Posterior distributions for the c, r, and u parameters for the Riefer et al. (2002) data set.

The code `MPT_1.m` or `MPT_1.R` applies the model to make inferences for the three trials. Figure 14.3 shows the posterior distributions for each of the c, r, and u parameters for each trial.

Exercises

Exercise 14.1.1 What do you conclude from the posterior distributions in Figure 14.3 about learning over the course of the trials?

Exercise 14.1.2 Because the u parameter corresponds to both the storage and retrieval of unclustered words, it is typically regarded as a nuisance parameter. In an approach to inference that is not fully Bayesian, the lack of interest in the posterior distribution of u might lead to the shortcut of a reasonable value being substituted, rather than assigning a prior distribution. Modify the graphical model so that u is set to a constant for each trial, given by the expected value of the posterior from the fully Bayesian analysis. How does this change affect the posterior distributions of c and r, the parameters of interest?

14.2 Latent-trait MPT model

The use of aggregated data relies on the assumption that subjects do not differ in terms of the psychological process used by the MPT model. Modeling individual subject data, and allowing the parameters for each subject to vary in some structured way, provides one powerful way to addresses this issue. Here we focus on a latent-trait approach developed by Klauer (2010), which not only allows for variation in parameters between individuals, but provides an explicit model of that variation.

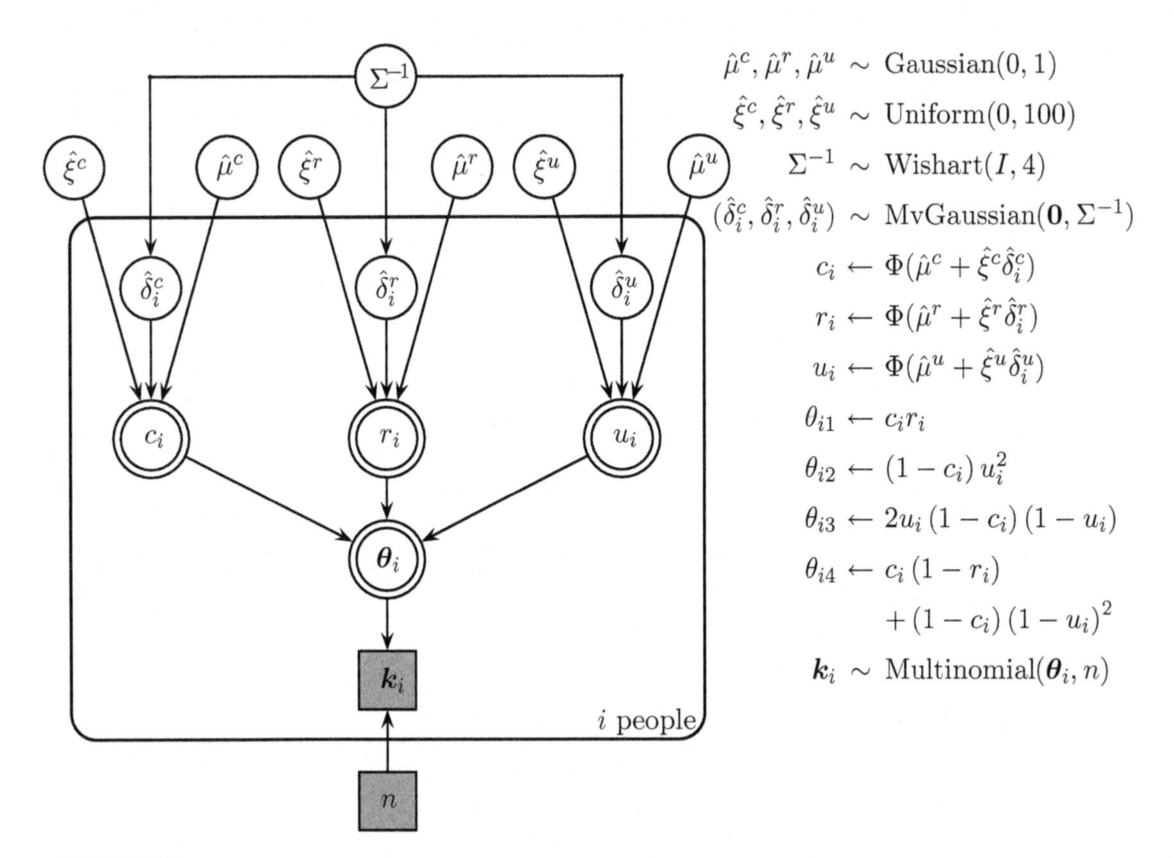

Fig. 14.4 Graphical model for the latent-trait pair-clustering model.

The latent-trait approach we adopt assumes that there are group means—μ_c, μ_r, and μ_u—for the three parameters. The values of these parameters for the ith subject—c_i, r_i, and u_i—are then modeled in terms of their displacement around these group means. The parameters that control the displacement come from a multivariate distribution, so that the variability and correlation between the displacements themselves are both modeled. The motivation for taking this additional modeling step is that the cognitive processes represented by the model parameters—the retrieval of clustered and unclustered word pairs, for example—may often be highly correlated, and part of the goal of modeling is to make inferences about these relationships from data.

Figure 14.4 presents the graphical model for the latent-trait pair-clustering model. The data $\boldsymbol{k}_i$ for the ith subject consist of counts of responses falling into the different response categories shown in Figure 14.1. There are now cluster-storage c_i, cluster-retrieval r_i, and unique-retrieval u_i parameters for the ith subject, which generate response probabilities $\boldsymbol{\theta}_i$ $(\theta_{i1}, \ldots, \theta_{i4})$, and $k_i \sim \text{Multinomial}(\boldsymbol{\theta}_i, n)$ as before, with n now being just the number of word pairs for a single subject.

Box 14.1 **Fundamental flaws**

"One argument for Bayesian inference is that it is better to wrestle with the practicalities of a method that is fundamentally sound, than to work with one having fundamental flaws." (O'Hagan & Forster, 2004, p. 17)

The individual subject parameters c_i, r_i, and u_i are all probabilities, as in the original model, but their variability is modeled in a probit-transformed space. The group means in the probit space are $\hat{\mu}^c$, $\hat{\mu}^r$ and $\hat{\mu}^u$, with, for example, $\mu^c = \Phi(\hat{\mu}^c)$. The priors on the group mean in the probit space are $\hat{\mu}^c, \hat{\mu}^r, \hat{\mu}^u \sim \text{Gaussian}(0,1)$ which corresponds in the probability space to $\mu^c, \mu^r, \mu^u \sim \text{Uniform}(0,1)$.

Individual differences come from displacement parameters $\hat{\delta}_i^c$, $\hat{\delta}_i^r$, and $\hat{\delta}_i^u$ for the ith subject. These are draws from a multivariate Gaussian distribution, with $(\delta_i^c, \delta_i^r, \delta_i^u) \sim \text{MvGaussian}(\mathbf{0}, \Sigma^{-1})$. The multivariate Gaussian[1] is zero-centered, with an unscaled covariance matrix Σ. The prior for the inverse covariance is $\Sigma^{-1} \sim \text{Wishart}(I, 4)$, where I is the 3×3 identity matrix, and there are 4 degrees of freedom. This is a standard prior that corresponds to a uniform distribution on the correlation between the model parameters (Gelman & Hill, 2007, pp. 284–287 and pp. 376–378; Klauer, 2010, pp. 77–78).

To improve the rate of convergence in MCMC sampling, a parameter expansion method is also used in the graphical model shown in Figure 14.4. This involves redundant multiplicative scale parameters $\hat{\xi}^c, \hat{\xi}^r, \hat{\xi}^u \sim \text{Uniform}(0, 100)$, that combine with the $\hat{\delta}_i^c$, $\hat{\delta}_i^r$, and $\hat{\delta}_i^u$ values to determine the offset in probit space each subject has from the group mean.

Putting all of this together, the cluster-storage parameter, for example, for the ith subject is given by $c_i \leftarrow \Phi(\hat{\mu}^c + \hat{\xi}^c \hat{\delta}_i^c)$. Exactly the same approach is used for the cluster-retrieval and unique-retrieval parameters.

The advantage of this latent-trait approach to individual differences—as compared, say, to simply drawing the individual subject parameters from independent group distributions for each parameter—is that the relationship between parameters is being modeled. The covariance matrix Σ contains the information needed to infer both the variance of the cluster-storage, cluster-retrieval, and unique-retrieval parameters, and the correlation between each pair of these parameters. The standard deviation for the cluster-storage parameter, for example, is given by $\sigma^c = |\hat{\xi}^c|\sqrt{\Sigma_{cc}}$, where Σ_{cc} is the diagonal element of the 3×3 covariance matrix Σ corresponding to the c parameter. The correlation between the cluster-storage and cluster-retrieval parameters, for example, is given by $\rho^{cr} = \hat{\xi}^c \hat{\xi}^r \Sigma_{cr} / \left(|\hat{\xi}^c|\sqrt{\Sigma_{cc}}|\hat{\xi}^r|\sqrt{\Sigma_{rr}}\right)$. With priors on the $\hat{\xi}$ parameters that allow only positive values, this simplifies to $\rho^{cr} = \Sigma_{cr} / \left(\sqrt{\Sigma_{cc}}\sqrt{\Sigma_{rr}}\right)$.

[1] Just as WinBUGS uses precisions instead of variances to parameterize the Gaussian distribution, it uses inverse covariance matrices instead of covariance matrices to parameterize the multivariate Gaussian distribution.

The script MPT_2.txt implements the graphical model in WinBUGS:

```
# Multinomial Processing Tree with Latent Traits
model{
  for (i in 1:nsubjs){
    # MPT Category Probabilities for Word Pairs
    theta[i,1] <- c[i] * r[i]
    theta[i,2] <- (1-c[i])*pow(u[i],2)
    theta[i,3] <- (1-c[i])*2*u[i]*(1-u[i])
    theta[i,4] <- c[i]*(1-r[i])+(1-c[i])*pow(1-u[i],2)
    # Data
    k[i,1:4] ~ dmulti(theta[i,1:4],n[i])
    # Probitize Parameters c, r, and u
    c[i] <- phi(muchat + xichat*deltachat[i])
    r[i] <- phi(murhat + xirhat*deltarhat[i])
    u[i] <- phi(muuhat + xiuhat*deltauhat[i])
    # Individual Effects
    deltahat[i,1:nparams] ~
       dmnorm(mudeltahat[1:nparams],SigmaInv[1:nparams,1:nparams])
    deltachat[i] <- deltahat[i,1]
    deltarhat[i] <- deltahat[i,2]
    deltauhat[i] <- deltahat[i,3]
  }
  # Priors
  mudeltahat[1] <- 0
  mudeltahat[2] <- 0
  mudeltahat[3] <- 0
  muchat ~ dnorm(0,1)
  murhat ~ dnorm(0,1)
  muuhat ~ dnorm(0,1)
  xichat ~ dunif(0,100)
  xirhat ~ dunif(0,100)
  xiuhat ~ dunif(0,100)
  df <- nparams+1
  SigmaInv[1:nparams,1:nparams] ~ dwish(I[1:nparams,1:nparams],df)
  # Post-Processing Means, Standard Deviations, Correlations
  muc <- phi(muchat)
  mur <- phi(murhat)
  muu <- phi(muuhat)
  Sigma[1:nparams,1:nparams] <- inverse(SigmaInv[1:nparams,1:nparams])
  sigmac <- xichat*sqrt(Sigma[1,1])
  sigmar <- xirhat*sqrt(Sigma[2,2])
  sigmau <- xiuhat*sqrt(Sigma[3,3])
  for (i1 in 1:nparams){
    for (i2 in 1:nparams){
      rho[i1,i2] <- Sigma[i1,i2]/sqrt(Sigma[i1,i1]*Sigma[i2,i2])
    }
  }
}
```

Note that the script, besides implementing the generative model for the data shown in Figure 14.4, also generates standard deviations and correlations in the sigma and rho variables. These could, in principle, be found by post-processing the posterior samples, but are implemented in the script for convenience.

The code MPT_2.m or MPT_2.R applies the model to make inferences for the three trials, based on individual subject data. Figure 14.5 shows the posterior distribu-

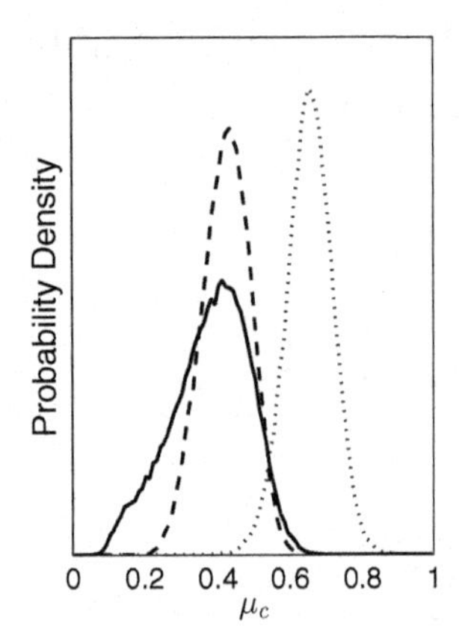

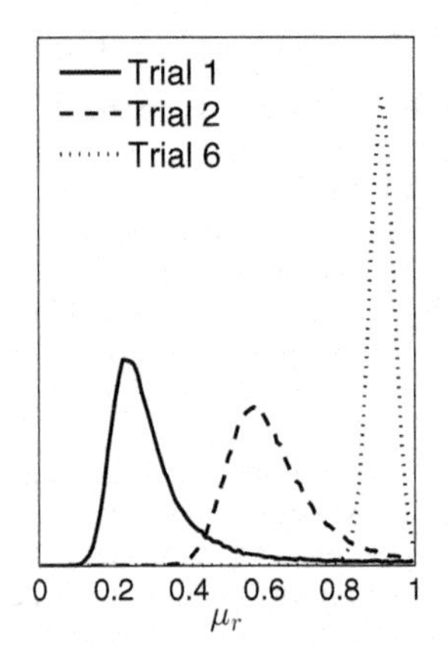

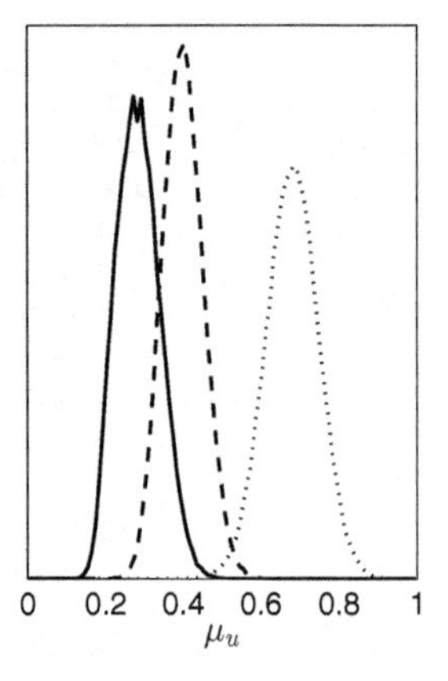

Fig. 14.5 Posterior distributions for group mean μ^c, μ^r, and μ^u parameters for the Riefer et al. (2002) data set, based on a latent-trait MPT model assuming individual differences.

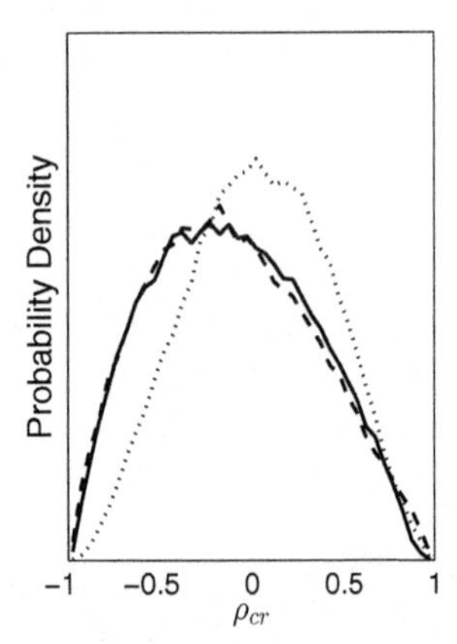

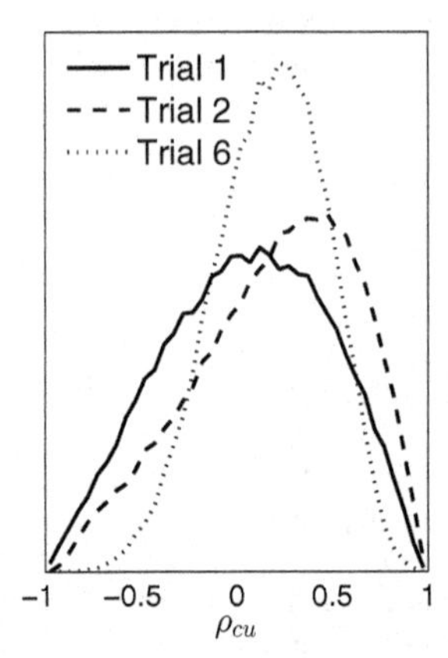

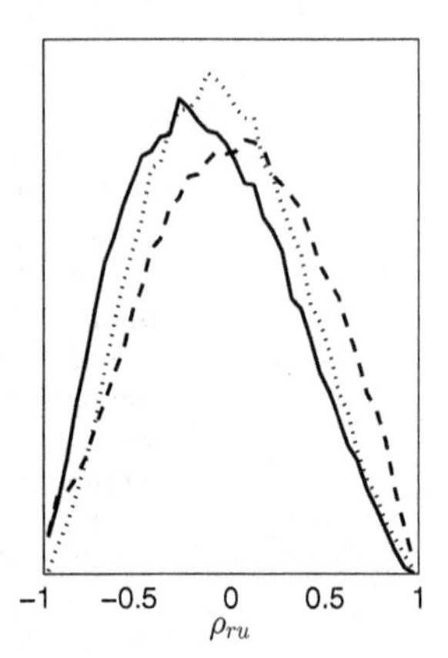

Fig. 14.6 Posterior distributions for the correlations ρ_{cr}, ρ_{cu}, and ρ_{cu} between parameters for the Riefer et al. (2002) data set, based on a latent-trait MPT model.

tions for the group means μ^c, μ^r, and μ^u for each trial. Figure 14.6 shows the posterior distributions for the correlations ρ_{cr}, ρ_{cu}, and ρ_{cu} between model parameters for each trial.

Exercises

Exercise 14.2.1 What do you conclude from the posterior distributions in Figure 14.5 about learning over the course of the trials? Compare your conclusions from the latent-trait model to the conclusions from the original MPT model.

Exercise 14.2.2 Extend the WinBUGS script to collect samples from the prior distributions for the standard deviation and correlation parameters. This will involve including variables `SigmaInvprior`, `Sigmaprior`, `rhoprior`, `sigmacprior`, `sigmarprior`, `sigmauprior`. Examine the prior and posterior distributions for the standard deviations and correlations. What can you conclude about the usefulness of including the correlation parameters in the latent-trait approach?

Exercise 14.2.3 The latent-trait approach deals with parameter heterogeneity as a result of individual differences between subjects, and relies on data that is aggregated over items. In many applications, however, it is reasonable to assume that the model parameters do not only differ between subjects but also between items. For example, it might be easier to cluster some pairs of semantically related words than others. This suggests using MPT models that incorporate both subject and item variability. Develop the graphical model that incorporates this extension. What is preventing the model from being applied to the current data?

15 The SIMPLE model of memory

15.1 The SIMPLE model

Brown, Neath, and Chater (2007) proposed the SIMPLE (Scale-Invariant Memory, Perception, and LEarning) model, which, among various applications, has been applied to the basic memory phenomenon of free recall. In this application, the SIMPLE model assumes memories are encoded by the time they were presented, but that the representations are logarithmically compressed, so that more temporally distant memories are more similar. It also assumes that distinctiveness plays a central role in performance on memory tasks, and that interference rather than decay is responsible for forgetting. Perhaps most importantly, the SIMPLE model assumes that the same memory processes operate at all time scales, unlike theories and models that assume different mechanisms for short-term and long-term memory.

The first application considered by Brown et al. (2007) involves seminal immediate free recall data reported by Murdock (1962). The data give the proportion of words correctly recalled averaged across participants, for lists of 10, 15, and 20 words presented at a rate of 2 seconds per word, and lists of 20, 30, and 40 words presented at a rate of 1 second per word.

Brown et al. (2007) make some reasonable assumptions about undocumented aspects of the task (e.g., the mean time of recall from the end-of-list presentation), to set the time T_i between learning and retrieval of the ith item. With these times established, the application of the SIMPLE model to the free recall data involves five stages, which are clearly described in Brown et al. (2007, Appendix).

First, the ith presented item, associated with time T_i, is represented in memory using logarithmic compression, given by $M_i = \log T_i$. Secondly, the similarity between each pair of items is calculated as $\eta_{ij} = \exp\left(-c\left|M_i - M_j\right|\right)$, where c is a parameter measuring the "distinctiveness" of memory. Thirdly, the discriminability of each pair of items is calculated as $d_{ij} = \eta_{ij} / \sum_k \eta_{ik}$. Fourthly, the retrieval probability of each pair of items is calculated as $r_{ij} = 1/\left(1 + \exp\left(-s\left(d_{ij} - t\right)\right)\right)$, where t is a threshold parameter and s is a threshold noise parameter. Finally, for free recall, the probability that the ith item in the presented sequence will be recalled is calculated as $\theta_i = \min\left(1, \sum_k r_{ik}\right)$.

These stages are implemented by the graphical model in Figure 15.1. The graphical model has variables corresponding to the observed times between learning and retrieval, T_i, and the observed number of correct responses y_{ix} for the ith item. The

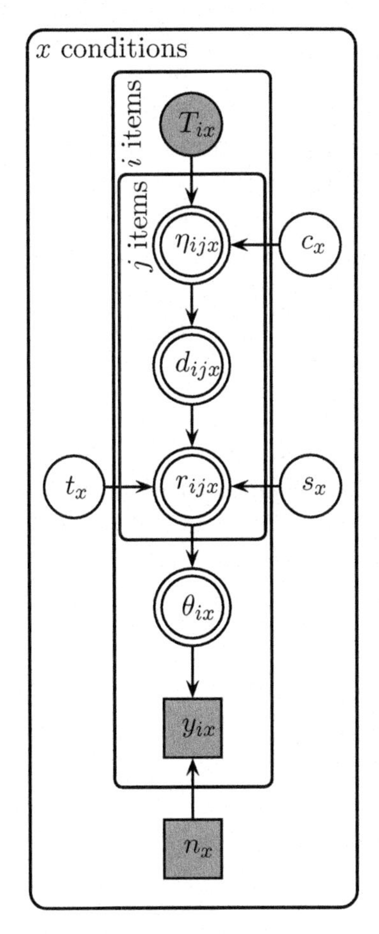

$$
\begin{aligned}
c_x &\sim \text{Uniform}(0, 100) \\
s_x &\sim \text{Uniform}(0, 100) \\
t_x &\sim \text{Uniform}(0, 1) \\
\eta_{ijx} &\leftarrow \exp\left(-c_x \left|\log T_{ix} - \log T_{jx}\right|\right) \\
d_{ijx} &\leftarrow \eta_{ijx} / \textstyle\sum_k \eta_{ikx} \\
r_{ijx} &\leftarrow 1/\left(1 + \exp\left(-s_x\left(d_{ijx} - t_x\right)\right)\right) \\
\theta_{ix} &\leftarrow \min\left(1, \textstyle\sum_k r_{ikx}\right) \\
y_{ix} &\sim \text{Binomial}(\theta_{ix}, n_x)
\end{aligned}
$$

Fig. 15.1 The SIMPLE model of free recall.

available data are aggregated across subjects, and the different word lists completed by subjects, and so n_x counts the total number of trials for the xth condition, which is the same for all words in that condition. The similarity η_{ijx}, discriminability d_{ijx}, retrieval r_{ijx}, and free recall probability θ_{ix} variables are deterministic. They implement the SIMPLE model in terms of its three parameters, linking the temporal representation of the items to the behavioral data, which is the accuracy of recalling the items.

In Figure 15.1 the times, responses, and free recall probabilities apply per item, and so are enclosed in a plate replicating over items. The similarity, discriminability, and retrieval measures apply to pairs of variables, and so involve an additional plate also replicating over items. We follow Brown et al. (2007) by fitting the c, t, and s parameters independently for each condition. This means the entire graphical model is also enclosed in a plate replicating over the 6 conditions in the Murdock (1962) data.

Box 15.1 Adhockery

"Faced with a new problem, a classical statistician is free to invent new estimators, confidence intervals or hypothesis tests ... In contrast, there is a unique Bayesian solution to any problem. That is the posterior distribution, which expresses the investigator's knowledge about θ after observing x. The Bayesian statistician's task is to identify the posterior distribution as accurately as possible, which usually entails identifying the prior distribution and the likelihood and then applying Bayes' theorem. There is no room for adhockery in Bayesian statistics." (O'Hagan & Forster, 2004, p. 19)

The script SIMPLE_1.txt implements the graphical model in WinBUGS. Note that the posterior predictive is calculated in some detail, leading up to pcpred, which is the posterior predicted proportion of correct recalls:

```
# SIMPLE Model
model{
  # Observed and Predicted Data
  for (x in 1:dsets){
    for (i in 1:listlength[x]){
      y[i,x] ~ dbin(theta[i,x],n[x])
      predy[i,x] ~ dbin(theta[i,x],n[x])
      predpc[i,x] <- predy[i,x]/n[x]
    }
  }
  # Similarities, Discriminabilities, and Response Probabilities
  for (x in 1:dsets){
    for (i in 1:listlength[x]){
      for (j in 1:listlength[x]){
        # Similarities
        sim[i,j,x] <- exp(-c[x]*abs(log(m[i,x])-log(m[j,x])))
        # Discriminabilities
        disc[i,j,x] <- sim[i,j,x]/sum(sim[i,1:listlength[x],x])
        # Response Probabilities
        resp[i,j,x] <- 1/(1+exp(-s[x]*(disc[i,j,x]-t[x])))
      }
      # Free Recall Overall Response Probability
      theta[i,x] <- min(1,sum(resp[i,1:listlength[x],x]))
    }
  }
  # Priors
  for (x in 1:dsets){
    c[x] ~ dunif(0,100)
    s[x] ~ dunif(0,100)
    t[x] ~ dbeta(1,1)
  }
}
```

The code Simple_1.m or Simple_1.R applies the model to the Murdock (1962) data. It involves some initial steps to get the data organized before passing it to

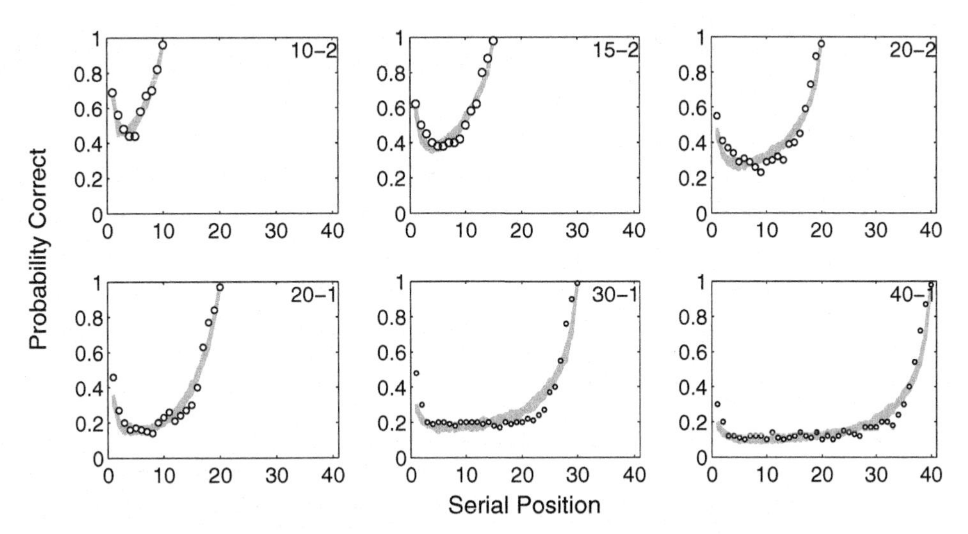

Fig. 15.2 Posterior prediction of the SIMPLE model for the six conditions of the Murdock (1962) immediate free recall data. The circles show the data, and the gray lines show the posterior predictive distribution.

WinBUGS. It also produces an analysis of the posterior predictions and the joint posterior parameter space.

The posterior predictive analysis for the six data sets is shown in Figure 15.2. The solid lines show the probability that the item in each serial position was correctly recalled. A total of 20 samples from the posterior predictive are shown for each serial position as gray points, making a gray area that spans the range in which the model expects the data to lie.

The posterior analysis is shown in Figure 15.3, which shows the joint posterior parameter distribution as a three-dimensional plot, with 20 posterior samples for each condition shown by different markers. The solid black dots projected onto the planes represent samples from the pairwise joint distributions of each possible combination of parameters, marginalized over the other parameter in each case. Finally, the marginal distributions for each parameter are shown along the three axes.

Figure 15.3 provides information about the distinctiveness, threshold, and threshold noise parameters, including information about variability and co-variation of the parameters across experimental conditions. This additional information is important to understanding how parameters should be interpreted, and for suggesting further model development. For example, the lack of overlap of the three-dimensional points for the six conditions suggests that there are important differences in model parameters for different item list lengths and presentation rates. In particular, it seems unlikely that an alternative approach to fitting the six conditions using a single discriminability level and threshold function will be adequate.

Box 15.2 Correcting the SIMPLE Model

Lee and Pooley (2013) point out that the part of the SIMPLE model that generates free recall probabilities is incorrect. The aim of this part of the model is to calculate the probability that an item will be recalled, and in free recall that is achieved if the word is recalled in any position in a recall sequence. Thus, all of the individual probabilities of recalling an item must be combined. The original formulation of the SIMPLE model does this additively, using $\theta_i = \min(1, \sum_x r_{ix})$, with the thresholding at 1 guaranteeing that a probability is produced.

But, as Lee and Pooley (2013) point out, if the probability of an event is 0.7, then the probability that event will occur at least once on two independent trials is not $0.7 + 0.7 = 1.4$, nor is it 1.4 thresholded to 1.0. The correct probability is naturally calculated by first finding the probability that the event occurs on neither trial as $(1 - 0.7) \times (1 - 0.7) = 0.09$, and taking the complement $1.0 - 0.09 = 0.91$ for an item. Thus, Lee and Pooley (2013) argue for $\theta_i = 1 - \prod_x (1 - r_{ix})$. Implementing this correction in WinBUGS requires only replacing `theta[i,x] <- min(1,sum(resp[i,1:listlength[x],x]))` with `theta[i,x] <- 1-prod(1-resp[i,1:listlength[x],x])`.

Another intuition, this time coming from the two-dimensional joint posteriors, is that there is a trade-off between the threshold and threshold noise parameters, since their joint distributions (shown by the solid black dots in the bottom plane) show a high level of negative correlation for all of the conditions. This means that the data in each condition are consistent with relatively high thresholds and relatively low levels of threshold noise, or with relatively low thresholds and relatively high levels of threshold noise. This is probably not ideal, since generally parameters are more easily interpreted and theoretically compelling if they operate independently of each other. In this way, the information in the joint parameter posterior suggests an area in which the model might need further development or refinement.

As a final example of the information in the joint posterior, note that the marginal distributions for the threshold parameter shown in Figure 15.3 seem to show a systematic relationship with item list length. In particular, the threshold decreases as the item list length increases from 10 to 40, with overlap between the two conditions with the most similar lengths (i.e., the 10–2 and 15–2 conditions, and the 20–2 and 20–1 conditions). This type of systematic relationship suggests that, rather than treating the threshold as a free parameter, it can be modeled in terms of the known item list length.

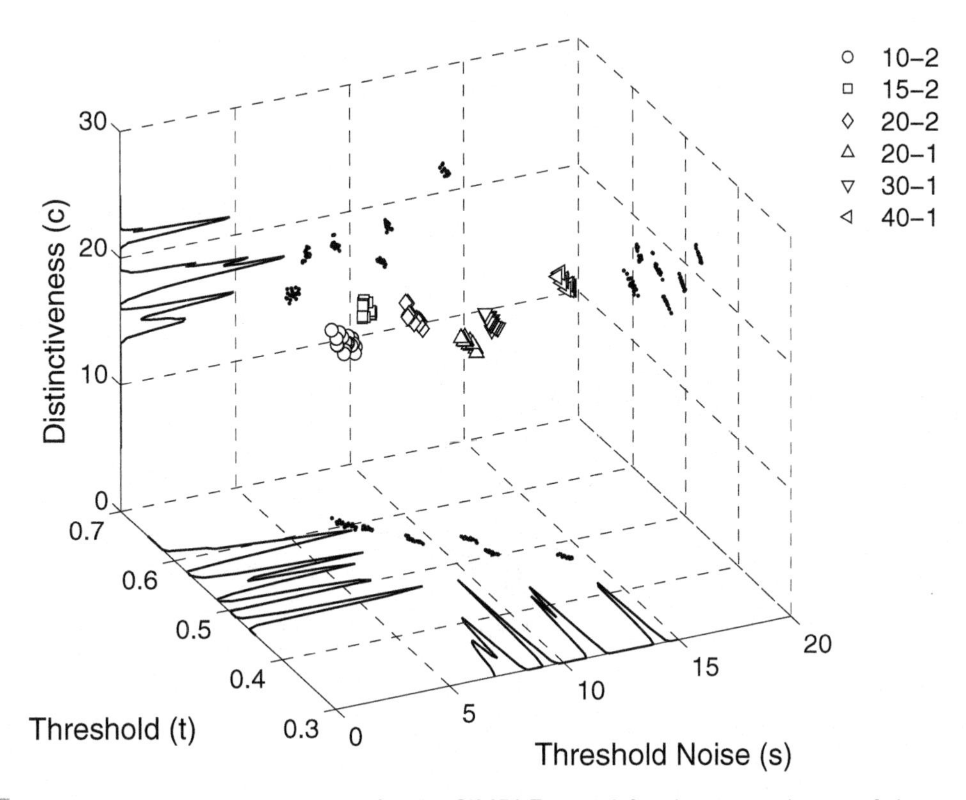

Fig. 15.3 Joint posterior parameter space for the SIMPLE model for the six conditions of the Murdock (1962) immediate free recall data.

Exercise

Exercise 15.1.1 Modify the graphical model so that the same parameter values are used to account for all of the data sets. You will also need to modify the Matlab or R code that produces the graphs.

15.2 A hierarchical extension of SIMPLE

We now consider how the possibility of a relationship between list length and thresholds can be implemented in a hierarchical extension to the SIMPLE model. The extended model is shown in Figure 15.4. There are two important changes. First, the distinctiveness c and threshold noise s parameters are now assumed to have the same value for all experimental conditions. In Figure 15.4, they are outside

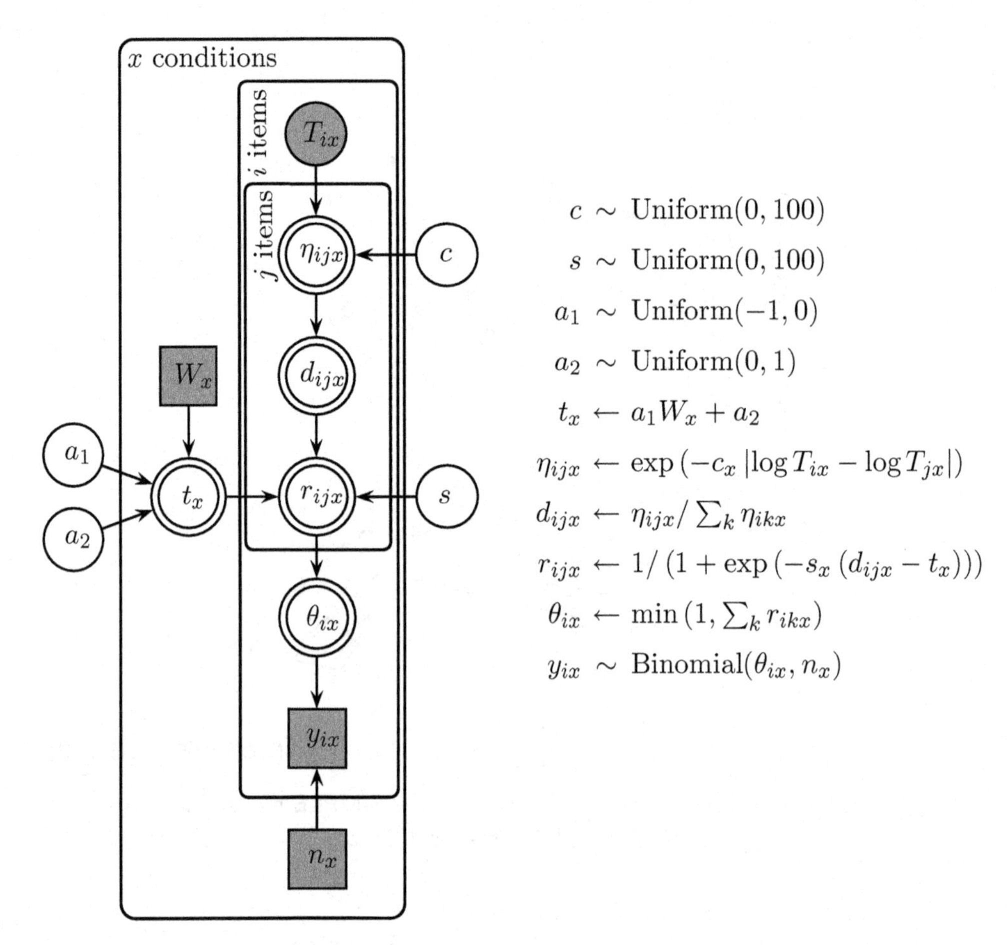

Fig. 15.4 Graphical model implementing a hierarchical extension to the SIMPLE model of memory.

the plate that replicates over the experimental conditions, and they are no longer indexed by x.[1]

The second change corresponds to the way the thresholds t_x are determined. Rather than being assumed to be independent, these thresholds now depend on the item list length, denoted W_x for the xth condition. The dependence is modeled as a linear function $t_x = a_1 W_x + a_2$, parameterized by the coefficients a_1 and a_2. Consistent with the intuitions gained from Figure 15.3, we make the assumption that the linear relationship expresses a decrease in threshold as item list length increases, by using the prior $a_1 \sim \text{Uniform}\big(-1, 0\big)$.

[1] This is probably not a theoretically realistic assumption—indeed, as we pointed out, the joint posterior in Figure 15.3 argues against it—but it is a simple assumption that makes it easy to focus on the hierarchical extension.

The goal of the hierarchical extensions is to move away from thinking of parameters as psychological variables that vary independently for every possible recall task. Rather, we now conceive of the parameters as psychological variables that themselves now need explanation, and attempt to model how they change in terms of more general parameters.

This approach not only forces theorizing and modeling to tackle new basic questions about how recall processes work, but also facilitates evaluation of the prediction and generalization capabilities of the basic model (Ahn, Busemeyer, Wagenmakers, & Stout, 2008). By making the threshold parameter depend on characteristics of the task—in this case, the number of words in the list—in a systematic ways, and by treating the other parameters as invariant, the hierarchical extension automatically allows the SIMPLE model to make predictions about other tasks.

The script `SIMPLE_2.txt` implements the graphical model in WinBUGS:

```
# Hierarchical SIMPLE Model
model{
  # Observed data
  for (x in 1:dsets){
    for (i in 1:listlength[x]){
      y[i,x] ~ dbin(theta[i,x],n[x])
    }
  }
  # Similarities, Discriminabilities, and Response Probabilities
  for (x in 1:gsets){
    t[x] <- max(0,min(1,a[1]*w[x]+a[2]))
    for (i in 1:listlength[x]){
      for (j in 1:listlength[x]){
        # Similarities
        sim[i,j,x] <- exp(-c*abs(log(m[i,x])-log(m[j,x])))
        # Discriminabilities
        disc[i,j,x] <- sim[i,j,x]/sum(sim[i,1:listlength[x],x])
        # Response Probabilities
        resp[i,j,x] <- 1/(1+exp(-s*(disc[i,j,x]-t[x])))
      }
      # Free Recall Overall Response Probability
      theta[i,x] <- min(1,sum(resp[i,1:listlength[x],x]))
    }
  }
  # Priors
  c ~ dunif(0,100)
  s ~ dunif(0,100)
  a[1] ~ dunif(-1,0)
  a[2] ~ dunif(0,1)
  # Predicted data
  for (x in 1:gsets){
    for (i in 1:listlength[x]){
      predy[i,x] ~ dbin(theta[i,x],n[x])
      predpc[i,x] <- predy[i,x]/n[x]
    }
  }
}
```

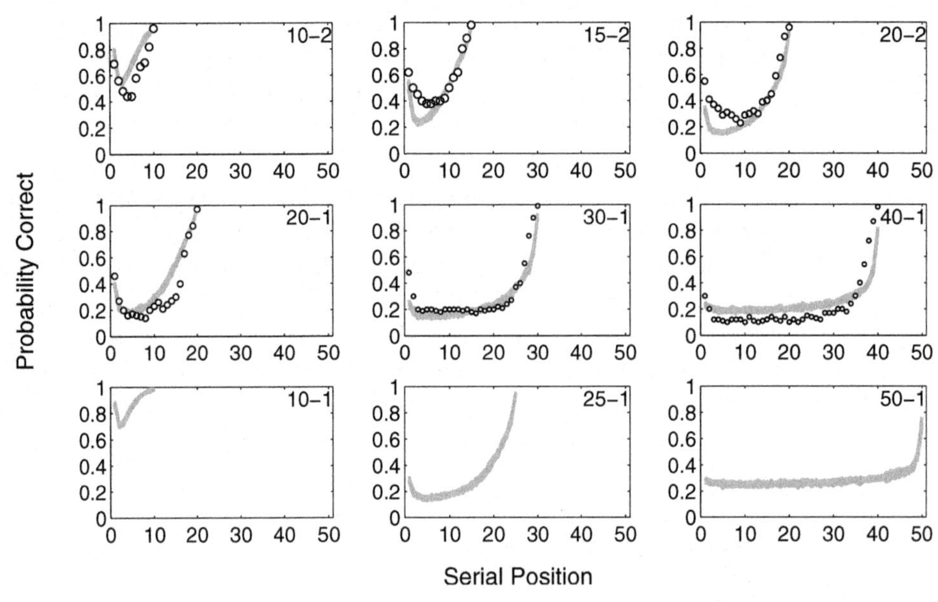

Fig. 15.5 Posterior prediction of the hierarchical extension of the SIMPLE model for the six conditions of the Murdock (1962) immediate free recall data, and in generalizing to three new conditions. The circles show the data, and the gray lines show the posterior predictive distribution.

The code `Simple_2.m` or `Simple_2.R` applies the hierarchical model to the Murdock (1962) conditions, and also to three other possible conditions, for which data are not available. These generalization conditions all involve presentation rates of 1 s per item, but with 10, 25, and 50 items, corresponding to both interpolations and extrapolations relative to the collected data.

The posterior predictive performance is shown in Figure 15.5. The top two rows show the Murdock (1962) conditions, while the bottom row shows the predictions the model makes about the generalization conditions. The model does not fit the data very well in the Murdock (1962) conditions, but our focus is on the possibility of the generalization predictions. Here, the hierarchical extension to the model allows it to predict serial recall curves for experimental conditions for which data are not available.

The posterior analysis in Figure 15.6 shows inferences about the threshold noise, distinctiveness, and threshold parameters. For the first two of these, the inferences take the form of single posterior distributions. For the threshold parameter, however, the posterior inference is now about its functional relationship to item list length. The posterior distribution for this function is represented in the right panel of Figure 15.6 by showing 50 posterior samples at each possible length $W = 1, \ldots, 50$. These posterior samples are found by taking joint posterior samples (a_1, a_2) and finding $t = a_1 W + a_2$ for all values of W.

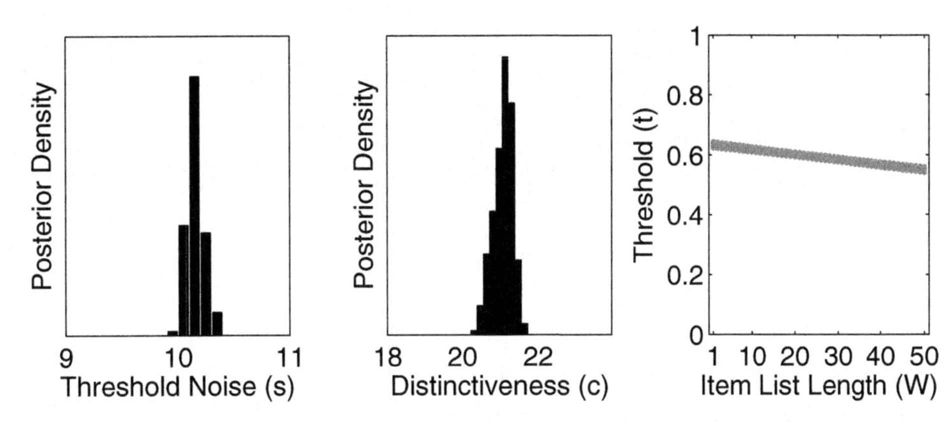

Fig. 15.6 Posteriors for the SIMPLE model parameters in its hierarchically extended form.

Exercise

Exercise 15.2.1 Why are empirical tests of generalization potentially more powerful or compelling evaluations of a model than fitting to existing data?

16 The BART model of risk taking

WITH DON VAN RAVENZWAAIJ

There is a large psychological literature studying individual differences in the judgments people make, and the preferences they have, in tasks involving risk. One controlled laboratory task developed for these purposes is the Balloon Analogue Risk Task (BART: Lejuez et al., 2002). Every trial in this task starts by showing a balloon representing a small monetary value, as in Figure 16.1. The subject can then either transfer the money to a virtual bank account, or choose to pump, which adds a small amount of air to the balloon, and increases its value. There is some probability, however, that pumping the balloon will cause it to burst, causing all the money to be lost. A trial finishes when either the subject has transferred the money, or the balloon has burst.

Fig. 16.1 The Balloon Analogue Risk Task (BART).

In the original version of the BART, the probability of the balloon bursting increased with every pump, but we consider a simplified version in which that probability is constant, and the expected gain of every decision to pump is zero. In the standard behavioral analysis of the BART, the risk propensity for a subject is measured as the average number of pumps for those balloons that did not burst. It is also possible to measure risk propensity using cognitive models of people's decisions on the BART (Rolison, Hanoch, & Wood, 2012; van Ravenzwaaij, Dutilh, & Wagenmakers, 2011; Wallsten, Pleskac, & Lejuez, 2005).

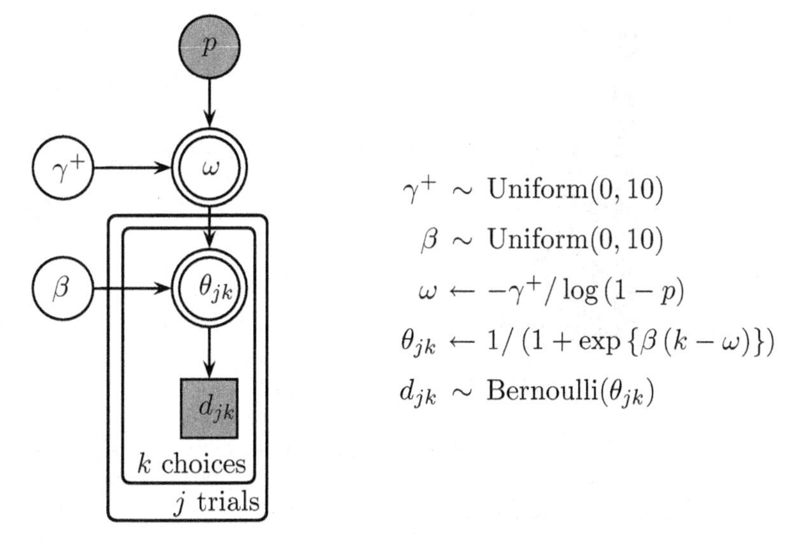

Fig. 16.2 Graphical model for the two-parameter BART model.

16.1 The BART model

We focus on a simple model using just two parameters, used by van Ravenzwaaij et al. (2011). One parameter, γ^+, controls risk taking and the other, β, controls behavioral consistency. It is assumed the subject knows the constant probability p that the balloon will burst any time it is pumped. The number of pumps the subject considers optimal, ω, depends on this probability, and on the propensity for risk taking, such that $\omega = -\gamma^+ / \log(1-p)$. Larger values of γ^+ lead to larger numbers of pumps being considered optimal, and so to greater risk seeking.

The probability that a subject chooses to pump on the kth opportunity within the jth trial depends on the number of pumps considered optimal, and on the behavioral consistency of the subject. These two factors are combined using the logistic function $\theta_{jk} = 1/\left(1+\exp\{\beta(k-\omega)\}\right)$. High values of β correspond to less variable responding. When $\beta = 0$, $\theta_{jk} = 0.5$, and both pumping and cashing in choices are always equally likely. As β becomes large, the choice becomes completely determined by whether or not k exceeds the number of pumps the subject considers optimal. Finally, the observed decision made on the kth choice within the jth trial simply follows the modeled choice, so that $d_{jk} \sim \text{Bernoulli}(\theta_{jk})$.

A graphical model that implements this account of decision-making on the BART is shown in Figure 16.2. Note that the probability θ_{jk} depends only on the choice being made within a trial (i.e., k) and not the trial itself (i.e., j), and so the model is more general than it needs to be.

The script `BART_1.txt` implements the graphical model in WinBUGS:

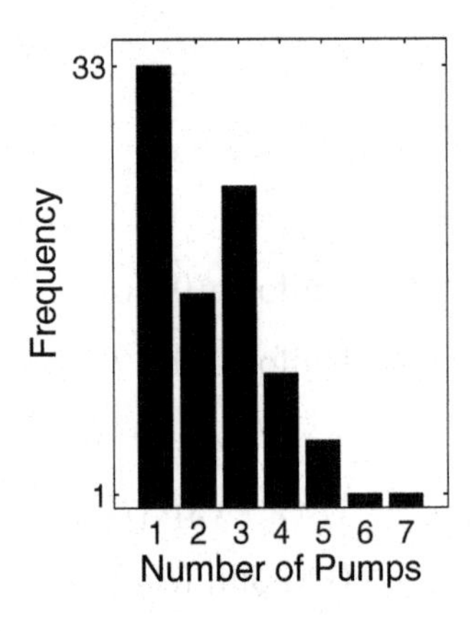

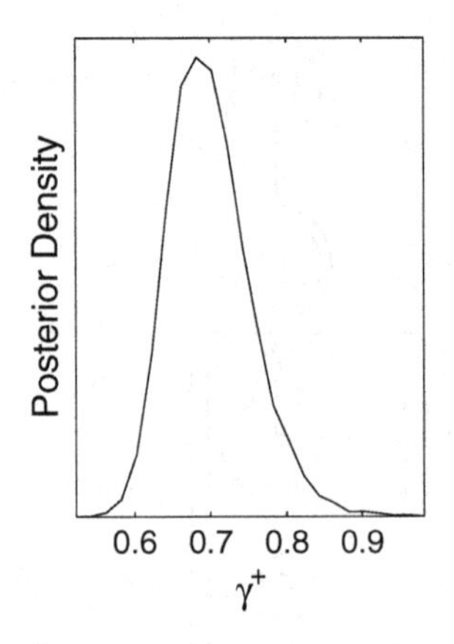

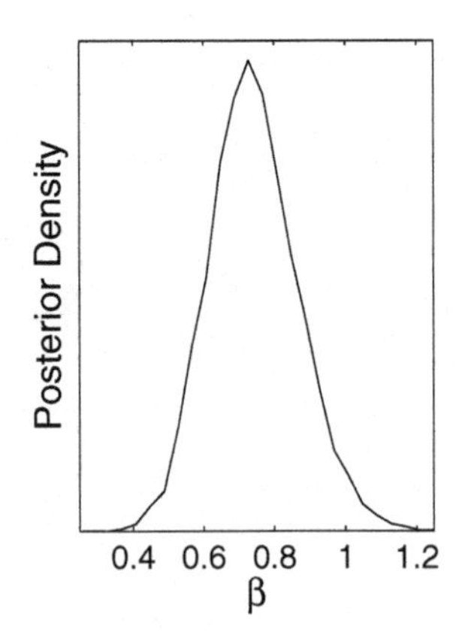

Fig. 16.3 The number of pump decisions (left panel), posterior distribution for risk propensity γ^+ (middle panel), and posterior distribution for behavioral consistency β (right panel) for subject George.

```
# BART Model of Risky Decision-Making
model{
  # Optimal Number of Pumps
  omega <- -gplus/log(1-p)
  # Choice Data
  for (j in 1:ntrials){
    for (k in 1:options[j]){
      theta[j,k] <- 1-(1/(1+max(-15,min(15,exp(beta*(k-omega))))))
      d[j,k] ~ dbern(theta[j,k])
    }
  }
  # Priors
  gplus ~ dunif(0,10)
  beta ~ dunif(0,10)
}
```

The code BART_1.m or BART_1.R applies the model to data from a single subject, known as George, provided in the file GeorgeSober.txt. Figure 16.3 shows the results that are produced. The left panel shows the empirical distribution of the number of pump decisions across all trials. The middle panel shows the posterior distribution for γ^+, George's propensity for risk. The right panel shows the posterior distribution for β, George's behavioral consistency.

Exercises

Exercise 16.1.1 Apply the model to data from a different subject, Bill, provided in the file BillSober.txt. Compare the estimated parameters for George and Bill. Who has the greater propensity for risk?

Exercise 16.1.2 What happens if two pumps are added to each trial for George's data? Make this change to the npumps variable in Matlab or R, and examine the new results. Which of the two parameters changed the most?

Box 16.1 Merit in application

"The merits of any statistical method are determined by the results it gives when applied to specific problems." (Jaynes, 1976, p. 178)

Exercise 16.1.3 Modify George's data in a different way to affect the behavioral consistency parameter.

16.2 A hierarchical extension of the BART model

Alcohol abuse can stimulate risk taking behavior. For example, alcohol abuse has been found to increase risk taking during driving (e.g., Burian, Liguori, & Robinson, 2002) and to increase participation in unsafe sex (e.g., McEwan, McCallum, Bhopal, & Madhok, 1992; but see Leigh & Stall, 1993). Examining performance on the BART, Lejuez et al. (2002) found that people with potentially problematic drinking habits also took more risks blowing up the balloons.

To examine the effects of alcohol on risk taking behavior more systematically, van Ravenzwaaij et al. (2011) carried out a within-subject manipulation, in which each subject completed the BART while sober, tipsy, and drunk. For a man weighing 70 kg, the blood alcohol concentration level for "drunk" required the consumption of 180 ml of vodka. We analyze here a subset of the data using a hierarchical extension of the individual two-parameter model.

The hierarchical extension is straightforward and requires only group-level distributions for the risk taking parameter γ^+ and behavioral consistency parameter β. We assume these parameters are drawn from Gaussian distributions, but are constrained to be positive. The graphical model now contains an extra plate that corresponds to the different levels or conditions of intoxication, and is shown in Figure 16.4.

The script `BART_2.txt` implements the graphical model in WinBUGS:

```
# Hierarchical BART Model of Risky Decision-Making
model{
  # Choice Data
  for (i in 1:nconds){
    gplus[i] ~ dnorm(mug,lambdag)I(0,)
    beta[i] ~ dnorm(mub,lambdab)I(0,)
    omega[i] <- -gplus[i]/log(1-p)
    for (j in 1:ntrials){
      for (k in 1:options[i,j]){
        theta[i,j,k] <- 1-(1/(1+max(-15,min(15,exp(beta[i]*(k-omega[i]))))))
        d[i,j,k] ~ dbern(theta[i,j,k])
      }
    }
  }
```

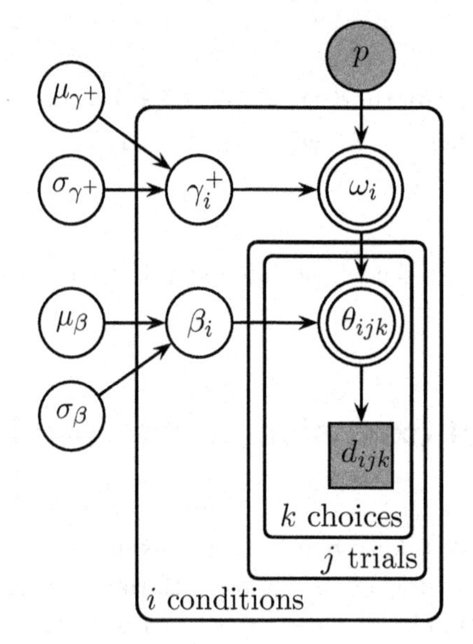

$$
\begin{aligned}
\mu_{\gamma^+} &\sim \text{Uniform}(0,10)\\
\sigma_{\gamma^+} &\sim \text{Uniform}(0,10)\\
\mu_{\beta} &\sim \text{Uniform}(0,10)\\
\sigma_{\beta} &\sim \text{Uniform}(0,10)\\
\gamma_i^+ &\sim \text{Gaussian}(\mu_{\gamma^+}, 1/\sigma_{\gamma^+}^2)\\
\beta_i &\sim \text{Gaussian}(\mu_{\beta}, 1/\sigma_{\beta}^2)\\
\omega_i &\leftarrow -\gamma_i^+/\log(1-p)\\
\theta_{ijk} &\leftarrow 1/\left(1+\exp\left\{\beta_i\left(k-\omega_i\right)\right\}\right)\\
d_{ijk} &\sim \text{Bernoulli}(\theta_{ijk})
\end{aligned}
$$

Fig. 16.4 Graphical model for the hierarchical two-parameter BART model.

```
  # Priors
  mug ~ dunif(0,10)
  sigmag ~ dunif(0,10)
  mub ~ dunif(0,10)
  sigmab ~ dunif(0,10)
  lambdag <- 1/pow(sigmag,2)
  lambdab <- 1/pow(sigmab,2)
}
```

The code `BART_2.m` or `BART_2.R` applies the model to George's data under different levels of intoxication, provided in the files `GeorgeSober.txt`, `GeorgeTipsy.txt`, and `GeorgeDrunk.txt`. It produces an analysis like that in Figure 16.5, showing the empirical distribution of the number of pumps, and posterior distributions for the two parameters, for all three intoxication conditions.

Exercises

Exercise 16.2.1 Apply the model to the data from the other subject, Bill. Does alcohol have the same effect on Bill as it did on George?

Exercise 16.2.2 Apply the non-hierarchical model in Figure 16.2 to each of the six data files independently. Compare the results for the two parameters to those obtained from the hierarchical model, and explain any differences.

Exercise 16.2.3 The hierarchical model in Figure 16.4 provides a structured relationship between the drinking conditions, but is still applied independently to each subject. Many of the applications of hierarchical modeling considered in our case studies, however, involve structured relationships between subjects, to capture individual differences. Develop a graphical model that extends Fig-

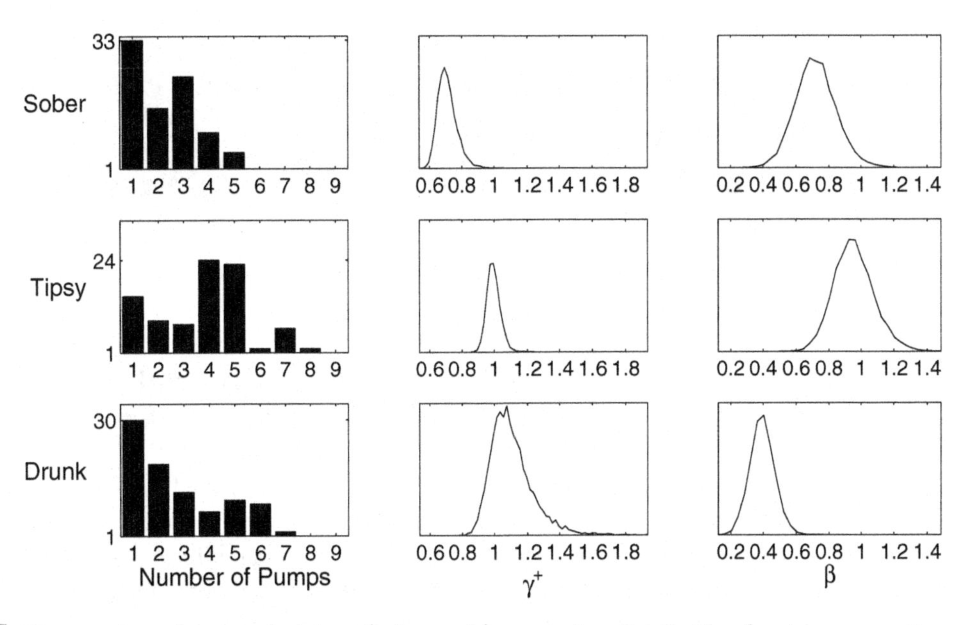

Fig. 16.5 The number of pump decisions (left panels), posterior distribution for risk propensity γ^+ (middle panels), and posterior distribution for behavioral consistency β (right panels) for subject George, in the sober (top row), tipsy (middle row), and drunk (bottom row) conditions.

ure 16.4 to incorporate hierarchical structure both for drinking conditions and subjects. How could interactions between these two factors be modeled?

17 The GCM model of categorization

WITH RUUD WETZELS

17.1 The GCM model

The Generalized Context Model (GCM: Nosofsky, 1984, 1986) is an influential and empirically successful model of categorization. It is intended to explain how people make categorization decisions in a task where stimuli are presented, one at a time, over a sequence of trials, and must be classified into one of a small number of categories (usually two) based on corrective feedback.

The GCM assumes that stimuli are stored as exemplars, using their values along underlying stimulus dimensions, which correspond to points in a multidimensional psychological space. The GCM then assumes people make similarity comparisons between the current stimulus and the exemplars, and base their decision on the overall similarities to each category.

A key theoretical component of the GCM involves selective attention. The basic idea is that, to learn a category structure, people selectively attend to those dimensions of the stimuli that are relevant to distinguishing the categories. Nosofsky (1984) showed that selective attention could help explain previously puzzling empirical regularities in the ease with which people learn different category structures (Shepard, Hovland, & Jenkins, 1961).

We consider category learning data from the "Condensation B" condition reported by Kruschke (1993).[1] This condition is shown in Figure 17.1, and involves eight stimuli—consisting of line drawings of boxes with different heights, with an interior line in different positions—divided into two groups of four, to make Category A and Category B stimuli. Kruschke (1993) collected data from 40 participants over 8 consecutive blocks of trials, within which each stimulus was presented once in a random order. These data can be summarized by y_{ik}, the number of times the ith stimulus was categorized as belonging to Category A by the kth participant, out of the $t = 8$ trials on which it was presented. In an analysis that does not consider individual differences, the data can be further summarized as $y_i = \sum_k y_{ik}$, the total number of times all participants categorized the ith stimulus into Category A, out of $t = 40 \times 8$ total presentations.

[1] The Kruschke (1993) category learning experiments involved corrective feedback after every trial. Usually, the GCM is not applied to this sort of task, but to categorization decisions made without feedback after a training period involving feedback.

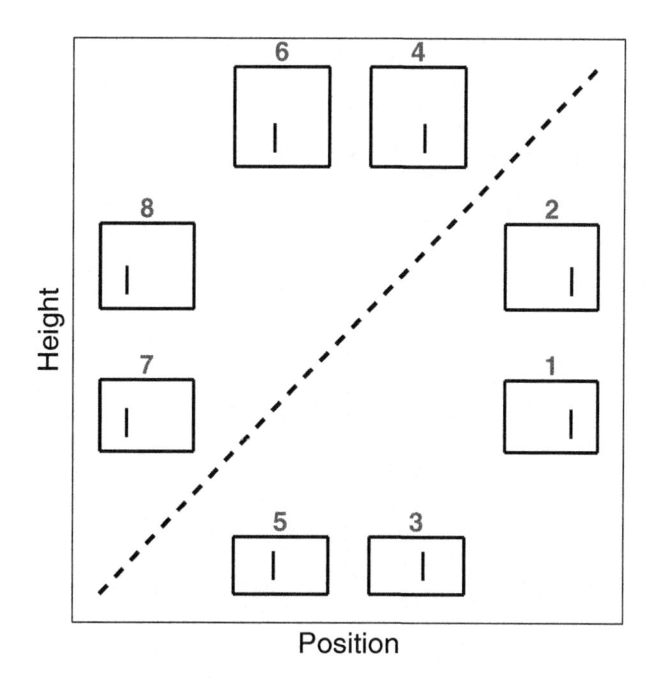

Fig. 17.1 Condensation category structure "B" from Kruschke (1993). Stimuli above the diagonal belong to category B and stimuli below the diagonal belong to category A.

A graphical model representation of the GCM, as applied to the group data, is shown in Figure 17.2. There are two stimulus dimensions, so the ith stimulus is represented by the point (p_{i1}, p_{i2}). The first dimension has attention weight w with $0 \leq w_d \leq 1$, and the second dimension then has attention weight $(1 - w)$. These weights act to "stretch" attended dimensions, and "shrink" unattended ones, so that the psychological distance between the ith and jth stimuli is $d_{ij} = w\,|p_{i1} - p_{j1}| + (1 - w)\,|p_{i2} - p_{j2}|$. The similarity between the ith and jth stimuli is $s_{ij} = \exp(-cd_{ij})$, where c is a generalization parameter. The overall similarity of the ith stimulus, when it is presented, to Category A is $s_{iA} = \sum_{j \in A} s_{ij}$. The graphical model uses these in an unbiased choice rule, so that the probability that the ith stimulus will be classified as belonging to Category A, rather than Category B, is $r_i = bs_{iA}/\,(bs_{iA} + (1 - b)\,s_{iB})$, with $b = 0.5$. The observed decisions themselves are then given by $y_i \sim \text{Binomial}\,(r_i, t)$.

The script `GCM_1.txt` implements the graphical model in WinBUGS. Note that it collects posterior predictive samples for the decision data in the variable `predy`:

```
# Generalized Context Model
model{
  # Decision Data
  for (i in 1:nstim){
    y[i] ~ dbin(r[i],t)
    predy[i] ~ dbin(r[i],t)
  }
```

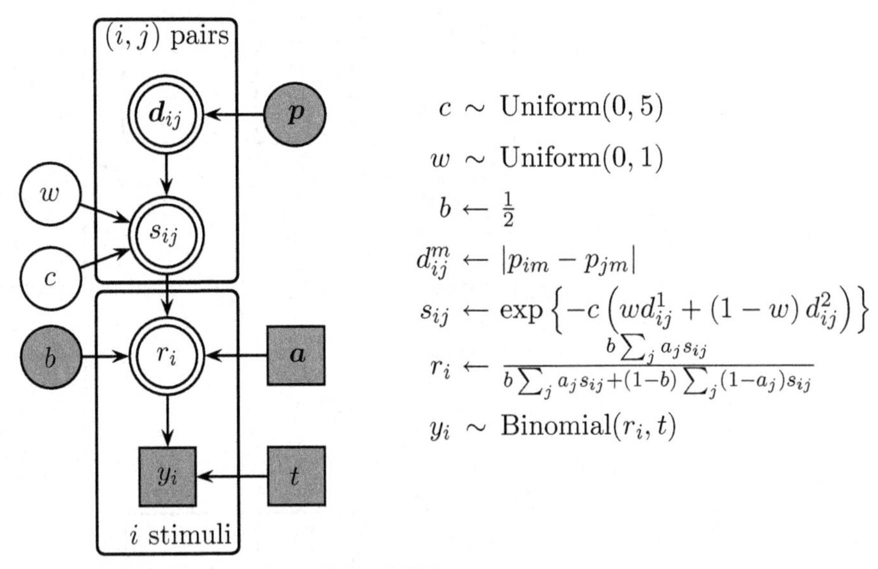

Fig. 17.2 Graphical model implementation of the GCM.

```
  # Decision Probabilities
  for (i in 1:nstim){
    r[i] <- sum(numerator[i,])/sum(denominator[i,])
    for (j in 1:nstim){
      tmp1[i,j,1] <- b*s[i,j]
      tmp1[i,j,2] <- 0
      tmp2[i,j,1] <- 0
      tmp2[i,j,2] <- (1-b)*s[i,j]
      numerator[i,j] <- tmp1[i,j,a[j]]
      denominator[i,j] <- tmp1[i,j,a[j]] + tmp2[i,j,a[j]]
    }
  }
  # Similarities
  for (i in 1:nstim){
    for (j in 1:nstim){
      s[i,j] <- exp(-c*(w*d1[i,j]+(1-w)*d2[i,j]))
    }
  }
  # Priors
  c  ~ dunif(0,5)
  w  ~ dbeta(1,1)
  b <- 0.5
}
```

The code `GCM_1.m` or `GCM_1.R` applies the model to the Kruschke (1993) data. The joint posterior distribution of the generalization parameter c and the attention weight parameter w is shown as a scatter-plot in Figure 17.3. The key result, in terms of the theory of category learning and selective attention, is that the attention parameter w lies between about 0.5 and 0.7. This can be interpreted as showing that

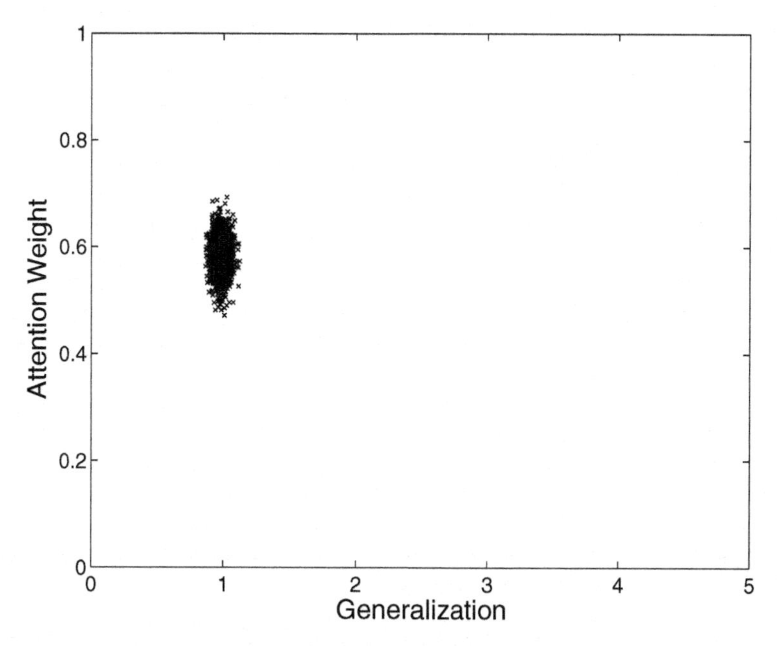

Fig. 17.3 Joint posterior distribution over attention w and generalization c parameters of the GCM, when applied to the Kruschke condensation data.

people give significant attention to both dimensions, although they are probably focusing a little more on the line position than the rectangle height.

This is consistent with the condensation task design of the category structure. It is clear from Figure 17.1 that both dimensions are relevant in determining to which category a stimulus belongs, and so the shared attention result makes sense.

Exercises

Exercise 17.1.1 Setting $b = 0.5$ to make the decision rule unbiased seems reasonable, since there are two alternatives with equal numbers of equally-often-presented stimuli. But, the assumption can be easily examined. Change the model so that the bias parameter b is given a uniform prior over the range 0 to 1, and is inferred from the data. Summarize the findings from this model, and compare them to the results from the original model.

Exercise 17.1.2 Figure 17.4 shows a posterior predictive analysis of the modeling. The average y_i counts are shown for each of the 8 stimuli, overlaid on gray violin plots showing the posterior predictive distributions. Also shown, by the broken lines, are individual participant data. These are linearly scaled from the individual count of 8, to the group count of 320, to allow visual comparison of the number of times. From this figure, what do you conclude about the ability of the GCM to describe the group data? What do you conclude about the adequacy of the group data as a summary of human performance?

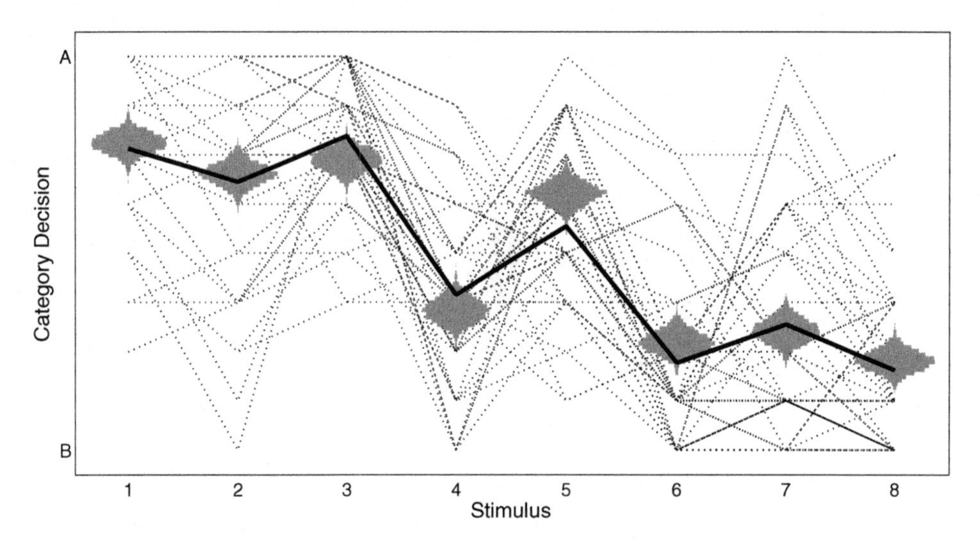

Fig. 17.4 Posterior predictive analysis of the GCM applied to the condensation data.

17.2 Individual differences in the GCM

Figure 17.4 suggests that there are significant individual differences in the categorization data. Figure 17.5 shows the individual data more clearly, with each subject's decisions about each stimulus presented in a separate panel.

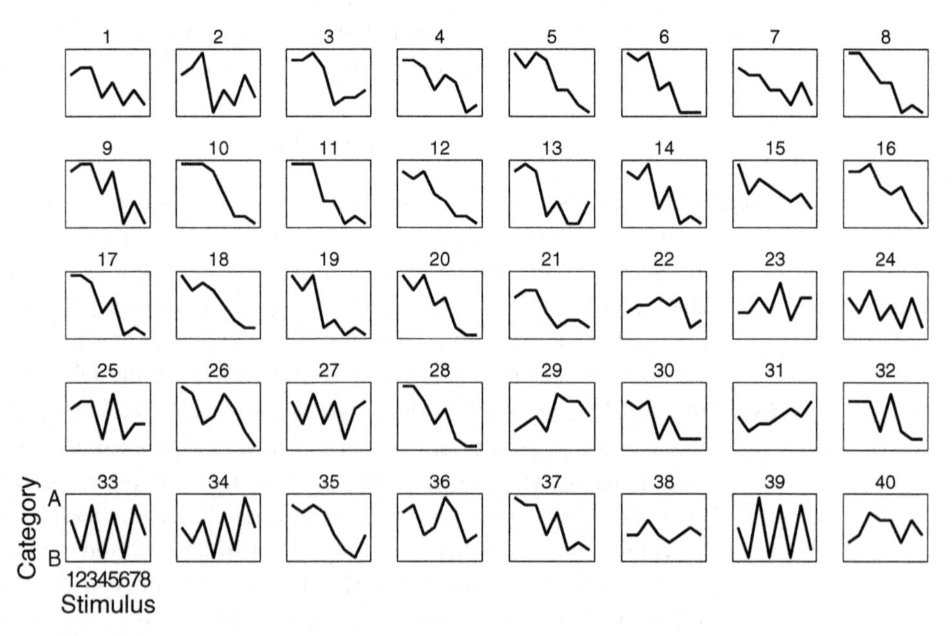

Fig. 17.5 Category decision data for all 40 subjects.

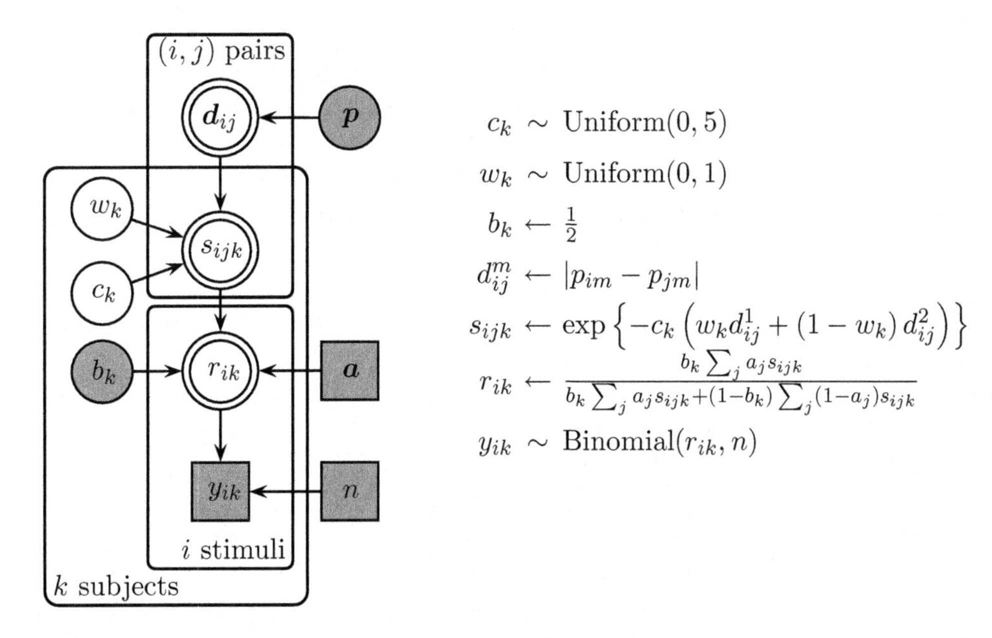

Fig. 17.6 Graphical model implementation of the GCM allowing for full individual differences in attention and generalization.

The simplest way to start to investigate these differences using the GCM is to infer parameter values for each subject independently. A graphical model to do this is presented in Figure 17.6. It simply allows the kth subject to have their own generalization and attention parameters c_k and w_k, and adds a plate replicating those parts of the GCM that depend upon the parameters. Notice that the extension to individual differences means the individual subject data, y_{ik}, for the ith stimulus and kth subject are now being modeled, rather than the summary data y_i.

The script GCM_2.txt implements the graphical model in WinBUGS:

```
# Generalized Context Model With Individual Differences
model{
  # Decision Data
  for (i in 1:nstim){
    for (k in 1:nsubj){
      y[i,k] ~ dbin(r[i,k],n)
      predy[i,k] ~ dbin(r[i,k],n)
    }
  }
  # Decision Probabilities
  for (i in 1:nstim){
    for (k in 1:nsubj){
      r[i,k] <- sum(numerator[i,k,])/sum(denominator[i,k,])
    }
  }
  # Base Decision Probabilities
  for (i in 1:nstim){
    for (j in 1:nstim){
```

```
      for (k in 1:nsubj){
        numerator[i,k,j] <- equals(a[j],1)*b*s[i,k,j]
        denominator[i,k,j] <- equals(a[j],1)*b*s[i,k,j]
                              + equals(a[j],2)*(1-b)*s[i,k,j]
      }
    }
  }
  # Similarities
  for (i in 1:nstim){
    for (j in 1:nstim){
      for (k in 1:nsubj){
        s[i,k,j] <- exp(-c[k]*(w[k]*d1[i,j]+(1-w[k])*d2[i,j]))
      }
    }
  }
  # Parameters and Priors
  for (k in 1:nsubj){
    c[k] ~ dunif(0,5)
    w[k] ~ dbeta(1,1)
  }
  b <- 0.5
}
```

The code `GCM_2.m` or `GCM_2.R` applies the model to the Kruschke (1993) data. The joint posterior distributions of the generalization parameter c and the attention weight parameter w, for all 40 subjects, are represented in Figure 17.7. The filled circles show the posterior mean for each individual subject, and the lines radiating from these points connect the mean to a small number of randomly selected samples of the joint posterior for that subject.

Exercises

Exercise 17.2.1 Compare the inferences about the attention parameter based on an individual subject analysis in Figure 17.7 with those based on a no individual differences analysis in Figure 17.3. Give a psychological interpretation of these differences, in terms of how the subjects selectively attended to the stimulus dimensions.

Exercise 17.2.2 Three of the individual subject joint posterior distributions—for Subjects 3, 31, and 33—in Figure 17.7 are labeled. These three subjects lie in different areas of the parameter space, and so possibly correspond to different sorts of categorization behavior. Look at the individual data in Figure 17.5 for these three subjects, and give a short description of the differences in their categorization decisions.

17.3 Latent groups in the GCM

The individual differences displayed in terms of basic behavior in Figure 17.5 and through inference about GCM parameters in Figure 17.7 provide information that

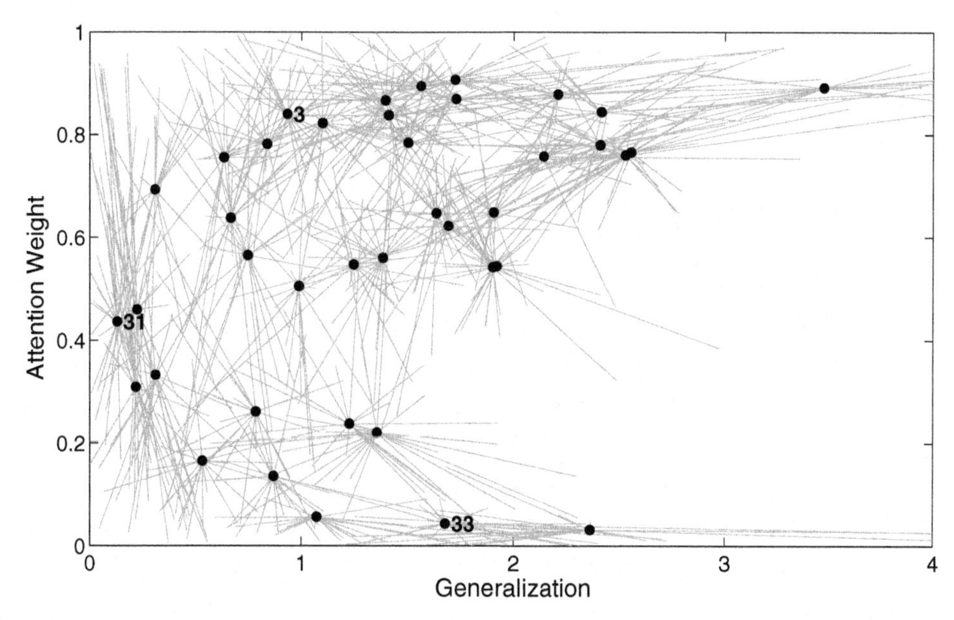

Fig. 17.7 Posterior distributions for each subject, for the attention w and generalization c parameters, inferred by the GCM with full individual differences.

can help formulate more constrained models of the variation in category learning in this data set. It is not the case that there is one correct model of these differences but, rather, that some models will be able to describe the variability better than others, and provide more useful insights into the underlying psychology.

One worthwhile model assumes there are three groups of subjects, and is shown as a graphical model in Figure 17.8. This model is based on the insight that there are three types of behavior, well exemplified by Subjects 3, 31, and 33. In the first group, subjects like Subject 31 seem to categorize each stimulus roughly equally often in each category, consistent with guessing behavior. These subjects might be interpreted as being contaminants, who are not attempting to do the task diligently. These subjects have GCM parameter posterior distributions with low generalization values, and large uncertainty about their attention values, as seen in Figure 17.7. This makes sense, since low values of the generalization parameter will lead to high similarity between all stimuli, and near-equal summed similarity for both category responses in this task. A much simpler way to account for the behavior of subjects like Subject 31, however, is not to use the GCM to account for their behavior at all, but to use a separate contaminant model that just sets their response probabilities to be $r_{ik} = \frac{1}{2}$ for all stimuli.

The groups of subjects represented by Subjects 3 and 33, meanwhile, show more thoughtful categorization behavior, but have different patterns of responding to some stimuli. Recall from Figure 17.1 that stimuli 1, 2, 3, and 5 belong to category A, and stimuli 4, 6, 7, and 8 belong to category B. Subjects like Subject 3 make

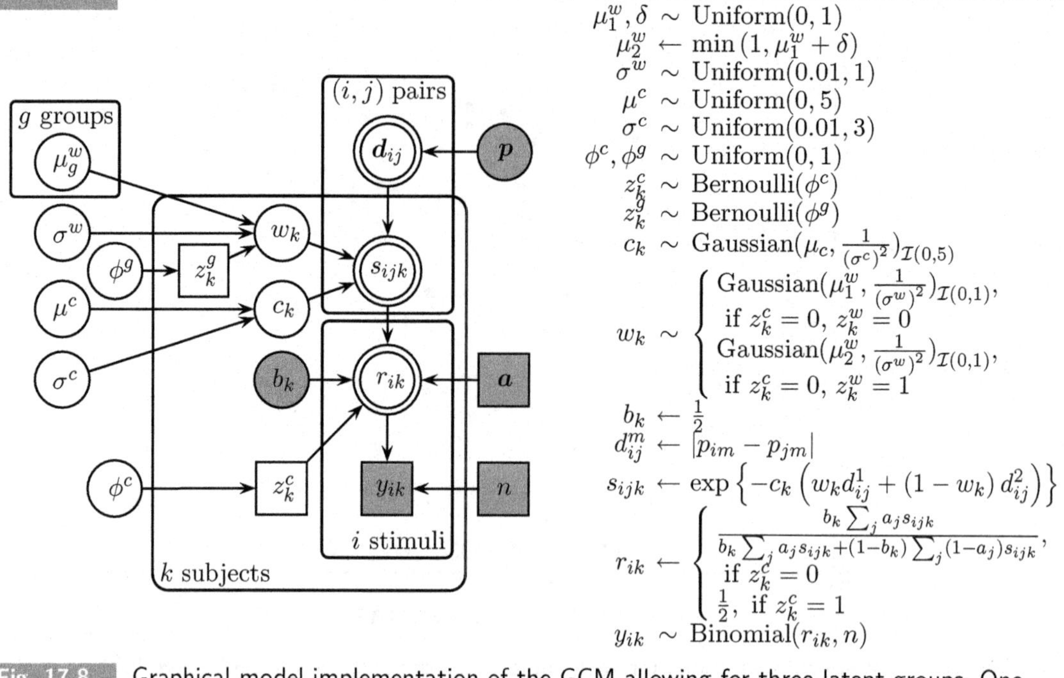

Fig. 17.8 Graphical model implementation of the GCM allowing for three latent groups. One group of subjects are contaminants, the other two groups use the GCM, but with different selective attention.

almost all of their errors on stimuli 4 and 5. This is consistent with a focus on the position dimension, corresponding to large values of the w attention parameter, which would assign these two stimuli incorrectly. Subjects like Subject 33, however, categorize stimuli 2 and 7 poorly. This is consistent with a focus on the height dimension, corresponding to low values of the attention parameter w.

The script `GCM_3.txt` implements the graphical model in WinBUGS. Note that posterior predictive distributions are generated at the group level, and that a three-valued classification variable **z** is constructed showing the group membership of each subject:

```
# Generalized Context Model With Contaminants and Two Attention Groups
model{
  # Decision Data
  for (i in 1:nstim){
    # Subjects
    for (k in 1:nsubj){
      y[i,k] ~ dbin(r[i,k],n)
    }
    # Groups
    for (g in 1:3){
      predyg[g,i] ~ dbin(rpredg[g,i],n)
    }
  }
  # Decision Probabilities
  for (i in 1:nstim){
    for (k in 1:nsubj){
```

```
    r[i,k] <- equals(zc[k],0)*sum(numerator[i,k,])/sum(denominator[i,k,])
              + equals(zc[k],1)*0.5
  }
  for (g in 1:2){
    rpredg[g,i] <- sum(numeratorpredg[g,i,])/sum(denominatorpredg[g,i,])
  }
  rpredg[3,i] <- 0.5
}
# Base Decision Probabilities
for (i in 1:nstim){
  for (j in 1:nstim){
    for (k in 1:nsubj){
      numerator[i,k,j] <- equals(a[j],1)*b*s[i,k,j]
      denominator[i,k,j] <- equals(a[j],1)*b*s[i,k,j]
                            + equals(a[j],2)*(1-b)*s[i,k,j]
    }
    for (g in 1:2){
      numeratorpredg[g,i,j] <- equals(a[j],1)*b*spredg[g,i,j]
      denominatorpredg[g,i,j] <- equals(a[j],1)*b*spredg[g,i,j]
                                 + equals(a[j],2)*(1-b)*spredg[g,i,j]
    }
  }
}
# Similarities
for (i in 1:nstim){
  for (j in 1:nstim) {
    for (k in 1:nsubj){
      s[i,k,j] <- exp(-c[k]*(w[k]*d1[i,j]+(1-w[k])*d2[i,j]))
    }
    for (g in 1:2){
      spredg[g,i,j] <- exp(-cpredg[g]*(wpredg[g]*d1[i,j]+(1-wpredg[g])*d2[i,j]))
    }
  }
}
# Subject Parameters
for (k in 1:nsubj){
  c[k] ~ dnorm(muc,lambdac)I(0,)
  w[k] ~ dnorm(muw[zg1[k]],lambdaw)I(0,1)
}
# Predicted Group Parameters
for (g in 1:2){
  wpredg[g] ~ dnorm(muw[g],lambdaw)I(0,1)
  cpredg[g] ~ dnorm(muc,lambdac)I(0,)
}
# Priors
b <- 0.5
# Latent Mixture
phic ~ dbeta(1,1)
phig ~ dbeta(1,1)
for (k in 1:nsubj){
  zc[k] ~ dbern(phic)
  zg[k] ~ dbern(phig)
  zg1[k] <- zg[k]+1
  z[k] <- equals(zc[k],0)*zg1[k] + 3*equals(zc[k],1)
}
# Mean Generalization
```

Box 17.1 Enduring knowledge

"[T]he most substantial and enduring advances [in cognitive science] have not been in the accumulation of empirical facts or the construction of models, but in the production of fruitful interactions between models and experimental research. Most experimental facts require continual reinterpretation and most models drop by the wayside like autumn leaves, but the results of interactions between models and experiments constitute most of our generalizable knowledge." (Estes, 2002, p. 3)

```
  muctmp ~ dbeta(1,1)
  muc <- 5*muctmp
  # Mean Attention
  muwtmp ~ dbeta(1,1)
  muw[1] <- muwtmp
  delta ~ dbeta(1,1)
  muw[2] <- min(1,delta+muw[1])
  # Standard Deviation Generalization
  sigmactmp ~ dbeta(1,1)
  sigmac <- max(.01,3*sigmactmp)
  # Standard Deviation Attention
  sigmawtmp ~ dbeta(1,1)
  sigmaw <- max(.01,sigmawtmp)
  # Precision
  lambdac <- 1/pow(sigmac,2)
  lambdaw <- 1/pow(sigmaw,2)
}
```

The code `GCM_3.m` or `GCM_3.R` applies the model to the Kruschke (1993) data, and generates two graphs that analyze the key results. The first of these is presented in Figure 17.9, and shows the latent assignment probabilities of each subject into the three possible groups. It is clear that almost all of the subjects are confidently classified into one of the three groups. This is a strong indication that the groups proposed by the model are useful characterizations of the data at the level of individual subjects. If a single group was not able to account well for individual data, the natural outcome of Bayesian inference is model averaging, in which a blend of different groups is used to describe the data, and there is significant posterior mass for more than one assignment.

Figure 17.10 shows a posterior predictive analysis for each of the three groups. Each panel corresponds to a group, with the squares showing the predictive distribution of the model. The thick lines show the average categorization behavior for subjects classified as belonging to the group, as determined from the mode of their posterior group membership shown in Figure 17.9. The thin lines show the individual behavior of these subjects. For each of the three groups, the qualitatively different pattern of categorization behavior seems well described by the posterior predictive distribution.

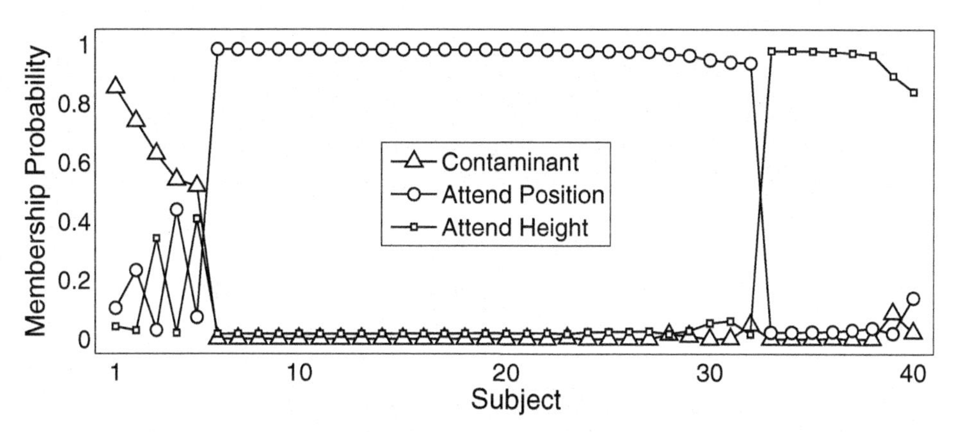

Fig. 17.9 The probability of assignment of each subject. Note that subjects have been ordered to make the information easy to parse, and this ordering is different from the subject ordering in the raw data, as shown in Figure 17.5.

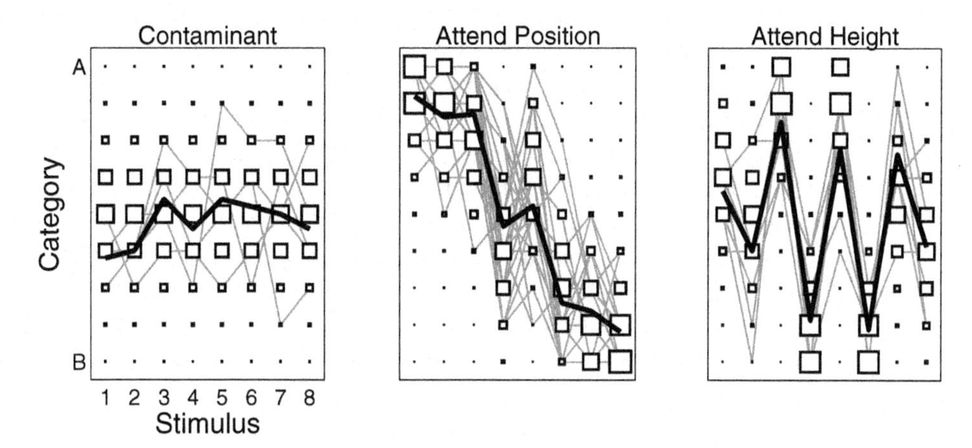

Fig. 17.10 Posterior predictive distributions for the latent, attend position and attend height groups, shown by squares, and the average and individual category decisions for subjects assigned to each group, shown by thick and thin lines, respectively.

Exercises

Exercise 17.3.1 The analyses presented focus on the inferred group membership, and the posterior predictive distributions for each group. These might be two of the most important inferences, but they are not the only available or useful ones. Extend the analysis by considering the posterior distributions of the inferred proportion of contaminant subjects ϕ^c, and the difference in group mean attention δ, giving psychological interpretations for both.

Exercise 17.3.2 Construct the posterior distribution for the probability that a subject is in the attend position group rather than the attend height group. This is not simply the posterior for the ϕ^g parameter.

18 Heuristic decision-making

18.1 Take-the-best

The take-the-best (TTB) model of decision-making (Gigerenzer & Goldstein, 1996) is a simple but influential account of how people choose between two stimuli on some criterion, and a good example of the general class of heuristic decision-making models (e.g., Gigerenzer & Todd, 1999; Gigerenzer & Gaissmaier, 2011; Payne, Bettman, & Johnson, 1990). TTB addresses decision tasks like "which of Frankfurt or Munich has the larger population?", "which of a catfish and a herring is more fertile?", and "which of these two professors has the higher salary?".

TTB assumes that all stimuli are represented in terms of the presence or absence of a common set of cues. In the well-studied German cities data set, this means cities are represented in terms of nine cues, including whether or not they have an international airport, whether they have hosted the Olympics, and whether they have a football team in the Bundesliga. Associated with each cue in TTB is a "cue validity." This validity measures the proportion of times that, for those pairs of stimuli where one has the cue and the other does not, the cue belongs to the stimulus that has the greater criterion value. For example, the cue "Is the city the national capital?" is highly valid because the capital city, Berlin, is also the most populous city.

The TTB model assumes that, when people decide which is the larger of two cities, they search the cues from highest to lowest validity, stopping as soon as a cue is found that one city has but the other does not. At this point, TTB says simply that people choose the city that has the cue. If all of the cues are exhausted, TTB assumes people guess.

The data we consider involve 20 subjects choosing between pairs of stimuli for the same 30 questions.[1] On each trial, a subject could search cues to find out whether or not the two stimuli had that cue. At any point during this search, the subject could choose one of the stimuli, on the basis of the available information. Thus, the experiment measures how people search for information, when they decide to stop that search, and what decision they make. We focus solely on the decision data collected during the experiment, with the goal of modeling the decisions made, and inferring from those decisions something about how people searched, and how they terminated their search.

[1] We thank Ben Newell for sharing these data.

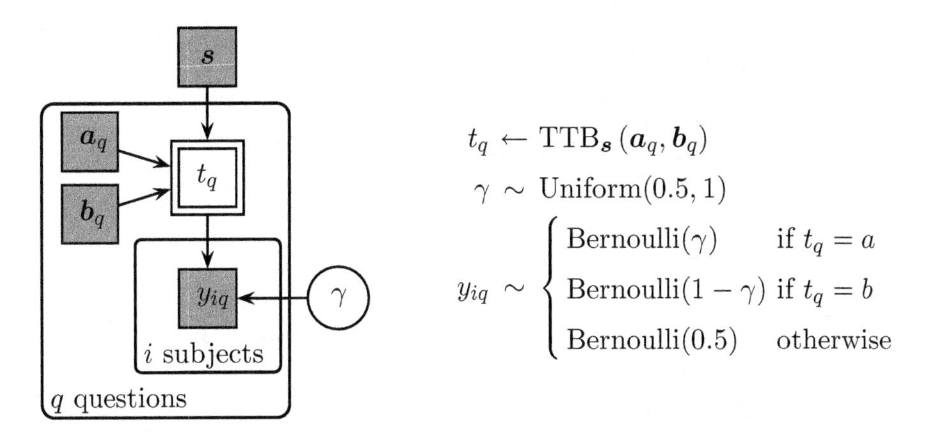

Fig. 18.1 Graphical model for the take-the-best model of heuristic decision-making.

A graphical model for implementing TTB is shown in Figure 18.1. The decision made by the ith subject on the qth question is $y_{iq} = 1$ if the first stimulus (stimulus 'a') is chosen, and $y_{ij} = 0$ if the second stimulus (stimulus 'b') is chosen. The cues for the stimuli in the jth question are given by the vectors $\boldsymbol{a}_q$ and $\boldsymbol{b}_q$, and the validity-based search order is given by $\boldsymbol{s}$.

To make the inherently deterministic TTB model a probabilistic model, a "responding rate" or "accuracy of execution" parameter γ is included, so that TTB decisions are followed with probability γ (Rieskamp, 2008). Thus, the TTB decision for the qth question is given by t_q, taking binary values corresponding to the choices "a" or "b." We can write this $t_q \leftarrow \text{TTB}_{\boldsymbol{s}}(\boldsymbol{a}_q, \boldsymbol{b}_q)$. The observed human decisions are distributed as

$$y_{iq} \sim \begin{cases} \text{Bernoulli}(\gamma) & \text{if } t_q = a \\ \text{Bernoulli}(1-\gamma) & \text{if } t_q = b \\ \text{Bernoulli}(0.5) & \text{otherwise.} \end{cases}$$

The probability TTB is followed is expected to be high, and certainly should not be below 0.5, which would reverse the decisions. Thus, the prior $\gamma \sim \text{Uniform}(0.5, 1)$ is used.

The script `TTB.txt` implements the graphical model in WinBUGS:

```
# Take The Best
model{
  # Data
  for (q in 1:nq){
    for (i in 1:ns){
      y[i,q] ~ dbern(ttb[t[q]])
      ypred[i,q] ~ dbern(ttb[t[q]])
    }
  }
  # TTB Model For Each Question
  for (q in 1:nq){
    # Add Cue Contributions To Mimic TTB Decision
```

```
      for (j in 1:nc){
        tmp1[q,j] <- (m[p[q,1],j]-m[p[q,2],j])*pow(2,s[j]-1)
      }
      # Find if Cue Favors First, Second, or Neither Stimulus
      tmp2[q] <- sum(tmp1[q,1:nc])
      tmp3[q] <- -1*step(-tmp2[q])+step(tmp2[q])
      t[q] <- tmp3[q]+2
    }
    # Cue Search Order Follows Validities
    for (j in 1:nc){
      s[j] <- rank(v[1:nc],j)
    }
    # Choose TTB Decision With Probability Gamma, or Guess
    ttb[1] <- 1-gamma
    ttb[2] <- 0.5
    ttb[3] <- gamma
    # Priors
    gamma ~ dunif(0.5,1)
}
```

The script simulates the TTB model by comparing all of the cues, but applying non-compensatory weights (the `pow(2,s[j]-1)` multiple) so that the first discriminating cue determines the decision. Note that the script determines the TTB validity-based search order from cues using the `rank` function.

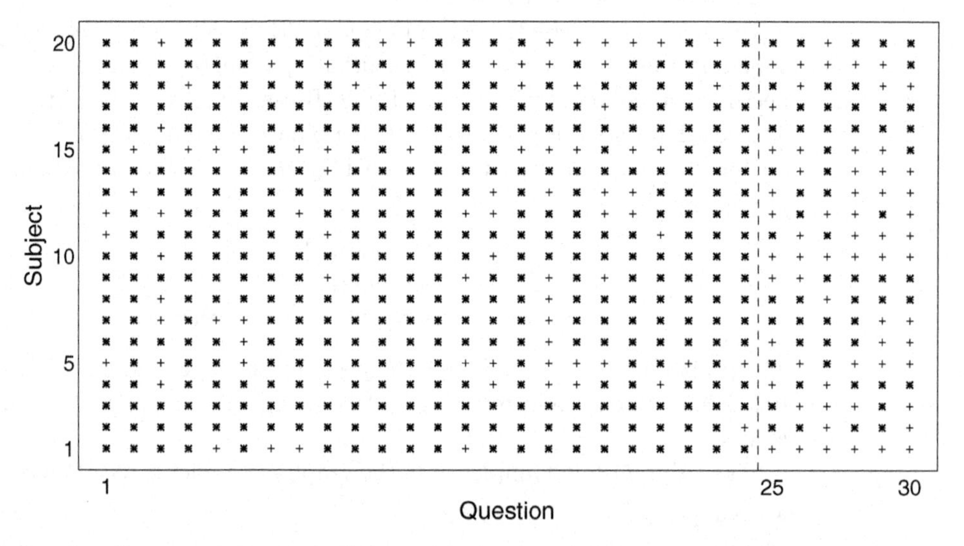

Fig. 18.2 Results of applying the probabilistic TTB model to the decision data.

The code `TTB.m` or `TTB.R` applies the model to the simulated data. Figure 18.2 summarizes the results, showing the agreement between subject decisions, and those produced by the TTB model. The x marker indicates that a subject chose stimulus "a," while the absence of a x marker indicates that they chose "b." A + marker is used to indicate the probability of the model choosing stimulus 'a." A full-size + marker means the model chooses "a" with probability one, while the complete absence of a + marker means the model choses "b" with probability one. Between these two extremes, the size of the + marker represents the probability of "a" being chosen by the model. Thus, agreement between the data and model is visually evident in a * marker, which is the combination of x marker and + marker, showing that both the subject and model chose "a," or the absence of any markers, showing that both chose "b."

The questions in Figure 18.2 are ordered, so that the first 24 are those in which the TTB model, and a more exhaustive model that examines all of the cues, lead to the same decision. The final 6 questions, however—separated by a broken line from the first 24—involve stimulus pairs where TTB makes a different decision. In the presentation of results, but not the presentation to subjects in the experiment, the stimulus pairs are also coded so that choice "a" always corresponds to the TTB choice.

Figure 18.2 shows, as expected, that the model always favors the TTB option. Especially for the first 24 questions, the subjects also often choose the TTB option, and the expected posterior agreement between the model and data over all decisions is calculated by the code as 63%. This is a measure of the descriptive adequacy of the model, and is often called "correspondence" in the judgment and decision-making literature (Dunwoody, 2009).

Exercise

Exercise 18.1.1 What is the posterior expectation of γ, and how can it be interpreted?

18.2 Stopping

A strong assumption of the TTB heuristic is that one discriminating cue is sufficient to trigger a decision. There is debate about this stopping rule, and heuristics that rely on one reason are often contrasted against an alternative model that assumes all of the cues are searched (e.g., Gigerenzer & Goldstein, 1996; Katsikopoulos, Schooler, & Hertwig, 2010).

A common comparison (e.g., Bergert & Nosofsky, 2007; Lee & Cummins, 2004) is between TTB and a model often called the Weighted ADDitive (WADD) model, which sums the evidence for both decision alternatives over all available cues, and chooses the one with the greatest evidence. The evidence provided by a discriminating cue can be determined from its validity, and is naturally measured on the log-odds scale as $x_k = \log \frac{v_k}{1-v_k}$. Thus, the WADD decision heuristic sums all the evidence values for cues that discriminate in favor of each alternative, and chooses the one with the most evidence.

A basic question these data were designed to address is whether subjects consistently used TTB or WADD and, if so, what proportion used each. This is naturally

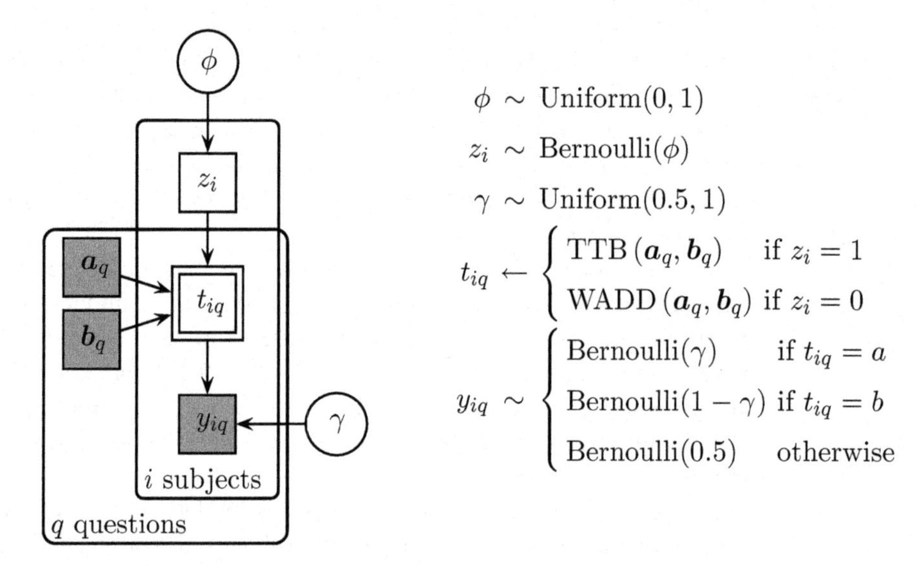

Fig. 18.3 Graphical model for a latent mixture of TTB and WADD stopping rules.

addressed using a latent-mixture model, where each subject either uses TTB or WADD to make decisions on every question. Figure 18.3 shows a graphical model that implements this approach. The z_i parameter functions as an indicator variable, with $z_i = 1$ if the ith subject uses TTB, and $z_i = 0$ if they use WADD. This indicator variable is distributed according to a base-rate of TTB subjects in the population, so that $z_i \sim \text{Bernoulli}(\phi)$. The deterministic node θ_{ij} for the ith subject is then given by

$$t_{iq} \leftarrow \begin{cases} \text{TTB}(\boldsymbol{a}_q, \boldsymbol{b}_q) & \text{if } z_i = 1 \\ \text{WADD}(\boldsymbol{a}_q, \boldsymbol{b}_q) & \text{if } z_i = 0 \end{cases}$$

with, as before

$$y_{iq} \sim \begin{cases} \text{Bernoulli}(\gamma) & \text{if } t_{iq} = a \\ \text{Bernoulli}(1-\gamma) & \text{if } t_{iq} = b \\ \text{Bernoulli}(0.5) & \text{otherwise.} \end{cases}$$

The script `Stop.txt` implements the graphical model in WinBUGS. The basic approach is to find the decision both TTB and WADD makes for each question, and use the indicator `z[i]` to determine how the ith subject makes their decisions:

```
# Stop
model{
  # Data
  for (i in 1:ns){
    for (q in 1:nq){
      y[i,q] ~ dbern(dec[t[i,q,z1[i]]])
      ypred[i,q] ~ dbern(dec[t[i,q,z1[i]]])
    }
  }
  # TTB Decision
  for (i in 1:ns){
```

```
    for (q in 1:nq){
      # Add Cue Contributions To Mimic TTB Decision
      for (j in 1:nc){
        tmp1[i,q,j] <- (m[p[q,1],j]-m[p[q,2],j])*pow(2,s[j]-1)
      }
      # Find if Cue Favors First, Second, or Neither Stimulus
      tmp2[i,q] <- sum(tmp1[i,q,1:nc])
      tmp3[i,q] <- -1*step(-tmp2[i,q])+step(tmp2[i,q])
      t[i,q,1] <- tmp3[i,q]+2
    }
  }
  # WADD Decision
  for (i in 1:ns){
    for (q in 1:nq){
      for (j in 1:nc){
        tmp4[i,q,j] <- (m[p[q,1],j]-m[p[q,2],j])*x[j]
      }
      # Find if Cue Favors First, Second, or Neither Stimulus
      tmp5[i,q] <- sum(tmp4[i,q,1:nc])
      tmp6[i,q] <- -1*step(-tmp5[i,q])+step(tmp5[i,q])
      t[i,q,2] <- tmp6[i,q]+2
    }
  }
  # Follow Decision With Probability Gamma, or Guess
  dec[1] <- 1-gamma
  dec[2] <- 0.5
  dec[3] <- gamma
  # Cue Search Order Follows Validities
  for (j in 1:nc){
    stmp[j] <- nc-j+1
    s[j] <- rank(v[1:nc],stmp[j])
  }
  # TTB and WADD Subjects in Latent Mixture
  for (i in 1:ns){
    z[i] ~ dbern(phi)
    z1[i] <- z[i]+1
  }
  # Priors
  gamma ~ dunif(0.5,1)
  phi ~ dbeta(1,1)
}
```

The code `Stop.m` or `Stop.R` applies the model to the same simulated data. Figure 18.4 summarizes the results, and calculates an improved 68% correspondence with the human decisions. Most of the improvement comes from the final 6 questions, where TTB and WADD make different decisions. Those people—like subjects 17, 18, 19, and 20—who did not make TTB-consistent decisions for the last 6 questions are better modeled.

Figure 18.5 highlights this point by showing the posterior expectation for the latent-mixture indicator variable for each subject. At least half the subjects—including subjects 17, 18, 19 and 20—are inferred, with high certainly, to belong to the WADD group.

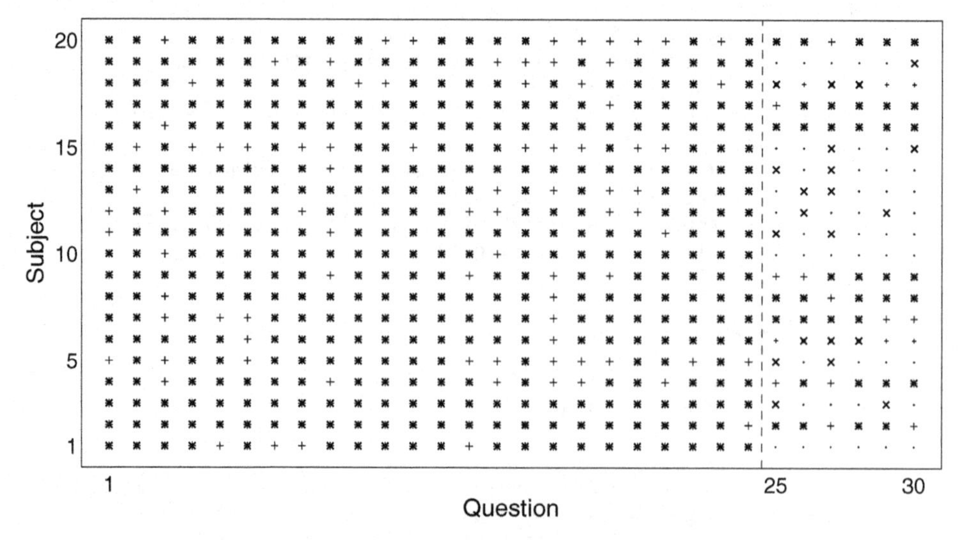

Fig. 18.4 Results of applying the latent mixture of the TTB and WADD models to the decision data.

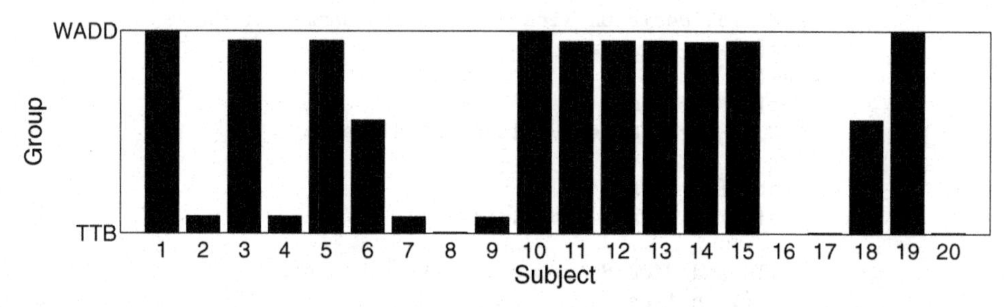

Fig. 18.5 Inferences about the use of the TTB and WADD models by each subject.

Exercise

Exercise 18.2.1 Plot and interpret the posterior distribution of ϕ. Approximate and interpret the Bayes factor comparing $\mathcal{H}_0 : \phi = 0$ versus $\mathcal{H}_1 : \phi \neq 0$.

18.3 Searching

Another way to extend the initial TTB account of the decision data is to preserve the one-reason stopping rule, but allow for flexibility in the search order. Most simply, it would be possible to consider a model in which everybody used a search order that did not follow cue validities. But, given the obvious individual differences in the data themselves, a more interesting model allows different people to have different search orders.

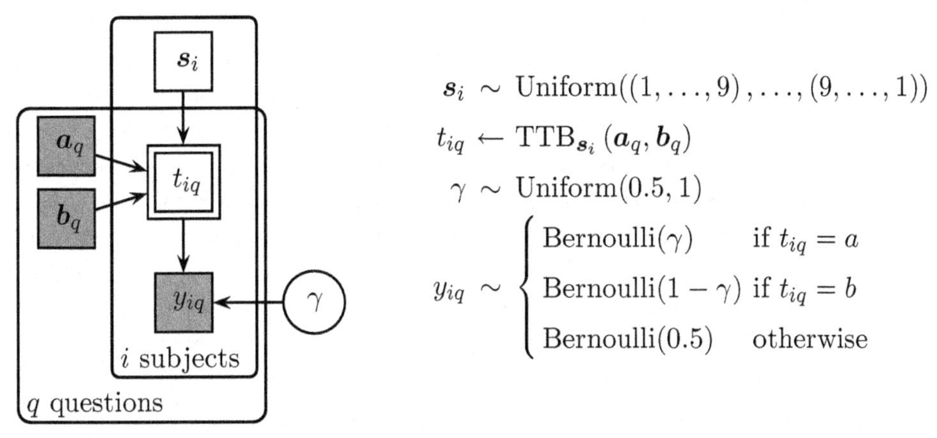

Fig. 18.6 Graphical model for individual differences in search orders.

The graphical model shown in Figure 18.6 implements this idea. The model allows each subject to have their own search order s_i, and treats these orders as latent parameters to be inferred from the data. Since the search orders are now model parameters, they must be given priors. In the graphical model, each of the $9! = 362,880$ search orders is made equally likely. That is, there is a uniform prior over all possible values of the search order parameters.

The script `Search.txt` implements the graphical model in WinBUGS:

```
# Individual Search Orders
model{
  # Data
  for (i in 1:ns){
    for (q in 1:nq){
      y[i,q] ~ dbern(ttb[t[i,q]])
      ypred[i,q] ~ dbern(ttb[t[i,q]])
    }
  }
  # One Reason Model, With Different Search Order Per Subject
  for (i in 1:ns){
    for (q in 1:nq){
      for (j in 1:nc){
        tmp1[i,q,j] <- (m[p[q,1],j]-m[p[q,2],j])*pow(2,s[i,j]-1)
      }
      # Find if Cue Favors First, Second, or Neither Stimulus
      tmp2[i,q] <- sum(tmp1[i,q,1:nc])
      tmp3[i,q] <- -1*step(-tmp2[i,q])+step(tmp2[i,q])
      t[i,q] <- tmp3[i,q]+2
    }
    # Cue Search Order From Ranking stmp
    for (j in 1:nc){
      s[i,j] <- rank(stmp[i,1:nc],j)
      stmp[i,j] ~ dnorm(0,1)I(0,)
    }
  }
  ttb[1] <- 1-gamma
  ttb[2] <- 0.5
  ttb[3] <- gamma
```

```
  # Prior
  gamma ~ dunif(0.5,1)
}
```

The generation of search orders has to be implemented in a low-level way, since WinBUGS has no standard distribution for placing a prior over orders. One approach would be to have a categorical variable that could take 9! values, each corresponding to a search order. Doing inference on this variable, however, would be hampered by the lack of structure in the relationships between the values. If two search orders are similar (say they have the same order, except for two neighboring cues, whose order in search are flipped) it would help if they had similar (i.e., neighboring) categorical numbers. But it is obviously impossible to map all the structure relationships between the orders onto the number line.

A better approach is to create an underlying continuous scale, and have each cue take a value on that scale. A search order can then be given simply by the order of those values, and changing the value for a cue in sampling-based inference would only change the order slightly, if at all. Thus, the `Search.txt` script uses underlying continuous positive variables for each of the cues, each with a Gaussian$(0, .001)$ prior.

The code `Search.m` or `Search.R` applies the model to the decision data. Figure 18.7 summarizes the new results, and calculates the correspondence to be 73%. The results show that this model is able to fit patterns in individual subject behavior that neither TTB, nor a mixture of TTB and WADD, could accommodate. Question 17 provides a good example, since about half the subjects choose one alternative, but half choose the other. This question requires choosing between an alternative that is defined by the presence of cue 2, and an alternative that is defined by the presence of cue 6. That is, both alternatives have only one cue, and it is cue 2 for one alternative, and cue 6 for the other. The first of these alternatives is chosen by TTB, since cue 2 has greater validity than cue 6. Thus, one explanation for the evident individual differences across subjects for this question is that some subjects search for cue 2 before 6, while others search for cue 6 before cue 2.

This speculation can be pursued by considering the posterior distribution over search orders inferred by the model. Conceptually, the posterior for an individual subject's search order is a distribution over the 9! possible orders. One way to display this is to list the orders in decreasing order of mass, as estimated from the number of times each is sampled. The code `Search.m` or `Search.R` displays how many unique orders were sampled, and then details the 10 specific orders with the highest posterior mass estimates.

To make this concrete, we consider the search orders inferred for subjects 12 and 13, who make different decisions for question 17. The summary of the posterior for the search patterns of subject 12 gives the following 10 orders.

```
Subject 12
There are 3709 search orders sampled in the posterior.
Order=(2  6  9  7  1  5  3  4  8), Estimated Mass=0.0013
```

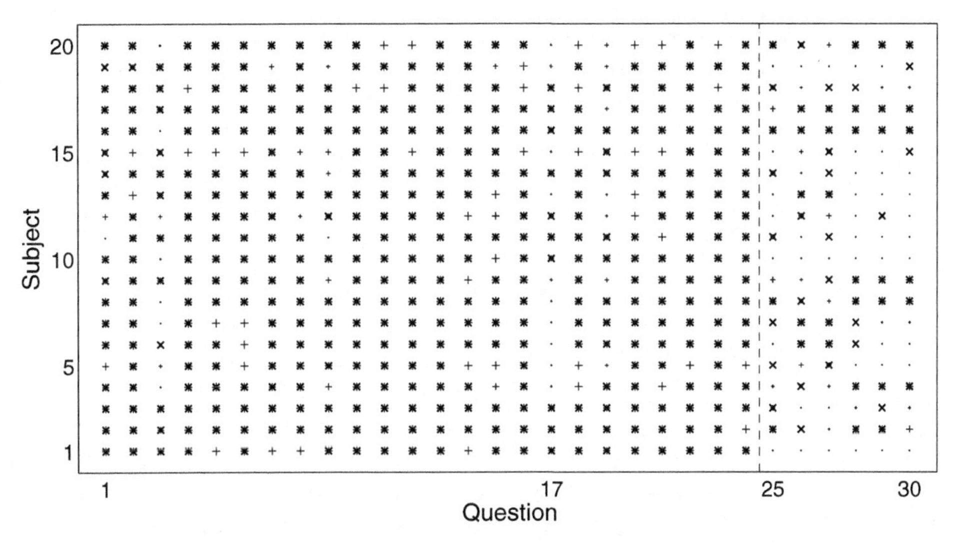

Fig. 18.7 Results of applying the individual search order model to the decision data.

```
Order=(2  6  5  9  7  1  8  3  4), Estimated Mass=0.0010
Order=(5  2  6  9  1  8  4  7  3), Estimated Mass=0.0010
Order=(2  6  5  9  1  8  7  3  4), Estimated Mass=0.0010
Order=(2  6  9  7  1  5  3  8  4), Estimated Mass=0.0010
Order=(2  6  9  5  1  3  4  7  8), Estimated Mass=0.0010
Order=(2  6  9  1  5  7  3  4  8), Estimated Mass=0.0010
Order=(2  6  5  1  3  9  7  4  8), Estimated Mass=0.0010
Order=(2  6  1  9  5  4  8  3  7), Estimated Mass=0.0010
Order=(2  1  6  9  7  5  4  8  3), Estimated Mass=0.0010
```

It is clear that, despite the uncertainty about the search order expressed by this posterior distribution, cue 2 is searched before cue 6. The summary of the posterior for the search patterns of subject 13 gives the following 10 orders.

```
Subject 13
There are 3350 search orders sampled in the posterior.
Order=(5  6  7  1  8  4  3  2  9), Estimated Mass=0.0018
Order=(5  4  6  1  7  3  2  8  9), Estimated Mass=0.0015
Order=(5  6  4  3  2  7  1  8  9), Estimated Mass=0.0013
Order=(5  6  7  8  1  9  4  3  2), Estimated Mass=0.0013
Order=(5  6  7  9  1  3  4  8  2), Estimated Mass=0.0013
Order=(5  6  4  7  1  9  2  8  3), Estimated Mass=0.0013
Order=(5  4  6  1  3  7  8  9  2), Estimated Mass=0.0013
Order=(5  6  4  8  7  1  2  9  3), Estimated Mass=0.0010
Order=(5  6  4  8  9  1  2  3  7), Estimated Mass=0.0010
Order=(5  4  6  9  7  1  2  3  8), Estimated Mass=0.0010
```

It seems clear for this subject that cue 6 is searched before cue 2. The difference in the order subjects 12 and 13 provides an explanation for their different decisions on question 17.

Exercises

Exercise 18.3.1 Look through the search order posterior distribution summaries for all of the subjects. How would you characterize the uncertainty they represent? Think about how many search orders are sampled, how many could be sampled, and how similar those sampled are to one another.

Exercise 18.3.2 Do you expect to be able to make inferences about the full search order if a subject is using a one-reason stopping rule like TTB? What consequences for analysis does this issue have?

Exercise 18.3.3 Are the inferred search orders for all of the subjects, and not just subjects 12 and 13, consistent with the individual differences observed in the answers to question 17?

18.4 Searching and stopping

The previous two models extended the original TTB model in two ways, to try and account for individual differences in the decisions. One model incorporated different stopping rules as the source of individual differences, while the other incorporates different search orders. The natural theoretical progression is to consider a model that allows for both different stopping rules and search orders.

Figure 18.8 presents a graphical model that combines searching and stopping. It is almost the logical combination of the stopping graphical model in Figure 18.3 and the searching graphical model in Figure 18.6, with one small adjustment. The stopping model considered each subject to use either the TTB or WADD stopping rule for all of their decisions, and inferred a base-rate over subjects. The model in Figure 18.8, however, assumes there is a base-rate ϕ_i over all of the questions with which the ith subject uses the WADD rather than the TTB stopping rule. Thus, there is an indicator z_{iq} corresponding to the stopping rule used by the ith subject on the qth question. Of course, if the WADD stopping rule is used, the search order does not matter, since all of the cues are examined. But, when the TTB one-reason stopping rule is used, the search order $\boldsymbol{s}_i$ for the ith subject influences the decision.

The script `SearchStop.txt` implements the graphical model in WinBUGS:

```
# Search and Stop
model{
  # Data
  for (i in 1:ns){
    for (q in 1:nq){
      y[i,q] ~ dbern(dec[t[i,q,z1[i,q]]])
      ypred[i,q] ~ dbern(dec[t[i,q,z1[i,q]]])
    }
```

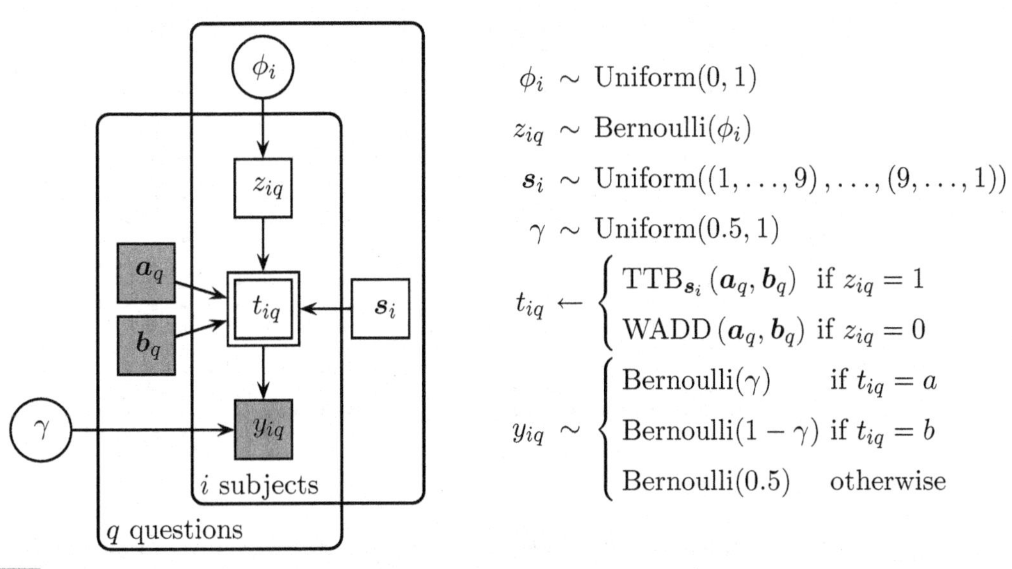

$$\phi_i \sim \text{Uniform}(0,1)$$
$$z_{iq} \sim \text{Bernoulli}(\phi_i)$$
$$\boldsymbol{s}_i \sim \text{Uniform}((1,\ldots,9),\ldots,(9,\ldots,1))$$
$$\gamma \sim \text{Uniform}(0.5,1)$$
$$t_{iq} \leftarrow \begin{cases} \text{TTB}_{\boldsymbol{s}_i}(\boldsymbol{a}_q,\boldsymbol{b}_q) & \text{if } z_{iq}=1 \\ \text{WADD}(\boldsymbol{a}_q,\boldsymbol{b}_q) & \text{if } z_{iq}=0 \end{cases}$$
$$y_{iq} \sim \begin{cases} \text{Bernoulli}(\gamma) & \text{if } t_{iq}=a \\ \text{Bernoulli}(1-\gamma) & \text{if } t_{iq}=b \\ \text{Bernoulli}(0.5) & \text{otherwise} \end{cases}$$

Fig. 18.8 Graphical model for inferring search orders and stopping rules.

```
}
# TTB Decision
for (i in 1:ns){
  for (q in 1:nq){
    for (j in 1:nc){
      tmp1[i,q,j] <- (m[p[q,1],j]-m[p[q,2],j])*pow(2,s[i,j]-1)
    }
    tmp2[i,q] <- sum(tmp1[i,q,1:nc])
    tmp3[i,q] <- -1*step(-tmp2[i,q])+step(tmp2[i,q])
    t[i,q,1] <- tmp3[i,q]+2
  }
}
# WADD Decision
for (i in 1:ns){
  for (q in 1:nq){
    for (j in 1:nc){
      tmp4[i,q,j] <- (m[p[q,1],j]-m[p[q,2],j])*x[j]
    }
    # Find if Cue Favors First, Second, or Neither Stimulus
    tmp5[i,q] <- sum(tmp4[i,q,1:nc])
    tmp6[i,q] <- -1*step(-tmp5[i,q])+step(tmp5[i,q])
    t[i,q,2] <- tmp6[i,q]+2
  }
}
# Follow Decision With Probability Gamma, or Guess
dec[1] <- 1-gamma
dec[2] <- 0.5
dec[3] <- gamma
# Cue Search Order From Ranking stmp
for (i in 1:ns){
  for (j in 1:nc){
    s[i,j] <- rank(stmp[i,1:nc],j)
```

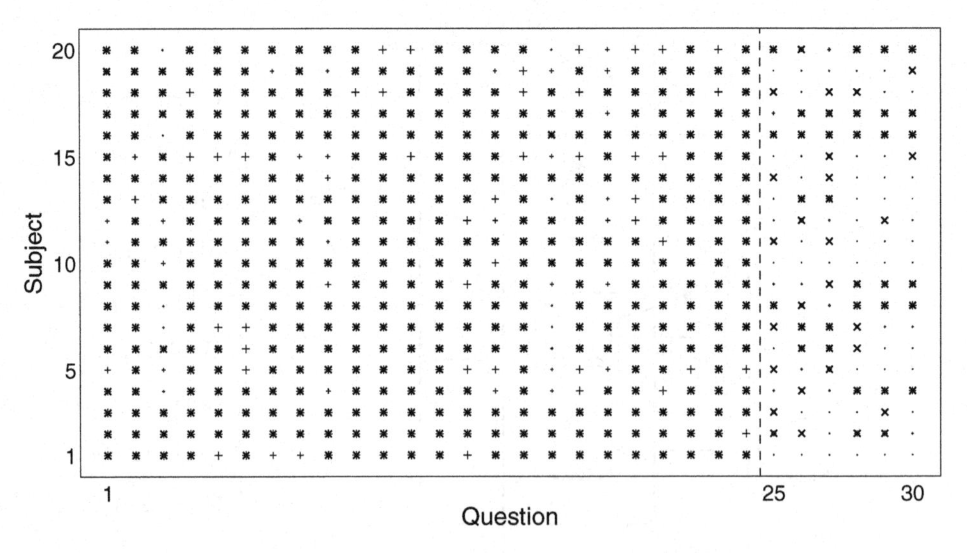

Fig. 18.9 Results of applying the flexible searching and stopping model to the decision data.

```
      stmp[i,j] ~ dnorm(0,1)I(0,)
    }
  }
  # TTB and WADD Rate Per Subject
  for (i in 1:ns){
    phi[i] ~ dbeta(1,1)
    for (q in 1:nq){
      z[i,q] ~ dbern(phi[i])
      z1[i,q] <- z[i,q]+1
    }
  }
  gamma ~ dunif(0.5,1)
}
```

The code `SearchStop.m` or `SearchStop.R` applies the model to the decision data. Figure 18.9 summarizes the results, and calculates the correspondence to be 74%. The incorporation of individual differences into the stopping rule does not appear to have had much impact on the posterior predicted decisions made by the search model in Figure 18.7.

Exercises

Exercise 18.4.1 How do the posterior expectations of the ϕ_i parameters for the searching and stopping model in Figure 18.8 compare to the posterior expectations of the z_i parameters for the stopping model in Figure 18.3? Interpret the similarities and differences.

Exercise 18.4.2 In what sense does the current model incorporate, and not incorporate, structured individual differences over searching and stopping? Suggest hierarchical extensions to the model in Figure 18.8 that could add structure to the individual differences in searching and stopping parameters.

19 Number concept development

A basic challenge in understanding human cognitive development is to understand how children acquire number concepts. Since the time of Piaget (1952), the concept of number has been one of the most active areas of research in the field. This chapter focuses on one prominent current theory about the origin of integer concepts, called the "knower-level" theory (Carey, 2001; Carey & Sarnecka, 2006; Wynn, 1990, 1992).

The knower-level theory asserts that children learn the exact cardinal meanings of the first three or four number words one-by-one and in order. That is, children begin by learning the meaning of "one" first, then "two," then "three," and then (for some children) "four," at which point they make an inductive leap, and infer the meanings of the rest of the words in their counting list. In the terminology of the theory, children start as PN-knowers (for "Pre-Number"), progress to one-knowers once they understand "one," through the two-knower, three-knower, and (for some children) four-knower levels, until they eventually become CP-knowers (for "Cardinal Principle").

There are at least two common behavioral tasks that are used to assess children's number knowledge. In the "Give-N" task, children are asked to give some number of objects, such as small toys, to the experimenter, or an experimenter substitute, such as a puppet (e.g., Frye, Braisby, Lowe, Maroudas, & Nicholls, 1989; Fuson, 1988; Schaeffer, Eggleston, & Scott, 1974; Wynn, 1990, 1992). In the "Fast-Cards" task, children are asked how many objects, such as pictures of animals, were displayed on a briefly presented card (e.g., Le Corre & Carey, 2007). In both tasks, the behavioral data are just a set of question–answer pairs, recording how many objects were present or asked for, and how many the child gave or answered.

The responses that children give in the Give-N and Fast-Cards tasks naturally depend on what numbers they understand, but also depend on properties of the tasks themselves. For example, if there are 15 toys available in a Give-N task, it is common for a child to give all 15 when asked for a number they do not understand. The same child, however, asked for the same number in a Fast-Cards task, would be very unlikely to answer 15, because there is nothing special about that answer in the different task setting. Thus, a good model needs to combine child-specific and task-specific components to capture the behavioral data.

We consider a knower-level model, developed by Lee and Sarnecka (2010), that incorporates task-specific components. We use a subset of 20 children from the data considered by Lee and Sarnecka (2010), first applying it to Give-N data, then to

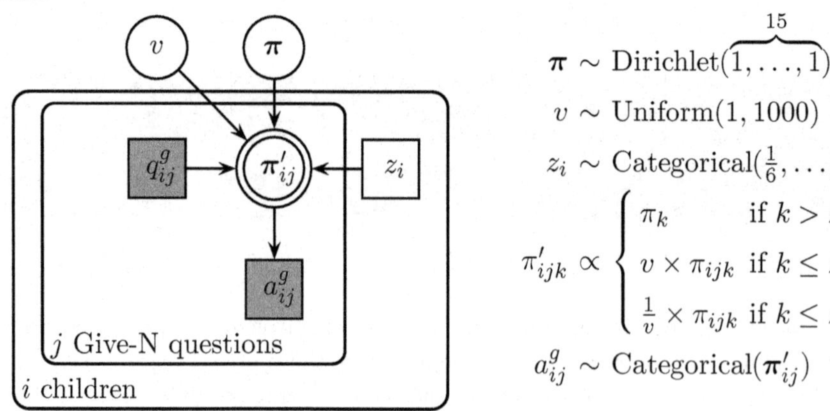

$$\boldsymbol{\pi} \sim \text{Dirichlet}(\overbrace{1,\ldots,1}^{15})$$

$$v \sim \text{Uniform}(1, 1000)$$

$$z_i \sim \text{Categorical}(\tfrac{1}{6},\ldots,\tfrac{1}{6})$$

$$\pi'_{ijk} \propto \begin{cases} \pi_k & \text{if } k > z_i \\ v \times \pi_{ijk} & \text{if } k \leq z_i \text{ and } k = q^g_{ij} \\ \frac{1}{v} \times \pi_{ijk} & \text{if } k \leq z_i \text{ and } k \neq q^g_{ij} \end{cases}$$

$$a^g_{ij} \sim \text{Categorical}(\boldsymbol{\pi}'_{ij})$$

Fig. 19.1 Graphical model for behavior on the Give-N task according to the Lee and Sarnecka (2010) knower-level model.

Fast-Cards data from the same children, and then to both sets of data simultaneously.

19.1 Knower-level model for Give-N

The basic idea behind the Lee and Sarnecka (2010) model is that a behavioral task has a base-rate for responding. This is simply a predisposition towards giving each possible answer, before any question has been asked. For the Give-N task, for example, we might expect one toy, two toys, a small handful of toys, or all of the available toys, to be possible answers with relatively high probabilities in the base-rate. The child's actual response comes from updating that base-rate when they are asked a question. The base-rate is task-specific, and the updating depends on the question and the child's number knowledge.

The nature of the updating in the model can be explained in terms of two cases. In the first case, the child is asked to give a number they understand. This means the probability of giving that number increases, and the probability of giving other numbers they understand but were not asked to give all decrease. The relative probabilities of giving other numbers in the base-rate are not changed, because the child does not know about them. So, for example, if a three-knower is asked to give two, their probability of giving 2 increases, and their probability of giving 1 or 3 decreases, relative to the base-rate. But the numbers 4 and above do not change in their relative probabilities.

In the second case, the child is asked to give a number they do not know. This means the probabilities of giving numbers they do know decrease, but all of the other numbers will retain the same relative probabilities. So, for example, if a three-knower is asked to give five, they become much less likely to give 1, 2, or 3, but equally relatively likely to give 4 and above.

Figure 19.1 presents a graphical model for this knower-level model. The data are the observed q^g_{ij} questions and a^g_{ij} answers for the ith child on their jth question. The task-specific base-rate probabilities are represented by the vector $\boldsymbol{\pi}$, so that π_k is the probability for giving k as an answer. The base-rate is updated to $\boldsymbol{\pi}'$. The updating occurs using the number asked for, the knower level z_i of the child, and an evidence value v that measures the strength of the updating. The updating logic explained by the two cases earlier can be formalized as

$$\pi'_{ijk} \propto \begin{cases} \pi_k & \text{if } k > z_i \\ v \times \pi_{ijk} & \text{if } k \leq z_i \text{ and } k = q^g_{ij} \\ \frac{1}{v} \times \pi_{ijk} & \text{if } k \leq z_i \text{ and } k \neq q^g_{ij}. \end{cases}$$

The actual answer produced by the child is then assumed to be sampled according to the probabilities for each possibility in the updated base-rate $\boldsymbol{\pi}'$. The categorical distribution, which can be thought of as the extension of the Bernoulli distribution to cases where there are more than two possible choices, expresses this naturally as

$$a^g_{ij} \sim \text{Categorical}\big(\boldsymbol{\pi}'\big).$$

The knower-level parameter z_i is given a categorical prior so that the six possibilities—PN-knower, one-knower, two-knower, three-knower, four-knower, and CP-knower—are given equal prior probability. The base-rate prior is given by a Dirichlet distribution that makes all possible distributions over the 15 toys equally likely. The Dirichlet distribution can be thought of as an extension of a beta distribution to cases where there are more than two possible alternatives, so $\text{Dirichlet}\big(1,\ldots,1\big)$ is naturally conceived as the extension of the uniform distribution $\text{Beta}\big(1,1\big)$.

The script `NumberConcept_1.txt` implements the graphical model in WinBUGS. Notice that the script collects posterior predictions for each individual child, as well as for each knower level as a group of children:

```
# Knower Level Model Applied to Give-N Data
model{
  # Data
  for (i in 1:ns){
    for (j in 1:gnq[i]){
      # Probability a z[i]-Knower Will Answer ga[i,j] to Question gq[i,j]
      # is a Categorical Draw From Their Distribution over the 1:gn Toys
      ga[i,j] ~ dcat(npiprime[z[i],gq[i,j],1:gn])
    }
    # Posterior Predictive
    for (j in 1:gn){
      predga[i,j] ~ dcat(npiprime[z[i],j,1:gn])
    }
  }
  # Model
  for (i in 1:nz){
    for (j in 1:gn){
```

```
      for (k in 1:gn){
        piprimetmp[i,j,k,1] <- pi[k]
        piprimetmp[i,j,k,2] <- 1/v*pi[k]
        piprimetmp[i,j,k,3] <- v*pi[k]
        # Will be 1 if Knower-Level (i.e, i-1) is Same or Greater than Answer
        ind1[i,j,k] <- step((i-1)-k)
        # Will be 1 for the Possible Answer that Matches the Question
        ind2[i,j,k] <- equals(k,j)
        # Will be 1 for 0-Knowers
        ind3[i,j,k] <- equals(i,1)
        # Will be 1 for HN-Knowers
        ind4[i,j,k] <- equals(i,nz)
        ind5[i,j,k] <- ind3[i,j,k]+ind4[i,j,k]*(2+ind2[i,j,k])
                       + (1-ind4[i,j,k])*(1-ind3[i,j,k])
                       * (ind1[i,j,k]+ind1[i,j,k]*ind2[i,j,k]+1)
        piprime[i,j,k] <- piprimetmp[i,j,k,ind5[i,j,k]]
        npiprime[i,j,k] <- piprime[i,j,k]/sum(piprime[i,j,1:gn])
      }
    }
  }
  # Posterior Prediction For Knower Levels
  for (i in 1:nz){
    for (j in 1:gn){
      predz[i,j] ~ dcat(npiprime[i,j,1:gn])
    }
  }
  # Base rate
  for (i in 1:gn){
    pitmp[i] ~ dunif(0,1)
    pi[i] <- pitmp[i]/sum(pitmp[1:gn])
  }
  predpi ~ dcat(pi[1:gn])
  # Priors
  v ~ dunif(1,1000)
  for (i in 1:ns) {
    z[i] ~ dcat(priorz[])
  }
  for (i in 1:nz){
    priorz[i] <- 1/6
  }
}
```

The code `NumberConcept_1.m` or `NumberConcept_1.R` applies the model to the Give-N data used by Lee and Sarnecka (2011), and produces several analyses.

Figure 19.2 shows the base-rate that is inferred, by showing the posterior distribution it generates over the 15 toys. It shows an intuitively reasonable result, giving high probability to small numbers of toys, as well as to the whole set of 15 toys. It is important to understand that this is a latent 15-dimensional parameter that was inferred from the data, and not something that was assumed to explain the data. This is a powerful piece of inference that is straightforward using the Bayesian approach, and makes it theoretically possible to propose a complicated construct like a base-rate as an important component of determining the observed data.

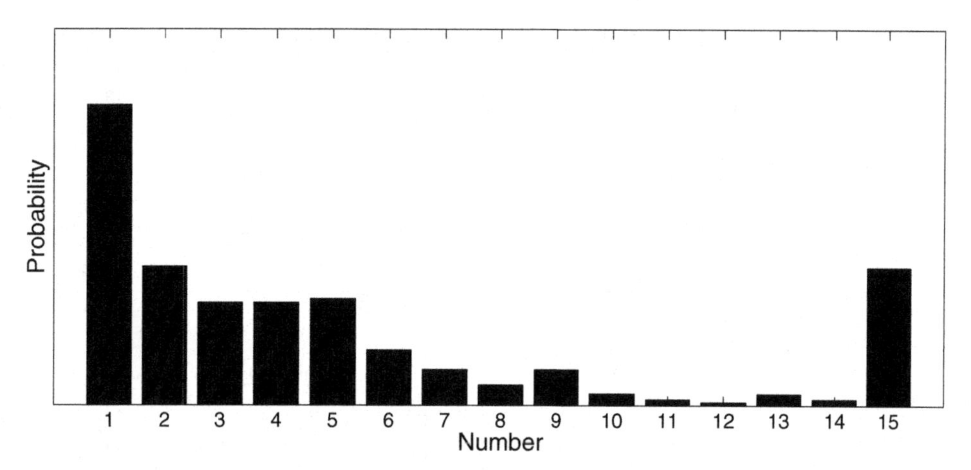

Fig. 19.2 Posterior base-rate for giving $1, \ldots, 15$ toys, inferred by applying the knower-level model to the Give-N data.

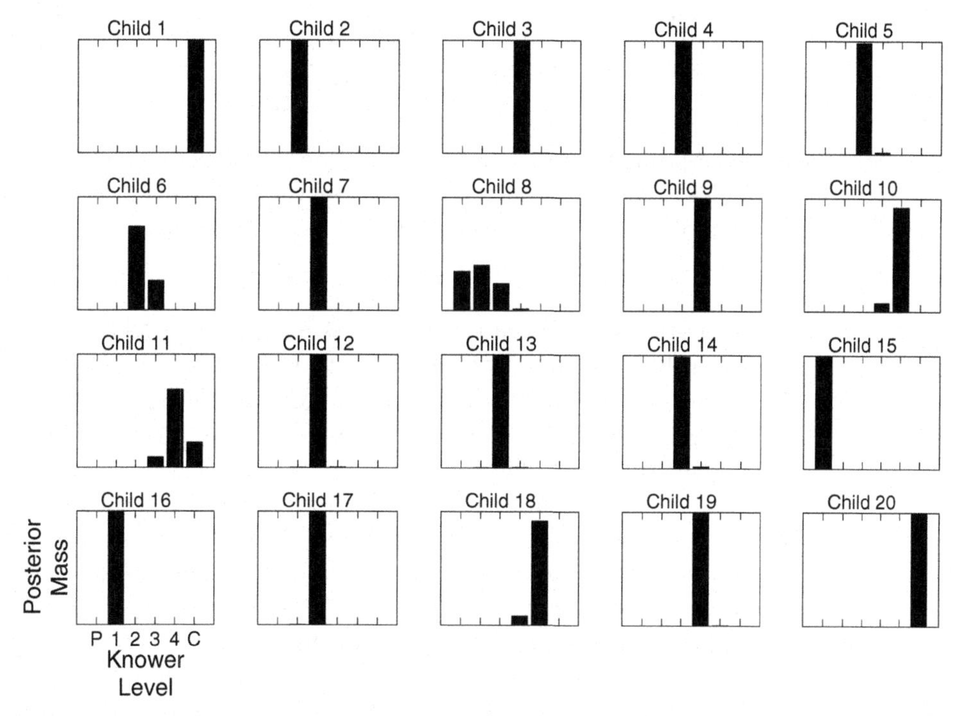

Fig. 19.3 Posterior over knower levels for the 20 children, based on the Give-N data. P="Pre-Number Knower" and C="Cardinality-Principle Knower".

Figure 19.3 shows the posterior distribution over the six possible knower levels for each child. The noteworthy feature of this result is that most of the children are classified with high certainty into a single knower level. There are exceptions, such as children 6, 8, and 11, but, for the most part, there is confidence in a

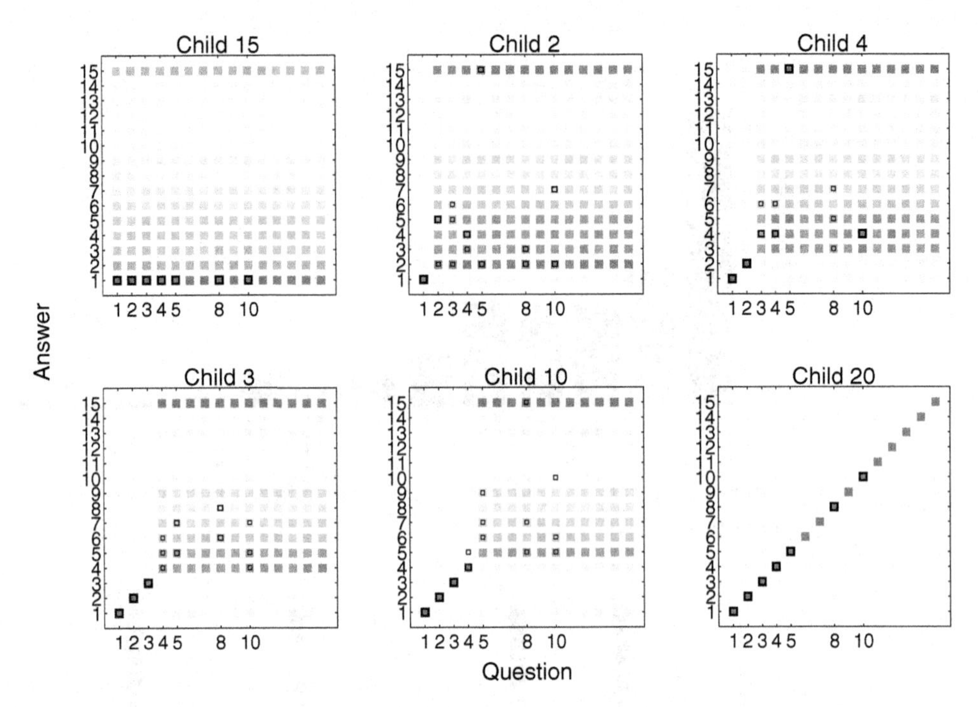

Fig. 19.4 Posterior prediction for six selected children on the Give-N data. The posterior predictive mass of each question and answer combination is shown by shading. The distribution of the child's observed data is shown by squares.

single classification. When inferring a discrete latent variable representing a class, highly-peaked posteriors like this are a good sign that the model is a useful one. When models are badly mis-specified, Bayesian inference naturally does "model averaging" blending over a mixture of possibilities to try and account for the data, making interpretation difficult.

Figure 19.4 shows the relationship between the posterior predictions of the model and the observed behaviors for six children, chosen to span a range of knower levels. Each panel corresponds to a child, with the x-axis giving the question asked and the y-axis giving the answer given. The darkness of the shading in each cell corresponds to the posterior probability that the child will give that many toys when asked that question. The squares show the distribution of the child's observed behavior.

Figure 19.5 shows the same sort of analysis for the posterior predictive for each knower level. The observed data now include every child inferred, using their maximum a posteriori knower level from Figure 19.3.

Exercises

Exercise 19.1.1 Report the posterior for the evidence parameter v.

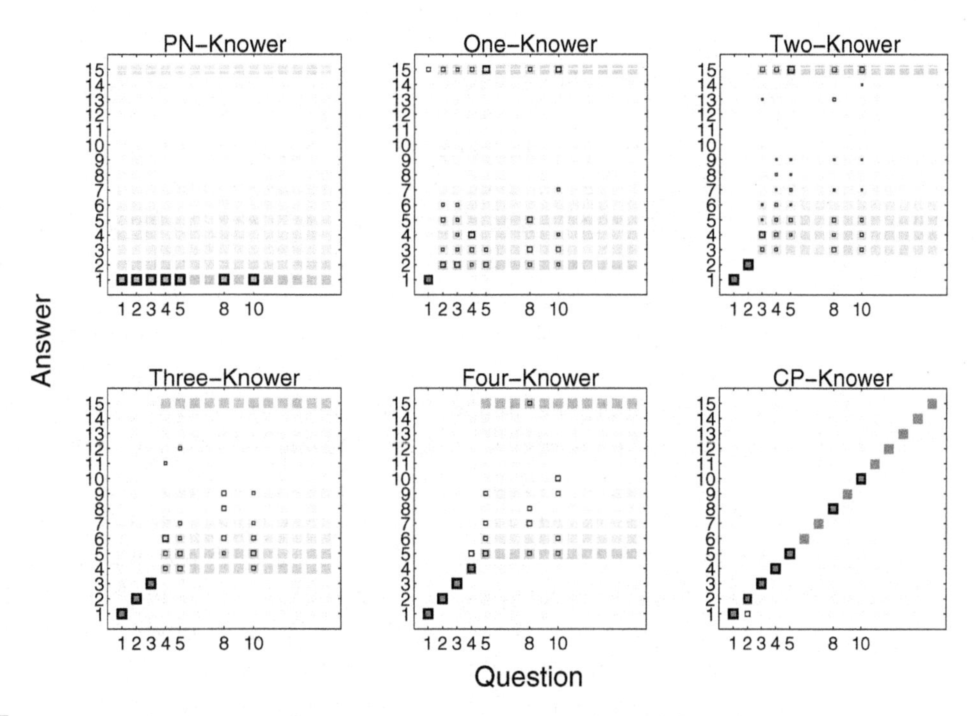

Fig. 19.5 Posterior prediction for the six knower levels on the Give-N data. The posterior predictive mass of each question-and-answer combination is shown by shading. The distribution of the observed data aggregated over every child classified as belonging to that knower level is shown by squares.

Exercise 19.1.2 Interpret the distinctive visual patterns of posterior prediction in Figure 19.4 for each child, explaining how they combine knower-level knowledge, and the task base-rate.

Exercise 19.1.3 What do you think of relying on the maximum a posteriori summary of the posterior uncertainty about knower levels to classify children? What might be a justifiable alternative?

Exercise 19.1.4 The model currently assumes that the final decision is sampled from the distribution over all possible responses, in proportion to the mass associated with each response. Does this probability-matching strategy seem psychologically plausible? What is an alternative model of this part of the decision-making process?

Exercise 19.1.5 Explain why the distribution shown in Figure 19.2 is not exactly the posterior distribution of the base-rate π. What distribution is shown, how is it related to the posterior of π, and what are the advantages of presenting the distribution in Figure 19.2?

Box 19.1 Bayesian statistical analysis of Bayesian cognitive models

One way to think about Bayesian methods is that they address the problem of drawing inferences over structured models—hierarchical models, mixture models, and so on—from sparse and noisy data. That is obviously a basic challenge for any method of statistical inference. It also, however, seems like a central challenge faced by the mind. We have structured mental representations, and we must deal with incomplete and inherently uncertain information. This analogy suggests that Bayesian statistics can be used not just as a method of analyzing models and data in the cognitive sciences, but also as a theoretical metaphor for developing models of cognition in the first place.

In fact, the "Bayes in the head" theoretical position, which assumes that the mind does Bayesian inference, is an important, and controversial, one in cognitive science (e.g., Chater et al., 2006; Griffiths et al., 2008; Jones & Love, 2011; Tenenbaum et al., 2011). The Bayesian cognitive models developed within this approach usually focus on providing "rational" accounts of psychological phenomena. This means they are mostly (see Sanborn et al., 2010, for an exception) pitched at the computational level within the three-level hierarchy described by Marr (1982), giving an account of why people behave as they do, without trying to account for the mechanisms, processes, or algorithms that produce the behavior, nor how those processes are implemented in neural hardware.

The model of number-development is probably unique among those considered in this book as being interpretable as a Bayesian model of cognition. The base-rate over responses is naturally interpreted as a prior, and the updating that takes place depending on the information presented and the child's knowledge defines a likelihood function. The logic of the model corresponds to applying Bayes' rule to produce a posterior distribution, and observed behavior is sampled from this posterior. This means we are making Bayesian inferences about a Bayesian model of cognition. Perhaps surprisingly, this is rarely done (see Kruschke, 2010b; Lee, 2011b, for critiques), even though all of the advantages Bayesian inference has for analyzing cognitive models still apply to Bayesian cognitive models. For example, it would be difficult to infer the base-rate distribution—that is, the prior the children bring to the task—from behavioral data without using Bayesian methods.

19.2 Knower-level model for Fast-Cards

Applying the graphical model in Figure 19.1 to Fast Cards data only requires changing the base-rate parameter to allow for answers greater than 15. In the data set we are considering, which now consists of the observed q^f_{ij} questions and a^f_{ij} answers for the ith child on their jth question, the largest number given as an answer by any child is 50.

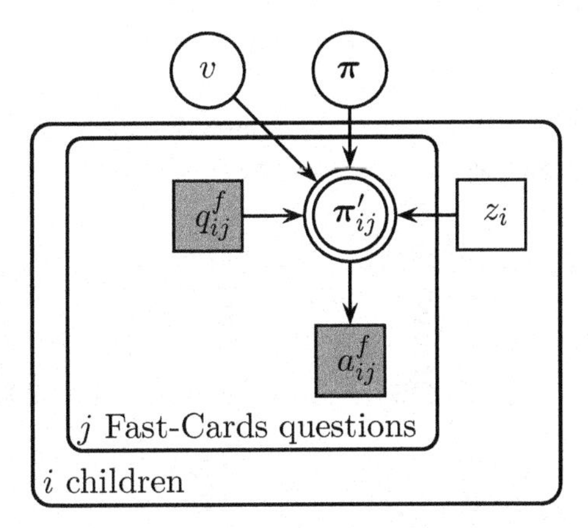

$$
\begin{aligned}
\boldsymbol{\pi} &\sim \text{Dirichlet}(\overbrace{1,\ldots,1}^{50})\\
v &\sim \text{Uniform}(1,1000)\\
z_i &\sim \text{Categorical}(\tfrac{1}{6},\ldots,\tfrac{1}{6})\\
\pi'_{ijk} &\propto \begin{cases} \pi_k & \text{if } k > z_i\\ v\times\pi_{ijk} & \text{if } k\le z_i \text{ and } k = q^f_{ij}\\ \frac{1}{v}\times\pi_{ijk} & \text{if } k\le z_i \text{ and } k\neq q^f_{ij}\end{cases}\\
a^f_{ij} &\sim \text{Categorical}(\boldsymbol{\pi}'_{ij})
\end{aligned}
$$

Fig. 19.6 Graphical model for behavior on the Fast-Cards task according to the Lee and Sarnecka (2010) knower-level model.

The revised graphical model is shown in Figure 19.6, and differs only in terms of the dimensionality of the base-rate parameter $\boldsymbol{\pi}$.

The script `NumberConcept_2.txt` makes this change to the original script. The code `NumberConcept_2.m` or `NumberConcept_2.R` applies the model to the Fast-Cards data. Note that the 50-element base-rate is implemented in an approximate but computationally efficient way, given its sparseness, by considering only those answers actually observed in the data.

Figure 19.7 shows the base-rate that is inferred. It is again intuitively reasonable, with most the greatest posterior mass given to numbers $1,\ldots,5$, which have high frequency in the child's environment. Relatively large mass is given to the remaining single-digit numbers, and most of the remaining mass occurs at so-called "prominent" numbers like 10, 20, 30, and 50 (Albers, 2001).

Figure 19.8 shows the posterior distribution over the six knower levels for each child.

Figure 19.9 shows the analysis for the posterior predictive for each knower level, based on the Fast-Cards data. As before, children have been classified into knower levels based on the mode of their posterior membership shown in Figure 19.8. Note that no child in this data set, for this task, is classified as being a PN-knower, but the model naturally still makes predictions about the behavior of children at this developmental stage.

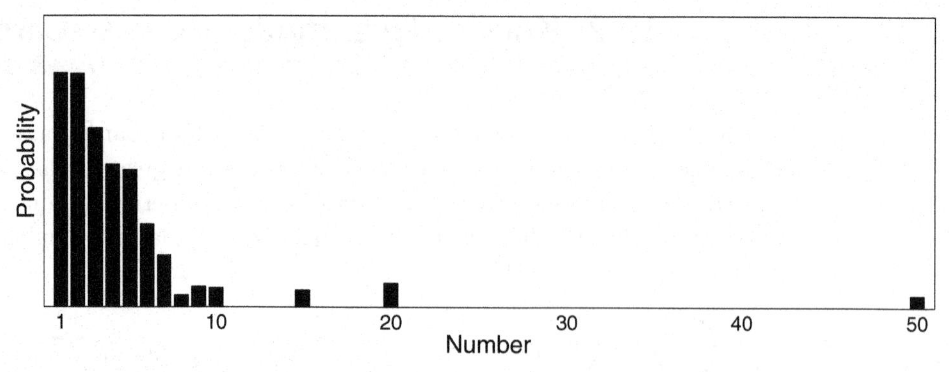

Fig. 19.7 Posterior base-rate for giving the answer $1, \ldots, 50$ toys, inferred by applying the knower-level model to the Fast-Cards data.

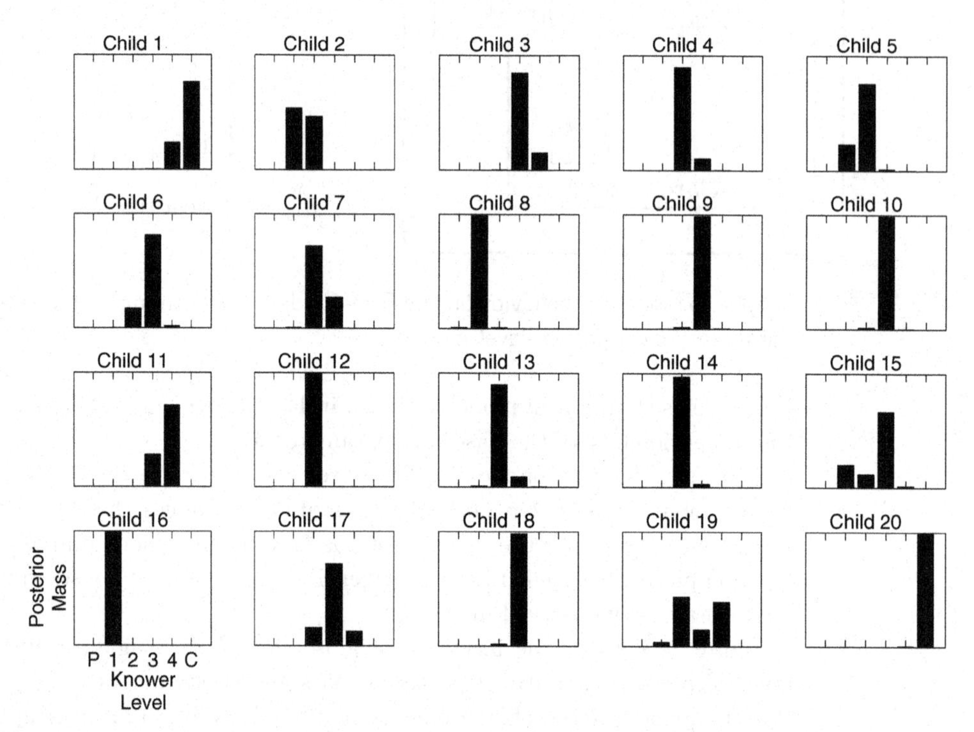

Fig. 19.8 Posterior over knower levels for 20 children, based on the Fast-Cards data. P="Pre-Number Knower" and C="Cardinality-Principle Knower."

Exercises

Exercise 19.2.1 Report the posterior for the evidence parameter v, and compare it to the value found in the Give-N analysis.

Exercise 19.2.2 Compare the posterior distributions over knower levels shown in Figure 19.8 with those inferred using the Give-N data in Figure 19.3.

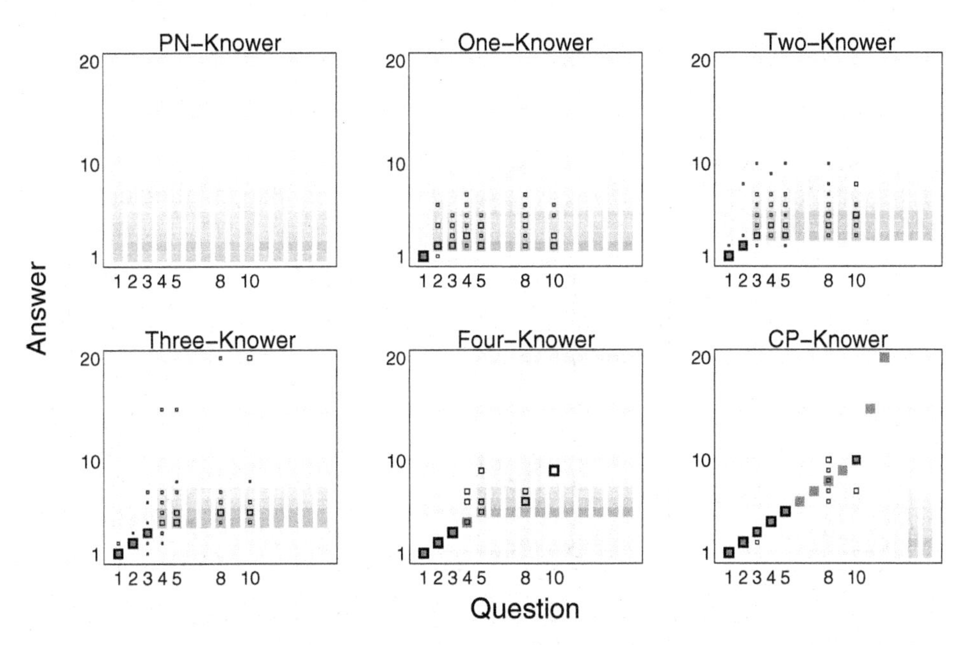

Fig. 19.9 Posterior prediction for the six knower levels on the Fast-Cards data. The posterior predictive mass of each question-and-answer combination is shown by shading. The distribution of the observed data aggregated over every child classified as belonging to that knower level is shown by squares.

Exercise 19.2.3 The uncertainty in the posterior distribution in Figure 19.8 always involves adjacent knower levels (e.g., it is uncertain whether child 2 is a one-knower or two-knower). Does this follow necessarily from the statistical definition of the z knower-level parameter? If so, how? If not, how do the patterns of uncertainty in Figure 19.8 arise?

19.3 Knower-level model for Give-N and Fast-Cards

Because of the within-subjects design of the data, in which each child did both the Give-N and Fast-Cards tasks, it is natural to think about combining both sources of behavioral data to infer knower levels. Conceptually, this is straightforward, with one underlying knower level generating both sets of behavioral data for any given child, according to the specific characteristics of each task.

The graphical model that integrates the two tasks in this way is shown in Figure 19.10. It is visually clear how it combines the graphical models in Figures 19.1 and 19.6, linking them through z_i, the common knower-level parameter. Notice that the base-rates and evidence-value parameters are allowed to be different for the two tasks. Thus, the way the data from the two tasks are modeled in Figure 19.10 corre-

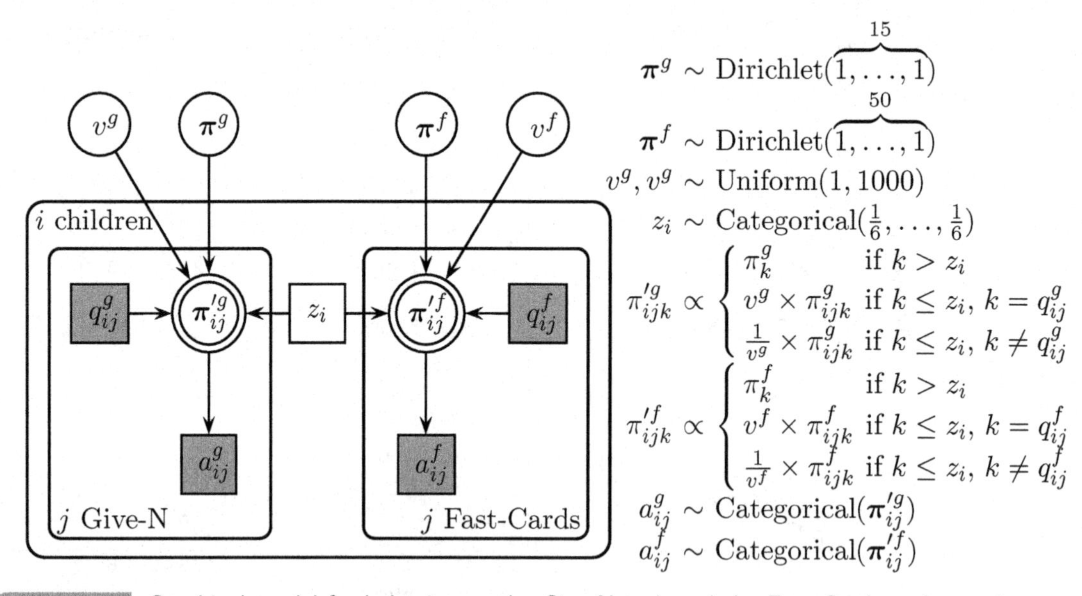

Fig. 19.10 Graphical model for behavior on the Give-N task and the Fast-Cards task, coming from each child's underlying knower level.

sponds to assuming that the latent psychological state of number knowledge is the same in both tasks, but that base-rate responding, and the evidence levels different stimuli provide, can vary in task-specific ways.

The script `NumberConcept_3.txt` implements the graphical model. The code `NumberConcept_3.m` or `NumberConcept_3.R` applies the model to the Given-N and Fast-Cards data:

```
# Knower Level Model Applied to Give-N and Fast-Cards Data
model{
  # Give-N Part
  # Data
  for (i in 1:ns){
    for (j in 1:gnq[i]){
      ga[i,j] ~ dcat(npiprime[z[i],gq[i,j],1:gn])
    }
    # Posterior Predictive
    for (j in 1:gn){
      predga[i,j] ~ dcat(npiprime[z[i],j,1:gn])
    }
  }
  # Model
  for (i in 1:nz){
    for (j in 1:gn){
      for (k in 1:gn){
        piprimetmp[i,j,k,1] <- pi[k]
        piprimetmp[i,j,k,2] <- 1/gv*pi[k]
        piprimetmp[i,j,k,3] <- gv*pi[k]
        ind1[i,j,k] <- step((i-1)-k)
        ind2[i,j,k] <- equals(k,j)
```

```
        ind3[i,j,k] <- equals(i,1)
        ind4[i,j,k] <- equals(i,nz)
        ind5[i,j,k] <- ind3[i,j,k]+ind4[i,j,k]*(2+ind2[i,j,k])
                       + (1-ind4[i,j,k])*(1-ind3[i,j,k])
                       * (ind1[i,j,k]+ind1[i,j,k]*ind2[i,j,k]+1)
        piprime[i,j,k] <- piprimetmp[i,j,k,ind5[i,j,k]]
        npiprime[i,j,k] <- piprime[i,j,k]/sum(piprime[i,j,1:gn])
      }
    }
}
# Fast-Cards Part
# Data
for (i in 1:ns){
  for (j in 1:fnq[i]){
    fa[i,j] ~ dcat(fnpiprime[z[i],fq[i,j],1:fn])
  }
  # Posterior Predictive
  for (j in 1:gn){
    predfa[i,j] ~ dcat(fnpiprime[z[i],j,1:fn])
  }
}
# Model
for (i in 1:nz){
  for (j in 1:gn){
    for (k in 1:fn){
      fpiprimetmp[i,j,k,1] <- fpi[k]
      fpiprimetmp[i,j,k,2] <- 1/fv*fpi[k]
      fpiprimetmp[i,j,k,3] <- fv*fpi[k]
      find1[i,j,k] <- step((i-1)-k)
      find2[i,j,k] <- equals(k,j)
      find3[i,j,k] <- equals(i,1)
      find4[i,j,k] <- equals(i,nz)
      find5[i,j,k] <- find3[i,j,k]+find4[i,j,k]*(2+find2[i,j,k])
                      + (1-find4[i,j,k])*(1-find3[i,j,k])
                      * (find1[i,j,k]+find1[i,j,k]*find2[i,j,k]+1)
      fpiprime[i,j,k]  <- fpiprimetmp[i,j,k,find5[i,j,k]]
      fnpiprime[i,j,k] <- fpiprime[i,j,k]/sum(fpiprime[i,j,1:fn])
    }
  }
}
# Posterior Prediction For Knower Levels
for (i in 1:nz){
  for (j in 1:gn){
    predgaz[i,j] ~ dcat(npiprime[i,j,1:gn])
    predfaz[i,j] ~ dcat(fnpiprime[i,j,1:fn])
  }
}
# Base rates
for (i in 1:fn){
  fpitmp[i] ~ dunif(0,1)
  fpi[i] <- fpitmp[i]/sum(fpitmp[1:fn])
}
for (i in 1:gn){
  pitmp[i] ~ dunif(0,1)
  pi[i] <- pitmp[i]/sum(pitmp[1:gn])
}
```

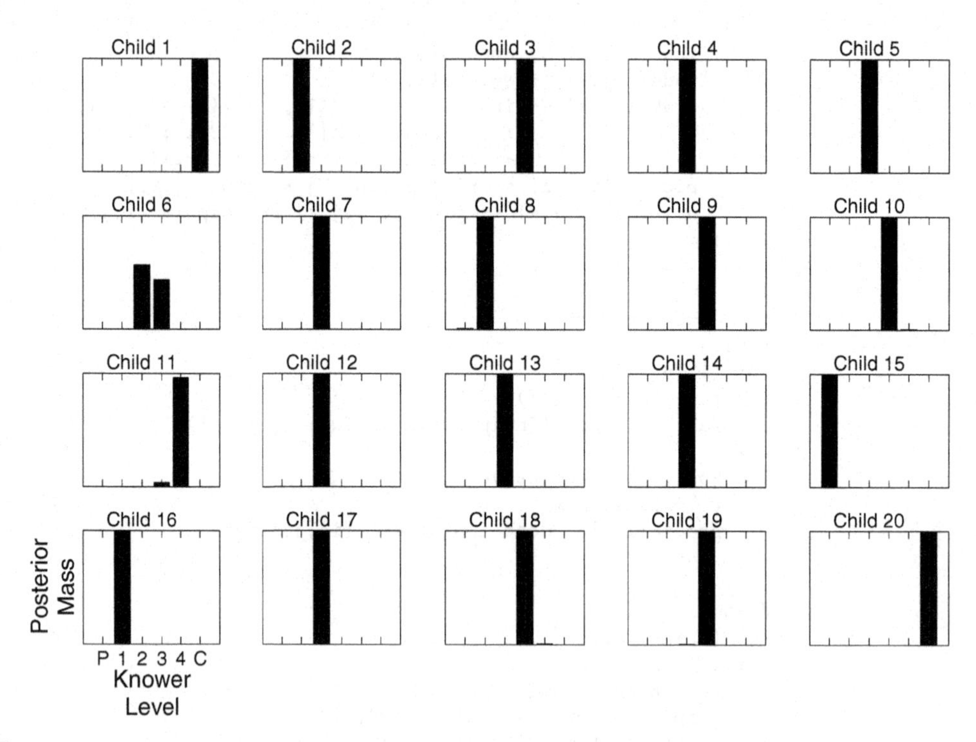

Fig. 19.11 Posterior over knower levels for 20 children, based on both the Give-N and Fast-Cards data. P="Pre-Number Knower" and C="Cardinality-Principle Knower."

```
  predpi  ~ dcat(pi[1:gn])
  predfpi ~ dcat(fpi[1:fn])
  # Priors
  gv ~ dunif(1,1000)
  fv ~ dunif(1,1000)
  for (i in 1:ns) {
    z[i] ~ dcat(priorz[])
  }
  for (i in 1:nz){
    priorz[i] <- 1/6
  }
}
```

Figure 19.11 shows the posterior distribution over the six knower levels for each child. For almost all of the children—with child 6 being the one exception—the posterior distributions have no uncertainty, and children are confidently inferred as belonging to a single knower level. The key point is that, by using both sources of empirical evidence simultaneously, clearer inferences are able to be made than were possible from either alone.

Exercise

Exercise 19.3.1 The behavioral data for Child 18 are detailed in Table 19.1, showing their answers to every question in both tasks. Explain why the inference based on only the Give-N data in Figure 19.3 has most posterior mass on four-knower, but there is some uncertainty, with three-knower also a possibility. Explain why the inference based on only the Fast-Cards data in Figure 19.8 has posterior mass almost entirely on the three-knower possibility. Explain why the combined inference in Figure 19.11 favors three-knower, and why the possible four-knower inference in the Give-N analysis could be viewed as arising from the nature of the task itself, rather than from the actual number knowledge of the child that is of primary interest developmentally.

Table 19.1 Behavior of Child 18 on Give-N and Fast-Cards tasks.

Question	Give-N Answers	Fast-Cards answers
1	1, 1, 1	1, 1, 2
2	2, 2, 2	2, 2, 2
3	3, 3, 3	3, 3, 3
4	4, 5, 4	50, 15, 3
5	9, 7, 6	15, 4, 4
8	7, 15, 5	20, 50, 5
10	10, 6, 5	20, 20, 20

References

Agresti, A. (1992). Modelling patterns of agreement and disagreement. *Statistical Methods in Medical Research, 1*, 201–218.

Ahn, W. Y., Busemeyer, J. R., Wagenmakers, E.-J., & Stout, J. C. (2008). Comparison of decision learning models using the generalization criterion method. *Cognitive Science, 32*, 1376–1402.

Albers, W. (2001). Prominence theory as a tool to model boundedly rational decisions. In G. Gigerenzer & R. Selten (Eds.), *Bounded Rtionality: The Adaptive Toolbox* (pp. 297–317). Cambridge, MA: MIT Press.

Andrich, D. (1988). *Rasch Models for Measurement.* London: Sage.

Andrieu, C., De Freitas, N., Doucet, A., & Jordan, M. I. (2003). An introduction to MCMC for machine learning. *Machine Learning, 50*, 5–43.

Angus, J. E. (1994). The probability integral transform and related results. *SIAM Review, 36*, 652–654.

Anscombe, F. J. (1963). Sequential medical trials. *Journal of the American Statistical Association, 58*, 365–383.

Banerjee, M., Capozzoli, M., McSweeney, L., & Sinha, D. (1999). Beyond kappa: A review of interrater agreement measures. *The Canadian Journal of Statistics, 27*, 3–23.

Bartlett, M. S. (1957). A comment on D. V. Lindley's statistical paradox. *Biometrika, 44*, 533–534.

Basu, S., Banerjee, M., & Sen, A. (2000). Bayesian inference for kappa from single and multiple studies. *Biometrics, 56*, 577–582.

Batchelder, W. H., & Riefer, D. M. (1980). Separation of storage and retrieval factors in free recall of clusterable pairs. *Psychological Review, 87*, 375–397.

Batchelder, W. H., & Riefer, D. M. (1986). The statistical analysis of a model for storage and retrieval processes in human memory. *British Journal of Mathematical and Statistical Psychology, 39*, 120–149.

Behseta, S., Berdyyeva, T., Olson, C. R., & Kass, R. E. (2009). Bayesian correction for attentuation of correlation in multi–trial spike count data. *Journal of Neurophysiology, 101*, 2186–2193.

Bem, D. J. (2011). Feeling the future: Experimental evidence for anomalous retroactive influences on cognition and affect. *Journal of Personality and Social Psychology, 100*, 407–425.

Berger, J. O. (1985). *Statistical Decision Theory and Bayesian Analysis* (2nd edn.). New York: Springer.

Berger, J. O., & Berry, D. A. (1988). The relevance of stopping rules in statistical inference. In S. S. Gupta & J. O. Berger (Eds.), *Statistical Decision Theory and Related Topics, Vol. 1* (pp. 29–72). New York: Springer Verlag.

Berger, J. O., & Delampady, M. (1987). Testing precise hypotheses. *Statistical Science, 2*, 317–352.

Berger, J. O., & Mortera, J. (1999). Default Bayes factors for nonnested hypothesis testing. *Journal of the American Statistical Association, 94*, 542–554.

Berger, J. O., & Pericchi, L. R. (1996). The intrinsic Bayes factor for model selection and prediction. *Journal of the American Statistical Association, 91*, 109–122.

Berger, J. O., & Wolpert, R. L. (1988). *The Likelihood Principle* (2nd edn.). Hayward, CA: Institute of Mathematical Statistics.

Bergert, F. B., & Nosofsky, R. M. (2007). A response-time approach to comparing generalized rational and take-the-best models of decision making. *Journal of Experimental Psychology: Learning, Memory, and Cognition, 33*, 107–129.

Besag, J. (1989). A candidate's formula: A curious result in Bayesian prediction. *Biometrika, 76*, 183.

Bowers, J. S., Vigliocco, G., & Haan, R. (1998). Orthographic, phonological, and articulatory contributions to masked letter and word priming. *Journal of Experimental Psychology: Human Perception and Performance, 24*, 1705–1719.

Broemeling, L. D. (2009). *Bayesian Methods for Measures of Agreement.* Boca Raton, FL: CRC Press.

Brooks, S. P., & Gelman, A. (1998). General methods for monitoring convergence of iterative simulations. *Journal of Computational and Graphical Statistics, 7*, 434–455.

Brown, G. D. A., Neath, I., & Chater, N. (2007). A temporal ratio model of memory. *Psychological Review, 114*, 539–576.

Brückner, H., & Bearman, P. (2005). After the promise: The STD consequences of adolescent virginity pledges. *Journal of Adolescent Health, 36*, 271–278.

Burian, S. E., Liguori, A., & Robinson, J. H. (2002). Effects of alcohol on risk-taking during simulated driving. *Human Psychopharmacology, 17*, 141–150.

Carey, S. (2001). Cognitive foundations of arithmetic: Evolutionary and ontogenetic. *Mind and Language, 16*, 37–55.

Carey, S., & Sarnecka, B. W. (2006). The development of human conceptual representations. In M. Johnson & Y. Munakata (Eds.), *Processes of Change in Brain and Cognitive Development: Attention and Performance XXI* (pp. 473–496). Oxford: Oxford University Press.

Carlin, B. P., & Chib, S. (1995). Bayesian model choice via Markov chain Monte Carlo methods. *Journal of the Royal Statistical Society, Series B, 57*, 473–484.

Chater, N., & Oaksford, M. (2000). The rational analysis of mind and behavior. *Synthese, 122*, 93–131.

Chater, N., Tenenbaum, J. B., & Yuille, A. (2006). Probabilistic models of cognition: Conceptual foundations. *Trends in Cognitive Sciences, 10*, 287–291.

Chechile, R. A. (1973). *The Relative Storage and Retrieval Losses in Short–Term Memory as a Function of the Similarity and Amount of Information Processing in the Interpolated Task.* Unpublished doctoral dissertation, University of Pittsburgh.

Chechile, R. A., & Meyer, D. L. (1976). A Bayesian procedure for separately estimating storage and retrieval components of forgetting. *Journal of Mathematical Psychology, 13*, 269–295.

Chen, M.-H. (2005). Computing marginal likelihoods from a single MCMC output. *Statistica Neerlandica, 59*, 16–29.

Chib, S. (1995). Marginal likelihood from the Gibbs output. *Journal of the American Statistical Association, 90*, 1313–1321.

Chib, S., & Jeliazkov, I. (2001). Marginal likelihood from the Metropolis–Hastings output. *Journal of the American Statistical Association, 96*, 270–281.

Cohen, J. (1960). A coefficient of agreement for nominal scales. *Educational and Psychological Measurement, 20*, 37–46.

Dawid, A. P. (2000). Comment on "The philosophy of statistics" by D. V. Lindley. *The Statistician, 49*, 325–326.

de Finetti, B. (1974). *Theory of Probability, Vol. 1 and 2.* New York: John Wiley.

Dennis, S. J., & Humphreys, M. S. (2001). A context noise model of episodic word recognition. *Psychological Review, 108*, 452–477.

Dennis, S. J., Lee, M. D., & Kinnell, A. (2008). Bayesian analysis of recognition memory: The case of the list-length effect. *Journal of Memory and Language, 59*, 361–376.

Dickey, J. M. (1971). The weighted likelihood ratio, linear hypotheses on normal location parameters. *The Annals of Mathematical Statistics, 42*, 204–223.

Dickey, J. M., & Lientz, B. P. (1970). The weighted likelihood ratio, sharp hypotheses about chances, the order of a Markov chain. *The Annals of Mathematical Statistics, 41*, 214–226.

Dienes, Z. (2011). Bayesian versus orthodox statistics: Which side are you on? *Perspectives on Psychological Science, 6*, 274–290.

Donner, A., & Wells, G. (1986). A comparison of confidence interval methods for the intraclass correlation coefficient. *Biometrics, 42*, 401–412.

Dunwoody, P. T. (2009). Introduction to the special issue: Coherence and correspondence in judgment and decision making. *Judgment and Decision Making, 4*, 113–115.

Edwards, W., Lindman, H., & Savage, L. J. (1963). Bayesian statistical inference for psychological research. *Psychological Review, 70*, 193–242.

Ernst, M. O. (2005). A Bayesian view on multimodal cue integration. In G. Knoblich, I. M. Thornton, J. Grosjean, & M. Shiffrar (Eds.), *Human Body Perception From the Inside Out* (pp. 105–131). New York: Oxford University Press.

Estes, W. K. (2002). Traps in the route to models of memory and decision. *Psychonomic Bulletin & Review, 9*, 3–25.

Evans, A. N., & Rooney, B. J. (2011). *Methods in Psychological Research* (2nd edn.). London: Sage.

Fleiss, J. L., Levin, B., & Paik, M. C. (2003). *Statistical Methods for Rates and Proportions* (3rd edn.). New York: John Wiley.

Forster, K. I., Mohan, K., & Hector, J. (2003). The mechanics of masked priming. In S. Kinoshita & S. J. Lupker (Eds.), *Masked Priming: The State of the Art* (pp. 3–38). New York, NY: Psychology Press.

Frye, D., Braisby, N., Lowe, J., Maroudas, C., & Nicholls, J. (1989). Young children's understanding of counting and cardinality. *Child Development, 60*, 1158–1171.

Fuson, K. C. (1988). *Children's Counting and Concepts of Number.* New York: Springer-Verlag.

Gallistel, C. R. (2009). The importance of proving the null. *Psychological Review, 116*, 439–453.

Gamerman, D., & Lopes, H. F. (2006). *Markov Chain Monte Carlo: Stochastic Simulation for Bayesian Inference.* Boca Raton, FL: Chapman & Hall/CRC.

Gelman, A. (1996). Inference and monitoring convergence. In W. R. Gilks, S. Richardson, & D. J. Spiegelhalter (Eds.), *Markov Chain Monte Carlo in Practice* (pp. 131–143). Boca Raton (FL): Chapman & Hall/CRC.

Gelman, A. (2004). Parameterization and Bayesian modeling. *Journal of the American Statistical Association, 99*, 537–545.

Gelman, A. (2006). Prior distributions for variance parameters in hierarchical models. *Bayesian Analysis, 1*, 515–534.

Gelman, A., & Hill, J. (2007). *Data Analysis Using Regression and Multilevel/Hierarchical Models.* Cambridge: Cambridge University Press.

Gelman, A., & Rubin, D. B. (1992). Inference from iterative simulation using multiple sequences (with discussion). *Statistical Science, 7*, 457–472.

Geurts, H. M., Verté, S., Oosterlaan, J., Roeyers, H., & Sergeant, J. A. (2004). How specific are executive functioning deficits in attention deficit hyperactivity disorder and autism? *Journal of Child Psychology and Psychiatry, 45*, 836–854.

Gigerenzer, G., & Gaissmaier, W. (2011). Heuristic decision making. *Annual Review of Psychology, 62*, 451-482.

Gigerenzer, G., & Goldstein, D. G. (1996). Reasoning the fast and frugal way: Models of bounded rationality. *Psychological Review, 103*, 650–669.

Gigerenzer, G., & Todd, P. M. (1999). *Simple Heuristics That Make Us Smart.* New York: Oxford University Press.

Gilks, W. R., Richardson, S., & Spiegelhalter, D. J. (1996). *Markov Chain Monte Carlo in Practice.* Boca Raton, FL: Chapman & Hall/CRC.

Gill, J. (2002). *Bayesian Methods: A Social and Behavioral Sciences Approach.* Boca Raton, FL: CRC Press.

Gönen, M., Johnson, W. O., Lu, Y., & Westfall, P. H. (2005). The Bayesian two-sample t test. *The American Statistician, 59*, 252–257.

Grant, J. A. (1974). Evaluation of a screening program. *American Journal of Public Health, 64*, 66–71.

Green, D. M., & Swets, J. A. (1966). *Signal detection theory and psychophysics.* New York: John Wiley.

Green, P. J. (1995). Reversible jump Markov chain Monte Carlo computation and Bayesian model determination. *Biometrika, 82*, 711–732.

Griffiths, T. L., Kemp, C., & Tenenbaum, J. B. (2008). Bayesian models of cognition. In R. Sun (Ed.), *Cambridge Handbook of Computational Cognitive Modeling* (pp. 59–100). Cambridge, MA: Cambridge University Press.

Heit, E. (2000). Properties of inductive reasoning. *Psychonomic Bulletin & Review, 7*, 569–592.

Heit, E., & Rotello, C. (2005). Are there two kinds of reasoning? In B. G. Bara, L. W. Barsalou, & M. Bucciarelli (Eds.), *Proceedings of the 27th Annual Conference of the Cognitive Science Society* (pp. 923–928). Mahwah, NJ: Erlbaum.

Hoijtink, H., Klugkist, I., & Boelen, P. (2008). *Bayesian Evaluation of Informative Hypotheses.* New York: Springer.

Hu, X., & Batchelder, W. H. (1994). The statistical analysis of general processing tree models with the EM algorithm. *Psychometrika, 59*, 21–47.

Hyman, R. (1985). The Ganzfeld psi experiment: A critical appraisal. *Journal of Parapsychology, 49*, 3–49.

Ivry, R. B. (1996). The representation of temporal information in perception and motor control. *Current Opinion in Neurobiology, 14*, 225–232.

Jaynes, E. T. (1976). Confidence intervals vs Bayesian intervals. In W. L. Harper & C. A. Hooker (Eds.), *Foundations of Probability Theory, Statistical Inference, and Statistical Theories of Science, Vol. II* (pp. 175–257). Dordrecht, Holland: D. Reidel Publishing Company.

Jaynes, E. T. (2003). *Probability Theory: The Logic of Science.* Cambridge, UK: Cambridge University Press.

Jefferys, W. H., & Berger, J. O. (1992). Ockham's razor and Bayesian analysis. *American Scientist, 80*, 64–72.

Jeffreys, H. (1961). *Theory of Probability* (3rd edn.). Oxford, UK: Oxford University Press.

Jones, M., & Love, B. (2011). Bayesian fundamentalism or enlightenment? On the explanatory status and theoretical contributions of Bayesian models of cognition. *Behavioral and Brain Sciences, 34*, 169–231.

Kass, R. E., & Raftery, A. E. (1995). Bayes factors. *Journal of the American Statistical Association, 90*, 377–395.

Kass, R. E., & Wasserman, L. (1995). A reference Bayesian test for nested hypotheses and its relationship to the Schwarz criterion. *Journal of the American Statistical Association, 90*, 928–934.

Kass, R. E., & Wasserman, L. (1996). The selection of prior distributions by formal rules. *Journal of the American Statistical Association, 91*, 1343–1370.

Katsikopoulos, K. V., Schooler, L. J., & Hertwig, R. (2010). The robust beauty of ordinary information. *Psychological Review, 117*, 1259-1266.

Klauer, K. (2010). Hierarchical multinomial processing tree models: A latent-trait approach. *Psychometrika, 75*, 70–98.

Kraemer, H. C. (1992). Measurement of reliability for categorical data in medical research. *Statistical Methods in Medical Research, 1*, 183–199.

Kraemer, H. C., Periyakoil, V. S., & Noda, A. (2004). Kappa coefficients in medical research. In R. B. D'Agostino (Ed.), *Tutorials in Biostatistics, Vol. 1: Statistical Methods in Clinical Studies.* New York: John Wiley.

Kruschke, J. K. (1993). Human category learning: Implications for backpropagation models. *Connection Science, 5*, 3–36.

Kruschke, J. K. (2010a). *Doing Bayesian Data Analysis: A Tutorial Introduction with R and BUGS.* Burlington, MA: Academic Press.

Kruschke, J. K. (2010b). What to believe: Bayesian methods for data analysis. *Trends in Cognitive Sciences, 14*, 293–300.

Kuss, M., Jäkel, F., & Wichmann, F. A. (2005). Bayesian inference for psychometric functions. *Journal of Vision, 5*, 478–492.

Landis, J. R., & Koch, G. G. (1977). The measurement of observer agreement for categorical data. *Biometrics, 33*, 159–174.

Lasserre, J., Bishop, C. M., & Minka, T. (2006). Principled hybrids of generative and discriminative models. In *Proceedings 2006 IEEE Conference on Computer Vision and Pattern Recognition (CVPR)* (pp. 87–94). Washington, DC: IEEE Computer Society.

Le Corre, M., & Carey, S. (2007). One, two, three, four, nothing more: An investigation of the conceptual sources of the verbal counting principles. *Cognition, 105*, 395–438.

Lee, M. D. (2011a). How cognitive modeling can benefit from hierarchical Bayesian models. *Journal of Mathematical Psychology, 55*, 1–7.

Lee, M. D. (2011b). In praise of ecumenical Bayes. *Behavioral and Brain Sciences, 34*, 206–207.

Lee, M. D., & Cummins, T. D. R. (2004). Evidence accumulation in decision making: Unifying the "take the best" and "rational" models. *Psychonomic Bulletin & Review, 11*, 343–352.

Lee, M. D., & Pooley, J. P. (2013). Correcting the SIMPLE model of free recall. *Psychological Review, 120*, 293–296.

Lee, M. D., & Sarnecka, B. W. (2010). A model of knower-level behavior in number concept development. *Cognitive Science, 34*, 51–67.

Lee, M. D., & Sarnecka, B. W. (2011). Number knower-levels in young children: Insights from a Bayesian model. *Cognition, 120*, 391–402.

Lehrner, J. P., Kryspin-Exner, I., & Vetter, N. (1995). Higher olfactory threshold and decreased odor identification ability in HIV-infected persons. *Chemical Senses, 20*, 325–328.

Leigh, B. C., & Stall, R. (1993). Substance use and risky sexual behavior for exposure to HIV: Issues in methodology, interpretation, and prevention. *American*

Psychologist, 48, 1035–1045.

Lejuez, C. W., Read, J. P., Kahler, C. W., Richards, J. B., Ramsey, S. E., Stuart, G. L., et al. (2002). Evaluation of a behavioral measure of risk taking: The balloon analogue risk task (BART). *Journal of Experimental Psychology: Applied, 8*, 75–84.

Leung, A. K.-Y., Kim, S., Polman, E., Ong, L. S., Qiu, L., Goncalo, J. A., et al. (2012). Embodied metaphors and creative "acts". *Psychological Science, 23*, 502–509.

Lewis, S. M., & Raftery, A. E. (1997). Estimating Bayes factors via posterior simulation with the Laplace–Metropolis estimator. *Journal of the American Statistical Association, 92*, 648–655.

Liang, F., Paulo, R., Molina, G., Clyde, M. A., & Berger, J. O. (2008). Mixtures of *g* priors for Bayesian variable selection. *Journal of the American Statistical Association, 103*, 410–423.

Lindley, D. V. (1972). *Bayesian Statistics, A Review.* Philadelphia, PA: SIAM.

Lindley, D. V. (1986). Comment on "Why Isn't Everyone a Bayesian?" by Bradley Efron. *The American Statistician, 40*, 6–7.

Lindley, D. V. (1993). The analysis of experimental data: The appreciation of tea and wine. *Teaching Statistics, 15*, 22–25.

Lindley, D. V. (2000). The philosophy of statistics. *The Statistician, 49*, 293–337.

Liu, C. C., & Aitkin, M. (2008). Bayes factors: Prior sensitivity and model generalizability. *Journal of Mathematical Psychology, 52*, 362–375.

Liu, J., & Wu, Y. (1999). Parameter expansion for data augmentation. *Journal of the American Statistical Association, 94*, 1264–1274.

Lodewyckx, T., Kim, W., Tuerlinckx, F., Kuppens, P., Lee, M. D., & Wagenmakers, E.-J. (2011). A tutorial on Bayes factor estimation with the product space method. *Journal of Mathematical Psychology, 55*, 331–347.

Lunn, D. J. (2003). WinBUGS development interface (WBDev). *ISBA Bulletin, 10*, 10–11.

Lunn, D. J., Spiegelhalter, D., Thomas, A., & Best, N. (2009). The BUGS project: Evolution, critique and future directions. *Statistics in Medicine, 28*, 3049–3067.

Lunn, D. J., Thomas, A., Best, N., & Spiegelhalter, D. (2000). WinBUGS – a Bayesian modelling framework: Concepts, structure, and extensibility. *Statistics and Computing, 10*, 325–337.

MacKay, D. J. C. (2003). *Information Theory, Inference, and Learning Algorithms.* Cambridge, MA: Cambridge University Press.

MacMillan, N., & Creelman, C. D. (2004). *Detection Theory: A User's Guide* (2nd edn.). Hillsdale, NJ: Erlbaum.

Marr, D. C. (1982). *Vision: A Computational Investigation into the Human Representation and Processing of Visual Information.* San Francisco, CA: W. H. Freeman.

Masson, M. E. J. (2011). A tutorial on a practical Bayesian alternative to null-hypothesis significance testing. *Behavior Reseach Methods, 43*, 679–690.

McEwan, R. T., McCallum, A., Bhopal, R. S., & Madhok, R. (1992). Sex and the risk of HIV infection: The role of alcohol. *British Journal of Addiction, 87*, 577–584.

Merkle, E. C., Smithson, M., & Verkuilen, J. (2011). Hierarchical models of simple mechanisms underlying confidence in decision making. *Journal of Mathematical Psychology, 55*, 57–67.

Murdock, B. B., Jr. (1962). The serial position effect in free recall. *Journal of Experimental Psychology, 64*, 482–488.

Myung, I. J. (2003). Tutorial on maximum likelihood estimation. *Journal of Mathematical Psychology, 47*, 90–100.

Myung, I. J., & Pitt, M. A. (1997). Applying Occam's razor in modeling cognition: A Bayesian approach. *Psychonomic Bulletin & Review, 4*, 79–95.

Nosofsky, R. M. (1984). Choice, similarity, and the context theory of classification. *Journal of Experimental Psychology: Learning, Memory, and Cognition, 10*, 104–114.

Nosofsky, R. M. (1986). Attention, similarity, and the identification–categorization relationship. *Journal of Experimental Psychology: General, 115*, 39–57.

Ntzoufras, I. (2009). *Bayesian Modeling Using WinBUGS.* Hoboken, NJ: John Wiley.

O'Hagan, A. (1995). Fractional Bayes factors for model comparison. *Journal of the Royal Statistical Society, Series B, 57*, 99–138.

O'Hagan, A., & Forster, J. (2004). *Kendall's Advanced Theory of Statistics, Vol. 2B: Bayesian Inference* (2nd edn.). London: Arnold.

Ortega, A., Wagenmakers, E.-J., Lee, M. D., Markowitsch, H. J., & Piefke, M. (2012). A Bayesian latent group analysis for detecting poor effort in the assessment of malingering. *Archives of Clinical Neuropsychology, 27*, 453–465.

Parsons, L. M., & Osherson, D. (2001). New evidence for distinct right and left brain systems for deductive and probabilistic reasoning. *Cerebral Cortex, 11*, 954–965.

Payne, J. W., Bettman, J. R., & Johnson, E. J. (1990). *The Adaptive Decision Maker.* New York: Cambridge University Press.

Piaget, J. (1952). *The Child's Conception of Number.* New York: Norton.

Plummer, M. (2003). JAGS: A program for analysis of Bayesian graphical models using Gibbs sampling. In K. Hornik, F. Leisch, & A. Zeileis (Eds.), *Proceedings of the 3rd International Workshop on Distributed Statistical Computing.* Vienna, Austria.

Poehling, K. A., Griffin, M. R., & Dittus, R. S. (2002). Bedside diagnosis of influenza virus infections in hospitalized children. *Pediatrics, 110*, 83–88.

Poirier, D. J. (2006). The growth of Bayesian methods in statistics and economics since 1970. *Bayesian Analysis, 1*, 969–980.

Ratcliff, R., & McKoon, G. (1997). A counter model for implicit priming in perceptual word identification. *Psychological Review, 104*, 319–343.

Riefer, D. M., Knapp, B. R., Batchelder, W. H., Bamber, D., & Manifold, V. (2002). Cognitive psychometrics: Assessing storage and retrieval deficits in special populations with multinomial processing tree models. *Psychological Assessment, 14*, 184–201.

Rieskamp, J. (2008). The probabilistic nature of preferential choice. *Journal of Experimental Psychology: Learning, Memory, and Cognition, 34*, 1446–1465.

Rips, L. J. (2001). Two kinds of reasoning. *Psychological Science, 12*, 129–134.

Rolison, J. J., Hanoch, Y., & Wood, S. (2012). Risky decision making in younger and older adults: The role of learning. *Psychology and Aging, 27*, 129–140.

Rouder, J. N., Speckman, P. L., Sun, D., Morey, R. D., & Iverson, G. (2009). Bayesian t tests for accepting and rejecting the null hypothesis. *Psychonomic Bulletin & Review, 16*, 225–237.

Rubin, D. C., Hinton, S., & Wenzel, A. (1999). The precise time course of retention. *Journal of Experimental Psychology: Learning, Memory, and Cognition, 25*, 1161–1176.

Rubin, D. C., & Wenzel, A. E. (1996). One hundred years of forgetting: A quantitative description of retention. *Psychological Review, 103*, 734–760.

Sanborn, A. N., Griffiths, T. L., & Shiffrin, R. M. (2010). Uncovering mental representations with Markov chain Monte Carlo. *Cognitive Psychology, 60*, 63–106.

Schaeffer, B., Eggleston, V. H., & Scott, J. L. (1974). Number development in young children. *Cognitive Psychology, 6*, 357–379.

Scheibehenne, B., Rieskamp, J., & Wagenmakers, E.-J. (2013). Testing adaptive toolbox models: A Bayesian hierarchical approach. *Psychological Review, 120*, 39–64.

Schwarz, G. (1978). Estimating the dimension of a model. *Annals of Statistics, 6*, 461–464.

Sellke, T., Bayarri, M. J., & Berger, J. O. (2001). Calibration of p values for testing precise null hypotheses. *The American Statistician, 55*, 62–71.

Shepard, R. N., Hovland, C. I., & Jenkins, H. M. (1961). Learning and memorization of classifications. *Psychological Monographs, 75*, 1–42.

Sheu, C.-F., & O'Curry, S. L. (1998). Simulation-based Bayesian inference using BUGS. *Behavior Research Methods, Instruments, & Computers, 30*, 232–237.

Shrout, P. E. (1998). Measurement reliability and agreement in psychiatry. *Statistical Methods in Medical Research, 7*, 301–317.

Sisson, S. A. (2005). Transdimensional Markov chains: A decade of progress and future perspectives. *Journal of the American Statistical Association, 100*, 1077–1089.

Sloman, S. A. (1998). Categorical inference is not a tree: The myth of inheritance hierarchies. *Cognitive Psychology, 35*, 1–33.

Smith, A. F. M., & Spiegelhalter, D. J. (1980). Bayes factors and choice criteria for linear models. *Journal of the Royal Statistical Society, Series B, 42*, 213–220.

Smith, J. B., & Batchelder, W. H. (2010). Beta-MPT: Multinomial processing tree models for addressing individual differences. *Journal of Mathematical*

Psychology, 54, 167–183.

Smithson, M. (2010). A review of six introductory texts on Bayesian methods. *Journal of Educational and Behavioral Statistics, 35*, 371–374.

Spiegelhalter, D. J., Best, N. G., Carlin, B. P., & Van Der Linde, A. (2002). Bayesian measures of model complexity and fit. *Journal of the Royal Statistical Society, Series B, 64*, 583–639.

Stan Development Team. (2013). *Stan: A C++ Library for Probability and Sampling, Version 1.3.*

Stone, C. J., Hansen, M. H., Kooperberg, C., & Truong, Y. K. (1997). Polynomial splines and their tensor products in extended linear modeling (with discussion). *The Annals of Statistics, 25*, 1371–1470.

Tenenbaum, J. B., Kemp, C., Griffiths, T. L., & Goodman, N. D. (2011). How to grow a mind: Statistics, structure, and abstraction. *Science, 331*, 1279–1285.

Townsend, J. T. (1975). The mind–body equation revisited. In C. Cheng (Ed.), *Philosophical Aspects of the Mind–Body Problem* (pp. 200–218). Honolulu, HI: Honolulu University Press.

Turing, A. M. (1950). Computing machinery and intelligence. *Mind, 59*, 433–460.

Uebersax, J. S. (1987). Diversity of decision-making models and the measurement of interrater agreement. *Psychological Bulletin, 101*, 140–146.

van Ravenzwaaij, D., Dutilh, G., & Wagenmakers, E.-J. (2011). Cognitive model decomposition of the BART: Assessment and application. *Journal of Mathematical Psychology, 55*, 94–105.

Vandekerckhove, J., Tuerlinckx, F., & Lee, M. D. (2011). Hierarchical diffusion models for two-choice response time. *Psychological Methods, 16*, 44–62.

Vanpaemel, W. (2010). Prior sensitivity in theory testing: An apologia for the Bayes factor. *Journal of Mathematical Psychology, 54*, 491–498.

Verdinelli, I., & Wasserman, L. (1995). Computing Bayes factors using a generalization of the Savage–Dickey density ratio. *Journal of the American Statistical Association, 90*, 614–618.

Wagenmakers, E.-J. (2007). A practical solution to the pervasive problems of p values. *Psychonomic Bulletin & Review, 14*, 779–804.

Wagenmakers, E.-J., Lee, M. D., Lodewyckx, T., & Iverson, G. (2008). Bayesian versus frequentist inference. In H. Hoijtink, I. Klugkist, & P. A. Boelen (Eds.), *Bayesian Evaluation of Informative Hypotheses* (pp. 181–207). New York: Springer Verlag.

Wagenmakers, E.-J., & Morey, R. D. (2013). Simple relation between one–sided and two–sided Bayesian point–null hypothesis tests. *Manuscript submitted for publication.*

Wagenmakers, E.-J., Wetzels, R., Borsboom, D., & van der Maas, H. L. J. (2011). Why psychologists must change the way they analyze their data: The case of psi. *Journal of Personality and Social Psychology, 100*, 426–432.

Wagenmakers, E.-J., Wetzels, R., Borsboom, D., van der Maas, H. L. J., & Kievit, R. A. (2012). An agenda for purely confirmatory research. *Perspectives on Psychological Science, 7*, 627–633.

Wallsten, T. S., Pleskac, T. J., & Lejuez, C. W. (2005). Modeling behavior in a clinically diagnostic sequential risk-taking task. *Psychological Review, 112*, 862–880.

Wetzels, R., Grasman, R. P. P. P., & Wagenmakers, E.-J. (2010). An encompassing prior generalization of the Savage–Dickey density ratio test. *Computational Statistics & Data Analysis, 54*, 2094–2102.

Wetzels, R., Lee, M. D., & Wagenmakers, E.-J. (2010). Bayesian inference using WBDev: A tutorial for social scientists. *Behavior Research Methods, 42*, 884–897.

Wetzels, R., Vandekerckhove, J., Tuerlinckx, F., & Wagenmakers, E. (2010). Bayesian parameter estimation in the Expectancy Valence model of the Iowa gambling task. *Journal of Mathematical Psychology, 54*, 14–27.

Wynn, K. (1990). Children's understanding of counting. *Cognition, 36*, 155–193.

Wynn, K. (1992). Children's acquisition of number words and the counting system. *Cognitive Psychology, 24*, 220–251.

Zeelenberg, R., Wagenmakers, E.-J., & Raaijmakers, J. G. W. (2002). Priming in implicit memory tasks: Prior study causes enhanced discriminability, not only bias. *Journal of Experimental Psychology: General, 131*, 38–47.

Zeigenfuse, M. D., & Lee, M. D. (2010). A general latent-assignment approach for modeling psychological contaminants. *Journal of Mathematical Psychology, 54*, 352–362.

Zellner, A., & Siow, A. (1980). Posterior odds ratios for selected regression hypotheses. In J. M. Bernardo, M. H. DeGroot, D. V. Lindley, & A. F. M. Smith (Eds.), *Bayesian Statistics* (pp. 585–603). Valencia: University Press.

Index

CPSIA information can be obtained
at www.ICGtesting.com
Printed in the USA
LVOW04s1001071017
551582LV00010B/183/P